However, ↑ solar energy also ↑ evapotranspiration, resulting in reductions in availab. of moisture for plants. Moist. is always an imp const. on agri in many areas and under climate warming bec. more crit. Of course, effects of these changes varies regionally with the moist. holding capacity of soils & other local conditions.

Global Climate Change

The Economic Costs of Mitigation and Adaptation

Global Climate Change

The Economic Costs of Mitigation and Adaptation

Proceedings of a Conference Sponsored by
Center for Environmental Information, Inc.,
46 Prince Street, Rochester, New York 14607-1016

Edited by
James C. White, PhD
Professor Emeritus, Cornell University

Associate Editors
William Wagner
Center for Environmental Information

Carole N. Beal
Center for Environmental Information

Elsevier
New York • Amsterdam • London • Tokyo

No responsibility is assumed by the publisher for any injury and/or damage to persons or property as a matter of products liability, negligence or from any use or operation of any methods, products, instructions, or ideas contained in material herein.

Elsevier Science Publishing Co., Inc.
655 Avenue of the Americas, New York, New York 10010

Sole distributors outside the United States and Canada:
Elsevier Applied Science Publishers Ltd.
Crown House, Linton Road, Barking, Essex IG11 8JU, England

© 1991 by Elsevier Science Publishing Co., Inc.

This book was printed on acid free paper.

This book has been registered with the Copyright Clearance Center, Inc. For further information please contact the Copyright Clearance Center, Inc., Salem, Massachusetts.

All inquiries regarding copyrighted material from this publication, other than reproduction through the Copyright Clearance Center, Inc., should be directed to: Rights and Permissions Department, Elsevier Science Publishing Co., Inc., 655 Avenue of the Americas, New York, New York 10010. FAX 212-633-3977.

ISBN 0-444-01647-3

Current printing (last digit):
10 9 8 7 6 5 4 3 2 1

Manufactured in the United States of America

Table of Contents

PREFACE

Bacteria in a test tube and humans on our planet follow the same population curve as time elapses. We on this earth, largely through control of disease, are rapidly approaching maximum population, and through energy demands are using up our resources and accumulating wastes.

We have come to realize the danger of a maturing system but have made essentially no impact in mitigating or preventing the problems we face. While we have been able to change the world we live in, most of our effects have been to degrade the planet.

We have nibbled away at the quality of the atmosphere for centuries and the accumulated damage has resulted in changing the world's climate at an ever-increasing rate. We now face the possibility of higher average temperatures over most of our planet, changing patterns of rainfall, rising sea levels, aggravated storm systems and a myriad of social effects.

We can prevent change to some degree and we can mitigate the effects in some ways, but how well and at what cost? Some things can be neither mitigated nor prevented. What are their costs in monetary and in social terms?

This book consists of papers presented at a meeting on *Global Climate Change: The Economic Costs of Mitigation and Adaptation*, held in Washington, D.C., in December 1990. It was sponsored by the Air Resources Information Clearinghouse (ARIC), a project of the Center for Environmental Information, Inc. (CEI), a nonprofit organization in Rochester, New York, and cosponsored by more than thirty-five United States, Canadian and international organizations and agencies.

The conference brought together a multidisciplinary group of economists, environmentalists, industry representatives, government agency staff, educators and scientists to attempt to answer the above questions. This distinguished group of presenters and panelists exchanged refreshing ideas and described original research on the past, present and future of the climate change problem, giving a variety of views on the costs of mitigation and prevention and on the public policies needed for coping with the future.

The conference made a special effort to consider the problems of the developing countries who fear that global control of emissions will stifle their efforts to improve their economic well-being. In particular, the discussion by the Chinese participant, Professor Yingzhong Lu, contains timely information and data on China's climate change.

Many people have contributed to the production of this volume. Special thanks are due to Debra Segura for producing the camera-ready copy and for her editorial assistance. William Wagner and Carole Beal, the associate editors, gave invaluable assistance in editing and producing this volume, and Linda Wall deserves special credit for her coordination and operation of the conference in Washington. The editor appreciates the office services provided by Cornell University.

It is inevitable that we learn to live with climate change. How much we can afford to spend in prevention and/or mitigation is a serious and engrossing problem of concern to the whole world. But spend it we will. This book gives a comprehensive view of various approaches to those costs and the factors which drive them; and it should prove useful to scientists, economists and those who establish environmental, energy and economic policy.

James C. White
Cornell University
May 1991

ABOUT THE CENTER FOR ENVIRONMENTAL INFORMATION

The Center for Environmental Information (CEI) was established in Rochester, New York, in 1974 as an answer to the growing dilemma of where to find timely, accurate and comprehensive information on environmental issues. To meet this need for current and comprehensive information, CEI has developed a multi-faceted program of publications, educational programs and information services. It is a private, nonprofit organization funded by membership dues, fees, contracts, grants and contributions.

The Center remains today a Rochester-based organization, but its services now reach far beyond the local community, reflecting the increasing number, scope and complexity of problems affecting the environment.

CEI acts as a catalyst to advance the public agenda toward soundly conceived environmental policies. CEI's communication network provides a link among the scientific community, educators, decision makers and the public, so that informed action follows the free interchange of information and ideas.

CONFERENCE STEERING COMMITTEE

J. Christopher Bernabo
Science & Policy Associates, Inc.
Rosina Bierbaum
Office of Technology Assessment
U.S. Congress
William Cline
Institute for International Economics
Alex Cristofaro
Air and Energy Studies
U.S. Environmental Protection Agency
Roger C. Dower
Climate, Energy and Pollution Program
World Resources Institute
Nelson E. Hay
American Gas Association
John Hughes
Global Climate Coalition
Gordon J. MacDonald
The MITRE Corporation
Alan Miller
Center for Global Change
University of Maryland
Richard Richels
Electric Power Research Institute
James C. White
Center for Environmental Research
Cornell University

CONFERENCE COSPONSORS

Funding Cosponsors:

American Gas Association
E.I. DuPont de Nemours & Co., Inc.
Electric Power Research Institute
Oak Ridge National Laboratory, Center for Global Environmental Studies
U.S. Environmental Protection Agency
World Resources Institute
Contributing Co-Sponsors:
Alliance for Responsible CFC Policy
American Fishing Tackle Manufacturers Association
American Petroleum Institute
Center for Global Change, University of Maryland
Environmental Defense Fund, Inc.
Motor Vehicle Manufacturers Association of the United States, Inc.
National Oceanic and Atmospheric Administration, Air Resources Laboratory
Ontario Ministry of the Environment
Resources for the Future
U.S. Council for Energy Awareness
U.S. Department of Agriculture, Economic Research Service
World Wildlife Fund and The Conservation Foundation

Cooperating Cosponsors:

American Chemical Society
American Council for an Energy Efficient Economy
Center for Environmental Research, Cornell University
Climate Institute
Edison Electric Institute
Environment Canada
Environmental and Energy Study Institute
Friends of the Earth, Environmental Policy Institute
Global Climate Coalition
International Institute for Applied Systems Analysis
Izaac Walton League of America
National Institute for Emerging Technology
Natural Resources Defense Council
Smithsonian Institution
World Meteorological Organization

MANAGING PLANET EARTH

William C. Clark

Center for International Affairs
John F. Kennedy School of Government
Harvard University
79 JFK Street
Cambridge, MA 02138

Every form of life continually faces the challenge of reconciling its innate capacity for growth with the opportunities and constraints that arise through its interactions with the natural environment. The remarkable success of our own species in meeting that challenge is only the beginning of the story.

As we seek to imagine different ways in which that story might unfold, analogies can be helpful. The global pattern of light created by today's civilizations is not unlike the pattern of exuberant growth that develops soon after bacteria are introduced to a nutrient-rich petri dish. In the limited world of the petri dish, such growth is not sustainable. Sooner or later, as the bacterial populations deplete available resources and submerge in their own wastes, their initial blossoming is replaced by stagnation or collapse.

The analogy breaks down in the fact that bacterial populations have no control over, and therefore no responsibility for, their ultimate collision with a finite environment. In contrast, the same wellsprings of human inventiveness and energy that are so transforming the earth have also given us an unprecedented understanding of how the planet works, how our present activities threaten its workings and how we can intervene to improve the prospects for its sustainable development. Our ability to look back on ourselves from outer space symbolizes the unique perspective we have on our environment and on where we are headed as a species. With this knowledge comes a responsibility not borne by the bacteria: the responsibility to manage the human use of planet earth.

At the individual level, people have begun to respond to increased awareness of global environmental change by altering their values, beliefs and actions. Changes in individual behavior are surely necessary but are not enough. It is as a global species that

Reprinted with permission. Copyright © 1989 by Scientific American, Inc. All rights reserved.

we are transforming the planet. It is only as a global species – pooling our knowledge, coordinating our actions and sharing what the planet has to offer – that we have any prospect for managing the planet's transformation along pathways of sustainable development. Self-conscious, intelligent management of the earth is one of the great challenges facing humanity as it approaches the 21st century.

Although efforts to manage the interactions between people and their environments are as old as human civilization, the management problem has been transformed today by unprecedented increases in the rate, scale and complexity of those interactions. What were once local incidents of pollution now involve several nations – witness the concern for acid deposition in Europe and in North America. What were once acute episodes of relatively reversible damage now affect multiple generations – witness the debates over chemical- and radioactive-waste disposal. What were once straightforward confrontations between ecological preservation and economic growth now involve multiple linkages – witness the feedbacks among energy consumption, agriculture and climatic change that are thought to enter into the greenhouse effect.

We have entered an era characterized by syndromes of global change that stem from the interdependence between human development and the environment. As we attempt to move from merely causing these syndromes to managing them consciously, two central questions must be addressed: What kind of planet do we want? What kind of planet can we get?

What kind of planet we want is ultimately a question of values. How much species diversity should be maintained in the world? Should the size or the growth rate of the human population be curtailed to protect the global environment? How much climatic change is acceptable? How much poverty? Should the deep ocean be considered an option for hazardous-waste disposal?

Science can illuminate these issues but cannot resolve them. The choice of answers is ours to make and our grandchildren's to live with. Because different people live in different circumstances and have different values, individual choices can be expected to vary enormously. As pointed out by Gro Harlem Brundtland, poor people and rich people are especially likely to place different values on economic growth and environmental conservation. Recently, however, the long-standing debate over growth versus environment has matured considerably. A broad consensus has begun to emerge that interactions between people and their environments should be managed with the goal of sustainable development.

The World Commission on Environment and Development (WCED), chaired by Prime Minister Brundtland, characterized sustainable development as paths of social, economic and political progress that meet "the needs of the present without compromising the ability of future generations to meet their own needs." Sustainable development thus reflects a choice of values for managing planet earth in which equity matters – equity among peoples around the world today, equity between parents and their grandchildren.

Managing the planet toward sustainable development is an undertaking made no less daunting by its urgency. The basic human dimensions of the task are explored by Nathan Keyfitz and by Jim MacNeill. The broad picture, although familiar, bears recounting. The planet today is inhabited by somewhat more than five billion people who each year appropriate 50 percent of the organic material fixed by photsynthesis on land, consume the equivalent of two tons of coal per person and produce an average of 150 kilograms of steel for each man, woman and child on the earth. The distribution of these people, their well-being and their impact on the environment vary significantly among countries.

At one extreme, the richest 15 percent of the world's population consumes more than one-third of the planet's fertilizer and more than half of its energy. At the other extreme, perhaps one-quarter of the world's population goes hungry during at least some seasons of the year. More than a third live in countries where the mortality for young children is greater than one in 10. The vast majority exist on per capita incomes below the official poverty level in the U.S.

As we look to the future, it is encouraging that the growth rate of the human population is declining virtually everywhere. Even if the trends responsible for the decline continue, however, the next century will probably see a doubling of the number of people trying to extract a living from planet earth. Nearly all of the increase will take place in today's poorer countries. According to the WCED, a fivefold to tenfold increase in world economic activity during the next 50 years will be required to meet the basic needs and aspirations of the future population. The implications of this desperately needed economic growth for the already stressed planetary environment are at least problematic and are potentially catastrophic.

Efforts to manage the sustainable development of the earth must therefore have three specific objectives. One is to disseminate the knowledge and the means necessary to control human population growth. The second is to facilitate sufficiently vigorous economic growth and equitable distribution of its benefits to meet the basic needs of the human population in this and subsequent generations. The third is to structure the growth in ways that keep its enormous potential for environmental transformation within safe limits – limits that are yet to be determined.

If the goals of sustainable development describe the type of planet people want, the second question still remains: What kind of planet can we actually get? When we address this question, the focus shifts from what we value to what we know.

In the end the strategies for sustainable development must translate into local action if they are to have any impact at all. As I have noted, however, many of today's most intractable challenges to sustainability involve time scales of decades or centuries and global spatial scales. Any significant improvements in our ability to manage planet earth will require that we learn how to relate local development action to a global environmental perspective.

Fortunately, understanding of global environmental change has been revolutionized in recent years. The revolution has its roots in the 1920s, with the Russian mineralogist Vladimir I. Vernadsky's seminal writings on the biosphere. It received important impetus from the International Geophysical Year of 1957 and is now being carried forward through a lively array of research and monitoring efforts around the world, capped by an ambitious new International Geosphere Biosphere Program. Although the "global change" revolution is far from complete, its broad outlines can be summarized.

The view of environmental change shows a planet dominated through decades and centuries by the interactions of climate and chemical flows of major elements, interactions that are woven together by the global hydrological cycle and are significantly influenced by the presence of life.

The climate system incorporates atmospheric and oceanic processes that govern the global distribution of wind, rainfall and temperature. Processes central to human transformation and management of planet earth include changes in concentrations of greenhouse gases and their impact on temperature; the effect of ocean circulation on the timing and distribution of climatic changes; and the role of vegetation in regulating the flux of water between land and atmosphere (see Schneider).

A second important component of the planet's environment is the global circulation and processing of major chemical elements such as carbon, oxygen, nitrogen, phosphorus and sulfur. These elements are the principal components of life. In chemical forms such as carbon dioxide, methane and nitrous oxide, they also exert a major influence on climate. Even in the absence of human influences, the earth's climate and chemistry have undergone abrupt and tightly linked changes. When added to these natural fluctuations, human activities have created disturbances in global chemical flows that manifest themselves as smog, acid precipitation, stratospheric ozone depletion and other problems (see Graedel and Crutzen).

The third component, the hydrological cycle, includes the processes of evaporation and precipitation, runoff and circulation. Water is a key agent of topographic change and an overall regulator of global chemistry and climate. As described by J.W. Maurits la Riviere, human impacts on the hydrological cycle that require attention include pollution of groundwater, surface waters and oceans, redistribution of water flows on the earth's surface and potential sea-level changes induced by global warming.

Life, the final component, has found the environment of planet earth to be replete with possibilities, resulting in the evolution of an astounding – but rapidly decreasing – degree of biological diversity (see Wilson). It has not been widely appreciated until recently that life is also a key player in conditioning and regulating the global environment, through its influence on the chemical and hydrological cycles. Finally, one form of life – the human species – has grown over the past several centuries from a position of negligible influence at the planetary scale to one of great significance as an agent of global change.

Although our knowledge of the earth system is quickly expanding, we do not yet know enough about it to say with any certainty how much change the system as a whole can tolerate or what its capacity may be for sustaining human development. We do, however, know a good deal about interactions between individual components of the global environment and specific human activities. This admittedly incomplete knowledge provides some useful perspectives on questions of planetary management.

Since the beginning of the 18th century, the human population has increased by a factor of eight; average life expectancy has at least doubled. During the same period human economic activity has become increasingly global, with demands for goods and services in one part of the planet being met with supplies from half a world away. The volume of goods exchanged in international trade has increased by a factor of 800 or more and now represents more than a third of the world's total economic product.

The three components of this growth and globalization of human activity that have had greatest impact on the environment are agriculture, energy and manufacturing. Agriculture has been the dominant agent of global land transformation; since the middle of the last century, nine million square kilometers of the earth's surface have been converted into permanent croplands (see Crosson and Rosenberg). Energy use has risen by a factor of 80 over the same period, with profound consequences for the planet's chemical flows of carbon, sulfur and nitrogen (see Gibbons, Blair and Gwin). Finally, the world's industrial production has increased more than 100-fold in 100 years, supported by long-term growth rates of more than 3 percent a year in the utilization of such basic metals as lead, copper and iron (see Frosch and Gallopoulos).

The transformation of the planetary environment induced by this explosion of human activity is particularly evident in changes to the physical landscape. Since the beginning of the 18th century, the planet has lost six million square kilometers of forest – an area larger than Europe. Land degradation has increased to a significant but uncertain degree. Sediment loads have risen threefold in major river systems and eightfold in smaller basins that support intense human activity; the resulting flow of carbon to the sea is between one and two billion tons a year. During the same period the amount of water humans withdraw from the hydrological cycle has increased from perhaps 100 to 3,600 cubic kilometers per year – a volume equivalent to that of Lake Huron.

Many substantial changes in the planet's other chemical flows have taken place. In the past 300 years agricultural and industrial development has doubled the amount of methane in the atmosphere and increased the concentration of carbon dioxide by 25 percent. The global flows of major elements such as sulfur and nitrogen that result from human activity are comparable to or greater than the natural flows of these elements. Among the trace metals, many of which are toxic to life, Jerome O. Nriagu of the Canadian National Water Research Institute and Jozef M. Pacyna of the Norwegian Institute for Air Pollution Research have shown that human emissions of lead, cadmium and zinc exceed the flux from natural sources by factors of 18, five and three, respectively. For several other metals, including arsenic, mercury, nickel and vanadium, the human contribution is now as much as two times that from natural sources. Finally, of the more than 70,000 chemicals synthesized by humans, a number – such as the chlorofluorocarbons and DDT – have been shown to affect the global environment significantly, even at very low concentrations.

Assessment of the prospects for sustainable development of the earth shows that the change in the rates at which human activities are transforming the planet may be as important as the absolute magnitudes involved. B.L. Turner, Robert W. Kates and I have analyzed historical transformation rates for several components of the global environmental system. For each component, we first characterized the recency of change – the date by which half of the total human transformation from prehistoric times to the present had taken place. Next, we assessed the acceleration of change by comparing the present rate of transformation with that of a generation ago. The dominant impression from this analysis is the relative recency of most global environmental change. None of the components we reviewed had reached 50 percent of its total transformation before the 19th century. Most passed the 50-percent level only in the second half of the 20th century.

Beyond this general conclusion, four broad patterns of transformation emerge. The first pattern, characterized by relatively long-established and still accelerating change, includes deforestation and soil erosion. The second, established relatively recently and still accelerating, includes the destruction of floral diversity, withdrawal of water from the hydrological cycle, sediment flows and human mobilization of carbon, nitrogen and phosphorus. There is little reason to believe that human society has yet learned to manage on a global scale any of these accelerating transformations of the environment.

More encouraging are two decelerating trends. Human-induced extinctions of terrestrial vertebrates reached half of their present total by the late 19th century and are apparently occurring more slowly today than they were a generation ago. The remaining group of transformations we examined – releases of sulfur, lead, radioactive fallout, a representative organic solvent and extinction of marine mammals – also represents primarily phenomena of the 20th century that are now decelerating.

The crude measure of long-term deceleration presented here gives no assurance that the declining transformation rates reflect increasing competence in planetary management. (Specific transformation rates could, for example, decline simply because there are no more species to exterminate or because we turn to cheaper fuels that happen to emit different pollutants.) Nevertheless, for most of the cases I have cited, at least some fraction of the deceleration can be attributed to deliberate large-scale, long-term efforts at environmental management.

The global patterns sketched so far provide a necessary but insufficient perspective from which to reflect on the prospects for improving the management of planet earth. Also needed is an appreciation of the regional faces of change. To analyze regional situations in any detail is beyond the scope of this essay; still, it will be helpful to recall the extraordinary range of local circumstances that will have to be dealt with if the human transformation of the planet is to be steered along paths of sustainable development.

Any classification of regional perspectives on sustainable development will inevitably oversimplify reality. But one of the most instructive simplifications distinguishes interactions between environment and development that are associated with affluence. Another distinguishes interactions involving low population densities from those with high population densities. Combining the two simplifications yields the classification illustrated in Figure 1.

Low-income, low-density areas such as Amazonia and Malaya-Borneo consititute settlement frontiers still available for use by people in the less developed countries. Until recently, such regions supported sparse populations, and intrusions from the industrialized world were confined to small plantation and mining sites. The situation has changed dramatically during the past 20 years as humans engaged in large-scale timber clearing and livestock raising have invaded these regions. The resulting mix of subsistence and commercial agriculture plus industrial resource extraction has led to a unique pattern of landscape transformation, the full implications of which cannot yet be assessed. Reduction of biological diversity and degradation of biological productivity nonetheless seem inevitable (see Wilson). The poverty of the landless farmers engaged in land clearing and the relative paucity of indigenous institutions that might guide the sustainable development of such regions will make them especially problematic components of any strategy for planetary management.

In contrast, regions with low population density but high investments in sophisticated technology are illustrated by the classic harsh environments of the earth. Such environments include the circumpolar arctic areas, deserts, mineral-extraction platforms and off-shore "fish factories." The large-scale transformation of these regions has become possible only within the past several decades as knowledge, prices and technology have converged to induce development.

Of the environmental changes associated with such development – oil spills, river diversions and landscape transformation – some have received widespread attention. Others, such as atmospheric pollution and cultural dislocation, have received less. The knowledge base for management remains poor. But since a relatively few, wealthy corporate actors seem likely to be involved in most transformations of consequence, the possibilities for institutionalizing sustainable-development strategies for such regions may be relatively good.

Typical of low-income, high-density regions are the Gangetic Plain of the Indian subcontinent and the Huang-Huai-Hai Plains of China. Here intensive agricultural development has been under way for centuries and has been joined in the past several decades by the rapid rise of industrial development in growing urban centers. Landscape degradation is the central problem as more and more people are employed on agricultural land that is already exploited to capacity. In addition, the rapid rise of heavy industry in such areas has led to pollution problems comparable to those that Europe faced several decades ago. The critical management challenge here is to provide employment that generates income and takes pressure off the land without aggravating urbanization problems or increasing regional competition for "smokestack" industries.

The greatest responsibility and the greatest immediate potential for the design of sustainable-development strategies may be in the high-income, high-density regions of the industrialized world. As is repeatedly stated in discussions of stratospheric ozone depletion and the greenhouse effect, advanced industrialized societies have been responsible for imposing a disproportionate share of global environmental burdens on the planet. Over the past several decades, however, places as different as Sweden, Japan and northeastern U.S. have all achieved significant improvement in numerous aspects of their regional environments. Forests have expanded, sulfur emissions have declined, locally extinct species have been successfully reintroduced. Some of these environmental victories are clearly the unintended by-product of unrelated economic changes. Others reflect the export of environmentally destructive activities to less fortunate parts of the world. Increasingly, however, such regions are benefiting from systematic strategies to mitigate the impacts of uncontrolled development and are beginning to design the kinds of environments in which their people want to live.

What kind of environments can such strategies attain? What kinds of development can they sustain? Apart from a basic knowledge of how the global environment works and how human development interacts with it, an understanding is also required of the impact policy can have on environmental change.

At the outset, it cannot be overemphasized that policy for managing planet earth must above all else be adaptive (see Ruckelshaus). Our understanding of the science behind global change is incomplete and will remain so into the foreseeable future. Surprises like the stratospheric ozone hole will continue to appear and will demand action well in advance of scientific certainty. Our understanding of the economic and social processes that contribute to global environmental change is even weaker. Conventional forecasts of population and energy growth could turn out to be conventional foolishness. Science can help, but it is our capacity to shape adaptive policies able to cope with surprises that will determine our effectiveness as managers of planet earth. Building such a capacity will require cultivation of leadership and of institutional competence in at least four areas.

The first requirement is to make the information on which individuals and institutions base their decisions more supportive of sustainable-development objectives. Part of the task, it cannot be said often enough, is simply to support the basic scientific research and planetary monitoring activities that underlie our knowledge of global change. Also essential is to improve the flow of information implicit in existing systems of prices, regulations and economic incentives. The failure of current economic accounts to track the real environmental costs of human activities encourages the inefficient use of resources. The artificially high prices maintained for many agricultural products have significantly exacerbated problems of land degradation and pollution in many parts of the world. Narrowly targeted government subsidies have been directly responsible for a significant fraction of today's global deforestation. All of these distorted information signals need to be addressed in designing adaptive policies for sustainable development.

A second requirement for adaptive planetary management is the invention and implementation of technologies for sustainable development. Such technologies will need to be resource-conserving, pollution-preventing or environment-restoring and at the same time economically sustainable. Significant technical progress has already been made toward delivering desired end-use services at significantly lower environmental costs. Surprisingly often, the economic costs of the "conserving technologies" also turn out to be lower; cost advantages – not environmental concerns – are responsible for halving the ratio of energy consumption to the gross national product in the U.S. since it peaked in the early 1920s.

Technologies for the restoration of environments degraded by salinization, acidification and mining have also been developed and are being effectively employed at a regional scale. The policy need is to tailor technological innovations to the specific local conditions encountered in various environment-development conflicts around the world.

A third requirement for adaptive planetary management is the construction of mechanisms at the national and international level to coordinate managerial activities. The need for formal international agreements in this area has been highlighted by the Montreal Protocol on Substances that Deplete the Ozone Layer and discussion of a possible international law of the atmosphere. In fact, a dozen or more global conventions for protection of the environment are now in effect.

Beneath this orderly surface, however, a large and rapidly growing number of non-governmental bodies, governmental agencies and international organizations are scrambling to play some part in the management of planet earth. Pluralism has much to recommend it. But are we not nearing a point of diminishing returns where too many meetings, too many declarations and too many visiting experts leave too few people with too few resources and too little time to actually do anything? The immediate need at the international level is for a forum in which ministerial-level coordination of environmental-management activities can be regularly discussed and implemented, much as is already done for international economic policy. As in the case of economic policy, the existence of such a formal, high-level governmental summit on global issues of environment and development could provide an occasion for parallel discussions involving non-government and private-sector interests.

Finally, building a capacity for adaptive management of planet earth will require a desire and an ability to reflect continually on the values and objectives that guide our efforts. In an important sense, there has turned out to be more to the notion of sustainable development than even the wise members of the World Commission intended. Individuals, organizations and entire nations have taken the concept as a point of departure for rethinking their interactions with the global environment.

In the Soviet Union, issues of ecological deterioration became a central point of debate in the first Congress of People's Deputies. In Kenya, an innovative project sponsored by the African Academy of Sciences has begun to explore and articulate alternative possibilities for the continent's development in the 21st century. In West Germany, a high-level commission representing all political parties and the scientific community evolved a consensual *Vorsorge*, or prevention, principle to guide the nation's environmental policies. In Sweden, a national best-seller and focal point for political debate emerged when environmental scientists and artist Gunnar Brusewitz collaborated in "painting the future" of Swedish landscapes under alternative paths of development.

The impact that these and similar explorations being conducted around the world will ultimately have in guiding the human transformation of the environment is far from clear. But there can be no question that, against all expectations, the explorations all reflect an emerging commitment to get on with the task of responsibly managing planet earth.

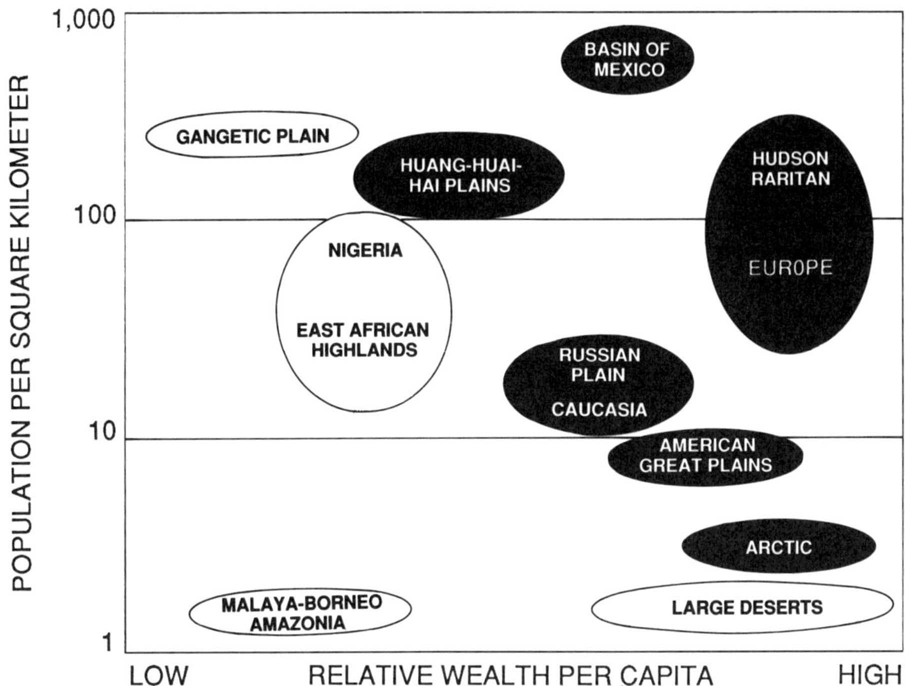

REGIONAL VARIETIES of environmental transformation can be visualized by plotting population density versus relative wealth. Regions with low density and low industrialization include many of the earth's remaining settlement frontiers where widespread agricultural development has only recently begun. In contrast, low-density areas with relatively high investments tend to be the harsh environments exploited by corporate developers of fuel and minerals. High-density, low-income regions have long histories of agricultural development; their challenge is to produce increasing food yields while relieving existing stresses on the land. The greatest responsibility for designing sustainable-development strategies lies with the high-density, wealthy regions that have imposed a disproportionate burden on the planet's environment. The figure is from work by B. L. Turner, Robert W. Kates and the author.

References

Scientific American, September, 1989:

Brundtland, Gro Harlem.

Crossen, Pierre R. and Rosenberg, Norman J., "Strategies for Agriculture," p. 128.

Frosch, Robert A. and Gallopoulos, Nicholas E., "Strategies for Manufacturing," p. 144.

Gibbons, John H., Blair, Peter D. and Gwin, Holly L., "Strategies for Energy Use," p. 136.

Graedel, Thomas E. and Crutzen, Paul J., "The Changing Atmosphere," p. 58.

Keyfitz, Nathan, "The Growing Human Population," p. 118.

MacNeill, Jim, "Strategies for Sustainable Economic Development," p. 154.

Maurits la Riviere, J.W., "Threats to the World's Water," p. 80.

Ruckelshaus, William D., "Toward A Sustainable World," p. 166.

Schneider, Stephen, H., "The Changing Climate," p. 70.

Wilson, Edward O., "Threats to Biodiversity," p. 108.

HOW SHOULD WE ADDRESS THE ECONOMIC COSTS OF CLIMATE CHANGE?

J. Christopher Bernabo

Science and Policy Associates, Inc.
1333 H Street, NW
Washington, DC 20005

The topic of how we should address the economic cost of climate change is not an open-and-shut one obviously; it's one on which there's a wide range of views, and we will try to reflect that diversity of views in this discussion. We do have a long list of speakers, people from the World Bank, environmental organizations, the Administration, and industry, and all presumably are going to give us different views on how to address this problem.

Clearly, as the issue of global climate change – which has been around for 100 years or more in a scientific sense – enters the present phase, much is known about the problem, but more needs to be known. As we determine what to do about it, policy and economics clearly are going to play a very important role in choosing and examining the options.

One of the challenges we address today is how to approach the problem. Do our standard ways of looking at cost-benefit analysis in the more limited scope of other problems apply here? Or are new techniques, new approaches needed? I think we're going to see a diversity of views and we encourage the audience to share in the discussion.

Published 1991 by Elsevier Science Publishing Company, Inc.
Global Climate Change: The Economic Costs of Mitigation and Adaptation
James C. White, Editor

SOME ECONOMICS OF GLOBAL CLIMATE CHANGE: THE VIEW FROM THE DEVELOPING COUNTRIES

Nancy Birdsall and John A. Dixon

World Bank
1818 H Street, NW
Washington, DC 20433

If global warming actually occurs it will have widespread economic effects. These effects will be different in different countries. Some countries will experience economic losses from inundated coastlines, changed and possibly more violent weather patterns, and increased climatic variability. Others may benefit from a milder climate and increased precipitation. As is obvious, there will be no necessary relation between a particular country's contribution to possible global warming in the form of greenhouse gas emissions (including loss of sequestered carbon, for example through forest burning) and the economic costs it bears or the economic benefits it enjoys as a result of global warming. The costs or benefits to a particular country that result, independent of its own contribution, constitute externalities (the effects of actions of other countries that those other countries do not take fully into account). These externalities occur because countries can pass on some of the costs of increased greenhouse gas emissions and cannot capture fully the benefits of prevention of those emissions.

The nations of the world vary greatly in their contribution (both historical and current) to the emission of CO_2 and other greenhouse gases, in their vulnerability to the possible effects of global warming, and, given differences in income and welfare, in their willingness and ability to take actions to reduce emissions of greenhouse gases. One obvious set of differences on all these counts is between the developed and the developing countries.

This paper examines these differences between developing and developed countries with the objective of deriving some initial answers to two apparently simple questions: Where should actions to reduce the possibility of global warming take place? Who should pay the costs of these actions? We refer to these two as the "siting" question and the "burden-sharing" question. The siting question does not exclude the possibility of world-wide actions, such as a global carbon tax, but allows for the possibility that even world-wide actions can be designed to affect different countries differentially. Addressing these questions in their fullest sense would require a discussion well beyond that presented here, raising issues of moral responsibility and intra- and intergenerational equity. Our

Copyright 1991 by Elsevier Science Publishing Company, Inc.
Global Climate Change: The Economic Costs of Mitigation and Adaptation
James C. White, Editor

goal is much more modest—to set out some of the facts and suggest some points that might guide a fuller discussion.

In the first section, we provide background information on the relative contributions of various groups of countries to current emissions and the accumulated stock of emissions. Later we take the perspective of the developing countries and consider the potential welfare costs in those countries of global warming and the nature of the costs those countries would face were they to undertake steps to reduce their own emissions. Finally, we conclude with preliminary answers to the siting and the burden-sharing questions. On the latter, our conclusion, surprising even to ourselves, is that there is currently little scope for the kind of directed burden-sharing that we expected.

In our discussion, we address qualitatively the costs and benefits of taking action at the national level from the point of view of developing vs. developed countries—in effect a crude benefit-cost analysis from the perspective of each set of countries. However, we do not address the question of how much, if anything, should be done by the world as a whole to reduce the potential threat of global climate change. We do not, in short, undertake a benefit-cost analysis of action at the global level.[1] In fact, given the poor basis for assessing the likelihood and timing of any climate change, the uncertainty about the potential costs of climate change, and the long gestation period for any interventions to make a difference, if there is any form of international cooperation it is likely to take the form of some agreement on a global target and on an accompanying schedule for reductions of emissions, perhaps allocated across countries or groups of countries. The search would then be for least-cost approaches to meeting the scheduled targets, with least-cost approaches being compared across countries as well as across policy or investment actions. Therefore, for the sake of this discussion, we assume that at the global level we are in a world of cost-effectiveness and thus least-cost analysis.

Contributions to Possible Climate Change Across Countries

Table 1 shows shares of population, GNP and carbon emissions grouped by developed, developing and oil-exporting countries. As the data on emissions cover carbon only, they probably understate the relative share in emissions of developed countries. Ninety-five percent of the carbon emissions from deforestation, omitted from this table, are from developing countries. Chlorofluorocarbons and other greenhouse gases, also omitted, come primarily from the developed countries. The basic point is obvious. With only 15-20% of world population, developed countries take up more than 50% of carbon emissions.

Table 2 shows the change in carbon emissions between 1960 and 1988 for various groups of countries. The top line includes the six largest emitters in 1960, excluding the Soviet Union. For these high-emitting developed countries, the share of world emissions declined over the period, from 54% to 39%. The share of the Soviet Union and Poland together increased almost 25%, from 17% to 21%. The share of the three large developing countries increased 50%, from 10% to 15%.

There are at least four explanations for the decline in the share of the developed countries. Population grew in those 6 countries about 25% over the period, while it doubled or more in the developing countries. In the developed countries, the shift from industry to services reduced energy use per dollar of GDP; in the developing countries, the shift out of agriculture into industry increased energy use per dollar of GDP. Across the countries shown, there were and are differences in fossil fuels used; China (as well as Poland and the Soviet Union) rely more heavily on coal which generates more emissions per unit of energy produced than do natural gas and oil. Finally, there are important differences across countries in energy efficiency, with the developing countries being less efficient in energy for given output. We return to this last issue below.

Table 3 shows the same data in absolute terms for selected countries. Though the U.S. share declined, its total emissions did increase. But much larger percentage increases occurred elsewhere; in Japan and Korea, countries whose economies grew rapidly, the increases were fourfold and tenfold respectively. In China and India, the tripling or more of emissions is associated with some economic growth combined with large absolute increases in population size.

Table 4 shows per capita emissions in 1960 and 1988 for selected countries. It captures the huge differences in per capita emissions between the developed and developing countries, especially in 1960—when there was for example a 50-fold difference between the U.S. and Korea. It also shows the generally much greater percentage increases in per capita emissions in the developing countries over the period. The 13-fold increase in per capita emissions in Korea suggests what success in economic growth can imply for increases in per capita carbon emissions. In the case of Korea, much of the rapid increase in carbon emissions is a direct result of the government's active promotion of public investment in a number of energy-intensive manufacturing sectors including petrochemicals, iron and steel, ship building and motor vehicle manufacturing.

Table 5, in the last column, shows emissions per dollar of GNP across selected countries. In general, the developing countries have higher emissions per GDP dollar. China is the most obvious, at over 2000 grams per dollar in 1987 compared to 276 in the U.S. and just 133 in France. Much of the difference across the two groups of countries is due to the greater efficiency of energy use in the developed countries.

Finally, we estimate very roughly that over the last thirty years the relative contribution of the developing countries to the total accumulated stock of greenhouse gases in the atmosphere is at most 20%. Greenhouse gases are retained in the atmosphere over long periods; annual emissions contribute relatively little to the accumulated stock, which is what matters for possible global warming.[2] This implies some logic to the idea of assigning rights across generations to use the earth's atmosphere as a sink for emissions, literally as a global commons. An obvious simple extension of the idea of allocating rights across generations is that of allocating rights across countries in the current generation. The facts above indicate that this would imply large immediate trading in such rights between developed and developing countries, with large transfers to the latter – whether those rights were assigned on a current flow basis or on the basis of accumulated deficits and surpluses.

In summary, it is clear that, though the developing countries contribute and have contributed relatively little to emissions, they are also the likely source of future growth in emissions – because their populations are growing more rapidly, their economies appear to be at an earlier, more energy-intensive stage of growth, their per capita emissions are still low, and they are not, at least presently, efficient in their use of energy.

The View from the Developing Countries

In this section we examine the relative costs of adapting to changes brought about by global warming in developed compared to developing countries, and the relative costs of unilateral preventive action in developing countries.

High Welfare Costs of Global Climate Change in Developing Countries

Global climate change is likely to produce some winners and some losers at the country level; the current state of scientific understanding is certainly not sufficient to establish how developing countries as a group might fare. (What is clear is that at least some countries will be worse off, e.g. those that will either literally go under water as a result of sea level rise such as some Pacific island nations, or that will lose a sizeable piece of their territory as in the case of Bangladesh.)

However, assuming that the net impact is negative at the global level, there is at least one reason to expect that developing countries will be worse off compared to developed countries. The developing countries have as a group less flexible, less responsive economies, making adaptation to any change more costly in welfare terms, i.e. in terms of the potential losses in human welfare.

Agriculture

The world rate of growth of agricultural production is as likely to increase as to decrease with global warming. It is possible that with increased variability of weather, the variance of production from harvest to harvest in particular settings will increase. It seems also almost certain that there will be changes in the location of agricultural production with changes in comparative advantage across space, with some settings having higher agricultural production and others lower production.

Table 6 shows the percent of the labor force in agriculture in 1980, percent of GDP of agriculture, and GNP per capita for selected developed and developing countries. Productivity per worker in agriculture is lower in developing countries. Any increase in local climate variability over time and over space implies costs of adaptation. In the absence of perfect capital markets (discussed further below), insurance costs to allow for variability over time in production will be high, either in terms of less risk-taking in agriculture or reduced consumption. If labor and capital must move to other sites, these adaptation costs will be relatively more costly in terms of effects on human welfare where productivity and thus income are lower. Even where mean production is stable, an increase in variance will raise costs – if there are longer periods of drought for example. Finally, it should be noted that developing countries are much more vulnerable to the possibility that net costs of climate change will be negative in agriculture, since a much higher proportion of not only their labor forces but of their economies is concentrated in agriculture and in economic activities linked to agriculture.

Poorly Developed Insurance Markets

Poorly developed insurance markets in most developing countries mean that risk-sharing is difficult to carry out and insurance is costly to purchase. Farmers and others are thus less able to weather increased climate-induced fluctuations than producers in more developed countries with better insurance markets (either private or government-run). Of course, improved insurance markets could develop over the lengthy period during which global climatic effects would begin to be felt. Nevertheless, the absence of such markets is likely to create at least major short-term problems. Moreover, with more of their economies in agriculture, risk-sharing may not be viable in many developing countries because of concentrated vulnerability (or covariance in exposure to risk). In the face of high-cost changes in the global climate, a world insurance market would almost surely need to be developed – and on grounds of vulnerability to risk, it is possible that developing countries would in principle face higher premiums.

Poorly Developed Capital Markets

Poor capital markets limit the ability of developing countries to raise and allocate the capital needed to adapt to economic changes induced by climate change. All other things the same, this raises the costs of investments, both to the public and the private sector. Typical costs of adaptation would be the construction of dikes or other safeguards in urban coastal areas, and the development of irrigation in agricultural areas no longer able to rely on rain-fed agriculture.

Weak Public Sectors

We can describe several types of policy failure in developing countries. In many developing countries, government has little capacity for effective intervention – these countries are what Myrdal referred to many years ago as "soft states." Rules and policies are generally ignored and government inaction is the norm. This is particularly true in much of Africa and parts of South Asia.

Public policy is driven less by public needs than by private interests, and rent-seeking is common.[3] Though this is also a problem in developed countries, in parts of Latin America and Asia the effect on policymaking is relatively greater in the absence of countervailing public interest groups.

Political regimes in many parts of the developing world are less stable than in most developed countries. This encourages short planning horizons on the part of those in power, vitiating their capacity to plan ahead in the event of foreseeable but distant problems such as might arise with global warming.

Other constraints, such as fear of inflation in much of Latin America, constrain policy choices in the short run. Fear of inflation makes raising energy prices more difficult than it would otherwise be. In fact, energy prices, to which we will return below, provide a straightforward example of the inability of governments in developing countries to take appropriate action – even when such action would reduce their fiscal problems as well as increase efficiency. (Though many governments in developed countries are equally ineffective in some policy areas, we are simply arguing here that in general, public policymaking is even less effective in developing countries.)

Finally, poverty itself assures that the welfare costs of adjusting to changes due to global warming will be relatively high. Assuming declining marginal utility of income as income rises, there are greater welfare losses from marginal reductions in income at low levels. In the extreme, the changes associated with global warming could increase mortality in developing countries – e.g. if urban sanitation collapsed due to changes in water availability – illustrating the point all too well.[4]

High Costs of Prevention in Developing Countries

Despite the high potential welfare costs of global warming, developing countries are unlikely to take unilateral action to reduce their own emissions. For a number of reasons, the costs to them of preventive policies are probably higher than the costs to developed countries.

It may or may not be true that economies go through various stages of development – and that the transition from agriculture to industry is inevitably energy-intensive. Developing countries need not necessarily follow as energy-intensive a path as did the now-industrialized countries if new, more efficient technologies are available. However, to the extent that it is more difficult to reduce energy use while industrializing, the apparent costs of measures to reduce emissions in developing countries could be high indeed in terms of foregone structural economic change.

There seems little doubt that developing countries have a higher opportunity cost of capital, and thus higher relative costs of investing in preventive action. For example, social rates of return to primary education are estimated to be as high as 24% in developing countries.[5]

The greater poverty that is the distinguishing characteristic of developing countries as a group implies that a higher discount rate is applied to all calculations of costs and benefits of various actions. This higher discount rate implies a higher relative cost to any preventive action that would be taken. At lower income levels, willingness to pay to preserve amenities and to reduce uncertainty is lower.

Finally, the weakness of the public sector, noted above, implies that the administrative and political costs of preventive action would be higher in developing countries. A limited supply of administrative and regulatory skills also implies higher costs of enforcing any policies that are adopted to reduce emissions.

So, despite relatively higher welfare losses that might be associated with global warming in developing countries, those countries have little incentive to act unilaterally to reduce sources of global warming. Without some explicit form of international cooperation, the developing countries are likely to be free-riders on any policy steps taken unilaterally in developed countries.

The Siting and Burden-Sharing Questions

The questions of siting and burden-sharing can of course be treated separately. The bulk of actions to reduce emissions could take place in developing countries, on the basis of financial transfers from developed countries. Does this approach make sense?

The Siting Question

It should be clear from the data in the first section that, if accumulated emissions are not to grow beyond, say, an additional 20%, there must be some reduction from the path of increasing emissions on which the developing countries are now set. The point is illustrated in Figure 1, which illustrates in highly notional terms the paths of total and per capita emissions for the two groups of countries. As the bottom half of the figure shows, it is unlikely that per capita emissions in developing countries will, even well into the next century, have risen to the projected per capita level for developed countries by that time, even assuming a decline in per capita emissions in the developed countries. However, even with the relatively low per capita emissions in the developing countries, because population size and growth are so much greater in those countries, achievement of a reduction in total emissions at the global level does depend heavily on whether their per capita emissions rise at a relatively slow or a relatively rapid rate.

The Burden-Sharing Question

On the one hand, there should be room for trade in global environmental services between developed and developing countries. Generally opportunities for trade in environmental services exist if the marginal costs of reducing emissions in one country are below the marginal benefits of such reduced emissions to another country, and below the marginal costs of the next cheapest solution within the second country.[6] There probably are at least some least-cost approaches to reducing emissions in developing countries – for example it can be shown that reduced burning of forests in the Amazon would reduce carbon emissions at much lower cost than reducing emissions in developed countries through a carbon tax.[7] At the same time, the demand for reduced emissions is much higher in developed countries, implying room for purchases of the global "service" of reduced emissions by the developed countries from the developing countries.

On the other hand, the scope for such trade, and thus for burden-sharing through mutually desirable "purchases" or "transfers," or through mutually agreed treaty arrangements, is limited. To understand why, consider a simple taxonomy of environmental policies that might reduce emissions of greenhouse gases in developing countries:

Type 1

These are "win-win" policies, i.e. policies that would yield a positive economic rate of return even without taking into account any national or global environmental benefits – and which also have direct environmental benefits at the national level. One example is increased fuel prices in Egypt, which would raise revenues relieving the fiscal burden, would reduce energy inefficiency with positive allocational effects as well as direct savings, and would simultaneously reduce emissions and local air pollution by some combination of increased energy efficiency and reduced overall energy consumption. A second example would be increased electricity prices in Brazil (where one estimate is that a $10 billion investment in end-use efficiency could offset demand for new capacity the investment cost of which would be $40 billion),[8] which would similarly have economic and environmental benefits at the national level. Another would be elimination of subsidies provided to some agricultural inputs (especially water, fertilizer and chemicals), which would discourage overuse of these inputs and reduce the financial drain on the economy (because of the subsidy), and reduce air, soil and water pollution. All of these policy changes would also have some benefits at the global level, by reducing consumption, production and thus emissions of greenhouse gases.

Presumably, political and institutional constraints and macroeconomic problems inhibit these win-win (even at the national level) policy changes. It is not at all clear that specific transfers under a treaty arrangement, in the manner of the Montreal Protocol Interim Multilateral Fund for financing reductions of ozone-depleting substances, could affect these constraints. Indeed, availability of such specific tied transfers could have a perverse effect – inducing countries to delay win-win reforms in the hope that such reforms would later be subsidized.[9] The existence of these constraints may, however, provide some logic for general development assistance tied to pricing and other policy reforms that are anyway in the interests of the countries concerned – much as is already done through the auspices of the World Bank and the bilateral donors.

Type 2

These are policies, also with global benefits, that yield a positive economic rate of return only once environmental externalities at the country level have been taken into account. Examples include pollution taxes (at some cost to economic growth) which have a positive return because they will reduce the local health costs of air pollution (note that the health benefits of reduced pollution would be calculated based on the local, not the global, demand for health), and policies to reduce deforestation which imply some cost in foregone income but an overall positive return due to watershed protection.

As is the case with type 1 policies, there is no obvious room for trade via "purchases" of services, nor a rationale for transfers from developed to developing countries specifically tied to reduced emissions of greenhouse gases under a negotiated treaty arrangement. Given that such policies are in the best interests of the countries anyway, the same issue of perverse incentives arises as with type 1 policies. There may be some rationale for including, in the envelope of development assistance, support for reform of national environmental policy and support for strengthening of environmental institutions in order to hasten the process of policy development and enforcement. Donors may see some merit in affording this kind of assistance higher priority than standard assistance, for example that aimed directly at the alleviation of poverty, because of their own direct interest in the reduction of emissions. However, the difficulty of implementing an approach to development assistance based on such overt preferences should be obvious from the comparison of aid for environment and aid for poverty alleviation; such preferences seem to undercut the fundamental basis for development assistance itself, which presumably includes improvement of the welfare of the citizens of poor countries.

Type 3

These are policies with benefits at the global level which yield a positive economic rate of return only when those global benefits are taken into account. At the national level, their economic returns, even taking into account environmental benefits at the national level, would be insufficient to warrant national-level policy change. Examples would be switching from coal to natural gas in Poland where coal is plentiful, the use of solar and biomass energy before these are economically viable, and carbon sequestration through reduced deforestation even when deforestation would be economic. In these examples, trade in environmental services is possible, and arrangements for international transfers on an agreed basis are analogously likely.

Indeed, such arrangements are already in place at a modest level in the form of the recently created Global Environmental Facility (GEF). This Facility is specifically directed to financing policies and programs in developing countries which would not otherwise be financed because their returns occur at the global level, i.e. they are not clearly in the interests of the recipient country itself relative to alternative investments.

Although there is some scope for burden-sharing, there is also a compelling argument for starting with the "win-win" policy changes described above. They make good economic sense, they help to change the pattern of resource use, they lead to more efficient use of available resources, and they can lead to measurable greenhouse gas emission reduction benefits. Not the least, these policy changes are probably implementable and will buy time while we improve understanding of the actual likely costs, if any, of global warming, and of the appropriate level of preventive effort.

In addition, there is an important contextual issue. In a world where many policy distortions already exist, it is necessary to lay a solid foundation before attempting to leap to a program of global siting decisions based on international burden sharing. At a minimum, therefore, countries should enact the first set of policy recommendations and begin to introduce the second set, those based on incorporation of national level externalities. All of these changes will mean increased private costs for energy and other forms of resource use but will begin to establish an economic framework within which global cooperation is both feasible and not unduly burdensome.

If the first set of "win-win" policy changes can be introduced, the second and third policy recommendations can be enacted more slowly, as increased information helps to build a global constituency for greenhouse gas reductions. This process will take time but some delay may well allow us to make better decisions and wiser investments.

Ironically, this approach to the matter seems to return us ineluctably to the more general issues of development, and the integration of environmental concerns into the development policies of developing countries – and thus to the difficult and messy areas of development assistance rather than the negotiation of burden-sharing agreements.

Table 1. Shares of Population, GNP and Carbon Emissions by Region, 1986 (in Percent)

	GNP	Population	Emissions
Developed	74.8	15.6	50.7
Developing	22.2	80.6	42.8
Oil Exporters	3.0	3.7	6.4

Source: Whalley and Wigle (1989)

Note: Per capita emissions in developed countries are about 6 times those of the rest of the world combined.

Table 2. Shares of Carbon Emissions, 1960, 1988 (In Percent)

	1960	1988
Developed countries a/	54	39
USA	31	22
Soviet Union, Poland	17	21
China, India, Mexico	10	15

Source: Boden, et al., 1990.

a/ In order of 1960 carbon emissions: United States, greater Germany, United Kingdom, France, Japan, Italy.

Table 3. Total Carbon Emissions,a/ Selected Countries, 1960, 1988

	1960	1988
United States	800	1,310
Greater Germany b/	221	272
United Kingdom	161	153
France	75	87
Japan	64	270
Canada	53	119
Italy	30	98
USSR	396	1,086
Poland	55	125
China	215	609
India	33	164
Mexico	17	84
Brazil	13	55
Korea	4	56
TOTAL GLOBAL	2,586	5,893

Source: Boden et al., 1990.
a/ Millions of metric tons of fossil fuels.
b/ The percentage increase was similar in East and West Germany.

Table 4. Absolute and Per Capita Carbon Emissions,a/ 1960, 1988

	Absolute a/		Per capita b/	
	1960	1988	1960	1988
United States	800	1,310	4.40	5.30
United Kingdom	161	153	3.10	2.70
West Germany	149	183	2.70	3.00
China	215	609	0.33	0.56
India	33	164	0.10	0.20
Brazil	13	55	0.20	0.40
Korea	4	56	0.10	1.30
TOTAL GLOBAL	2,586	5,893	0.90	1.20

Source: Boden, et al., 1990.
a/ Millions of metric tons.
b/ Metric tons.

Table 5. **FOSSIL FUEL CARBON EMISSIONS, 1987**

	TONS/YEAR PER CAPITA	GRAMS PER $GNP
U.S.	5.03	276
CANADA	4.24	247
AUSTRALIA	4.00	320
SOVIET UNION	3.68	436
POLAND	3.38	492
W. GERMANY	2.98	223
JAPAN	2.12	156
FRANCE	1.70	133
S. KOREA	1.14	374
MEXICO	0.96	609
CHINA	0.56	2,024
EGYPT	0.41	801
BRAZIL	0.38	170
INDIA	0.19	655
NIGERIA	0.09	359

Source: Flavin (1990).

Table 6. Agriculture Share in Labor Force (1980), Agricultural Share in GDP (1988) and Per Capita GNP (1988), Selected Countries

	Agriculture Labor Force Share (%)	Agricultural Share of GDP (%)	Per Capita GNP ($)
United States	4	2	19,840
West Germany	6	2	18,480
United Kingdom	3	2	12,810
France	9	4	16,090
Japan	11	3	21,020
Canada	5	4	16,960
Italy	12	4	13,330
Poland	29	--	1,860
China	74	32	330
India	70	32	340
Mexico	37	9	1,760
Brazil	31	9	2,160
Korea	36	11	3,600

Source: World Development Report 1988, 1990.

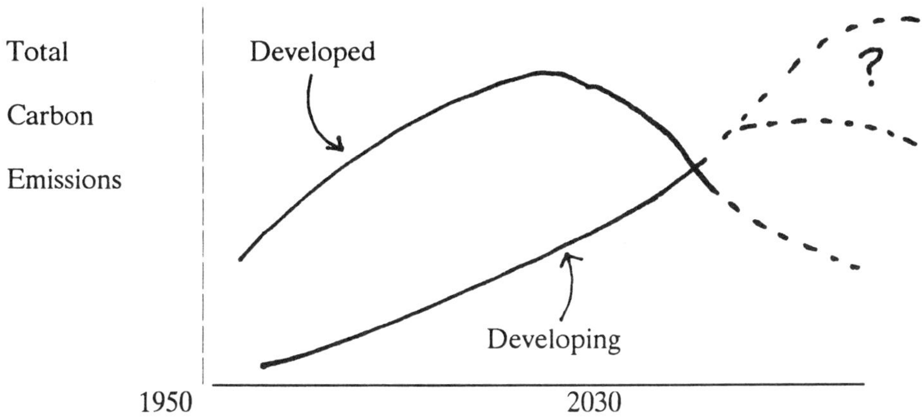

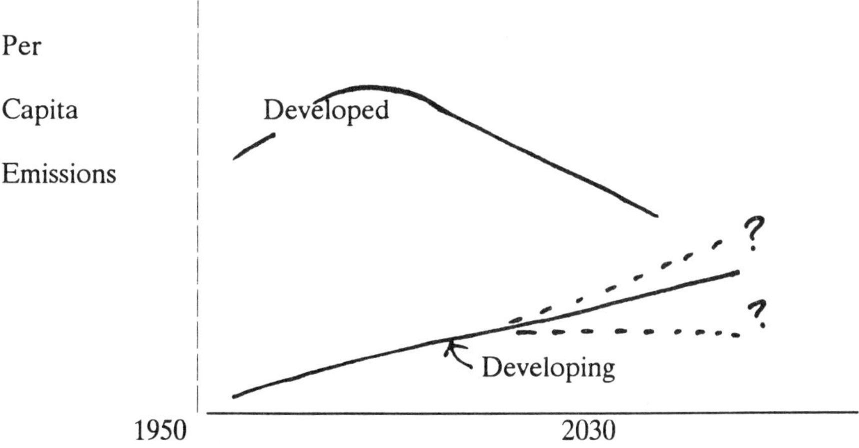

Figure 1.

Notional Total and Per Capita Carbon Emissions, Developed and Developing Countries, 1950 to 2030

Notes

1. But see, for example, Nordhaus, 1990.

2. Eckaus (1990) points out that because stock matters and accumulates slowly, the conventional procedure of computing present discounted values to assess the present significance of a future event makes little sense. Any positive discount rate would imply that global warming is not a problem. However, that implies putting off action to reduce emissions until a time when that action would make no difference (and might be difficult and much more costly to undertake). To get around this problem, some analysts have set some fixed future constraint on accumulated stock and then modeled how fast the stock should grow given that constraint and other factors (e.g. Nordhaus, 1977).

3. Krueger (1974) has characterized the problem as it affects development. For a discussion of effects on social programs in developing countries, see Birdsall and James (1990).

4. The recent rapid spread of cholera in Peru and neighboring countries is an example of the vulnerability of poor populations to external shocks – in this case the introduction of a new pathogen.

5. See Psacharopoulos, Tan and Jimenez, 1986.

6. Oates (1990) sets out this point formally.

7. Schneider (1991) uses Nordhaus's estimate that a reasonable expenditure to reduce carbon emissions, given the costs in economic growth, is $3-$13 per ton to show that the equivalent reduction in emissions can be obtained at much lower cost by preventing forest burning in the Amazon. Nordhaus (1990) also concludes that reducing deforestation is among the lowest-cost options at the global level for reducing emissions.

8. Arrhenius and Waltz, 1990.

9. We are grateful to our colleague Charles Blitzer for pointing this out.

Acknowledgments

The authors are, respectively, Chief and Economist, Environment Division, Latin America and Caribbean Region, World Bank. The views in this paper do not necessarily reflect the views of the World Bank. The authors are grateful to Robert Anderson and Charles Blitzer for comments on an earlier draft and to participants, particularly William Clark, in this conference and to participants at a presentation at Tulane University in April, 1991.

References

Arrhenius, E. and T.W. Waltz (1990). *The Greenhouse Effect: Implications for Economic Development*, World Bank Discussion paper No. 78. Washington, DC: The World Bank.

Birdsall, N. and E. James (1990). *Efficiency and Equity in Social Spending: How and Why Governments Misbehave*, revised version of World Bank Working Paper Series No. 294. Washington, DC: The World Bank.

Boden, T. A., P. Kanciruk, and M. P. Farrell (1990). *Trends 90: A Compendium of Data on Global Change*, Carbon Dioxide Information Analysis Center. Oak Ridge, TN: Oak Ridge National Laboratory.

Eckaus, R. S. (1991). *Economic Issues in Greenhouse Warming*, Center for Energy Policy Research. Cambridge, MA: M.I.T.

Flavin, C. (1990). "Slowing Global Warming" in L.R. Brown, ed. *State of the World 1990*. New York: W.W. Norton and Co.

Krueger, A. (1974). "The Political Economy of the Rent-Seeking Society," *American Economic Review*, 64, 3 (June), pp. 291-303.

Nordhaus, W.D. (1977). "Economic Growth and Climate: The Carbon Dioxide Problem," *American Economic Review*, 67, 1 (February), pp. 341-346.

Nordhaus, W. D. (1990). "Economic Approaches to Greenhouse Warming," paper prepared for a conference on *Economic Policy Response to Global Warming*, 4-6 October, Rome.

Oates, W.E. (1990). *Reflections on an "Optimal Treaty" for the Global Commons Involving Industrialized and Developing Nations*, mimeo. College Park, MD: University of Maryland.

Psacharopoulos, G., J.P. Tan and E. Jimenez (1986). *Financing Education in Developing Countries*. Washington, DC: The World Bank.

Schneider, R. (1991). *Brazil: An Economic Analysis of Environmental Problems in the Amazon*, draft report, Country Department 1. Washington, DC: The World Bank.

World Development Report, 1988 and *1990*. Washington, DC: The World Bank.

Whalley, J and R. Wigle (1989). "Cutting CO_2 Emissions: The Effects of Alternative Policy Approaches," background paper for the *World Bank Seminar on Global Warming*, 25 May. Washington, DC: The World Bank.

WHY CLIMATE CHANGE IS NOT A COST/BENEFIT PROBLEM

Peter G. Brown

University of Maryland
School of Public Affairs
College Park, Maryland 20742

Introduction

It is widely assumed by economists that the appropriate way to think about policies with respect to global climate change is through the application of cost/benefit techniques. Economists differ, however, on the feasibility of such analysis. Some, for example, William Nordhaus, think that it is possible to come up with net benefit calculations that are meaningful in policy discussions.[1] Others, such as Manne and Richels, emphasize how hard it is to come up with numerical estimates of benefits, and as a consequence concentrate solely on the costs of various measures to retard the release of heat-trapping gases.[2] Epstein and Gupta go further and assert that while, in principle, the problem can be framed in cost/benefit terms, "for the moment, it is not feasible, because of profound uncertainties surrounding the MB (marginal benefits) and, for different reasons, the MC (marginal cost) curves."[3]

In contrast, I believe that attempting to derive principles for planetary management—with implications for centuries to come—with cost/benefit techniques is infeasible and inappropriate. My argument has three parts. First, I agree with those who contend that the techniques are too ambiguous and imprecise at this time for use in deciding the issues associated with global climate change. This is due to problems of execution. Second, I argue that there are at least five reasons of principle that should lead us to distrust this framework. These I call problems of conception. Third, I argue that the problem should be thought of in terms of a planetary trust. I defend a version of this alternative by drawing on some of the moral implications of Darwinism.

I am not opposed to using cost/benefit analysis to assist many decisions, especially those in the private sector business context. It can be an appropriate tool when the consequences of the decision play out in a relatively short time, say 20 to 25 years. Even here, however, I hold that the "rules of the game" should frequently be decided on other grounds. In these cases, as in the case of stabilizing climate, the tools of the economist would be put to most fruitful use by assessing options to find the least-cost approach to meet objectives that have been otherwise determined.

Copyright 1991 by Elsevier Science Publishing Company, Inc.
Global Climate Change: The Economic Costs of Mitigation and Adaptation
James C. White, Editor

Problems of Execution

The first problem with cost/benefit analysis is deciding what to measure. It is fashionable to use changes in GNP as a measure of costs and benefits; but is GNP an appropriate measure of, or surrogate for, human well-being? A large and growing literature argues that GNP and similar current measures of economic activity, such as national income accounts, leave out a whole spectrum of welfare values. For example, *Blueprint for a Green Economy* (often referred to as the "Pearce report"), argues that conventional marginal cost pricing undervalues both the welfare externalities involved and the depletion of the resource in question.[4] A great variety of techniques are discussed that would improve valuation, none of which is news to economists. The range and plausibility of these alternatives do lend support to those who question the feasibility of deriving clear thinking about large-scale environmental problems on the basis of net weighted costs alone. At the least, pending a consensus regarding a proper measure of well-being, the issue should be held open.[5]

A second problem is created by the possibility of "surprises." Natural systems are subject to step-function changes in the behavior and interactions of their components. Were we to assume perfect definitions and measurements of costs and benefits, we would still have no way of knowing whether or not climate change would trigger some sharp change in temperature, rainfall or ocean currents that would result in very adverse ecological and economic consequences. We cannot physically experiment with the earth's systems so our probability estimates of these catastrophic (or perhaps partially beneficial) events are only as good as the models used today—which have a wide band of uncertainty. Of course, these models are likely to improve somewhat over time. Whether they will ever be good enough to legitimate playing Russian roulette with the climate remains to be seen. These uncertainties push many observers to advocate buying "greenhouse insurance." For example, we might hedge on the likelihood of climate related catastrophe by reducing heat-trapping gases. We will return to this subject later.

A third and related problem has to do with time horizons. The discussion of global climate change is often framed in terms of the next 35 to 100 years. Sometime during this period, the atmosphere will contain double the amount of gas, in carbon-dioxide equivalents, as it did at the beginning of the Industrial Revolution. Given the massive reserves of fossil fuels, principally coal, and the rapid increase in the human population that is all but certain to occur throughout this period, this "doubling" convention is seriously short-sighted. The discussions seem to imply, by almost universal failure to examine a longer time period, that the climate will somehow stabilize in the middle of the next century or that doubling will be a *de facto* ceiling. As William Cline has pointed out, it is far more fitting to think of the effects of atmospheric gases on the climate in terms of 200 or 300 years.[6] Over this longer term the accumulation of gases, absent policy-induced limitations, could lead to temperature increases of ten degrees centigrade, or even more. The magnitude of such a change could affect agriculture, ocean currents and levels, and so on, in innumerable ways that are not envisioned in the smooth transition to a slightly warmer climate that the shorter-term scenarios would have us accept.

A fourth problem of execution concerns technological change. How fast, for example, will alternatives to fossil fuel sources of energy be developed and at what cost? Because of the long lag time associated with climate change we need to know answers to this kind of question over very long periods in order for C/B techniques to work. If the technological optimists are right, our ability to design low-carbon and carbon-free ways of meeting our energy needs should be virtually assured. As a result, the linkage between GNP and fossil fuels could be very considerably weakened; and in principle it could be eliminated. Whether we will ever have the capacity to compute these figures in the future, or even chart trends, remains to be seen. No one, to my knowledge, argues that we know this now.

The long time horizon feature of climate change has had a very unfortunate impact on certain aspects of the debate. The costs of action occur in the present, while the benefits of various policies are spread out in the future. This makes the benefits harder to estimate than the costs. Some analysts, like Manne and Richels for example, don't even try to estimate the benefits. Costs are not then offset against benefits. It is then reported by the press that the costs of averting climate change are "staggering."[7] They are compared then with some other very large number like national defense and the inference is drawn that this is simply too much to pay, and hence is politically infeasible. This is policymaking by *big* numbers. But the opposite inference is more reasonable. In my argument that follows I will show that what these studies demonstrate is that the costs of averting and/or slowing climate change are quite small – and easily within our reach. All we really lack in the world is will and, in the United States, leadership.

Problems of Conception

I doubt that the problems of executing cost/benefit analysis that I have identified can be resolved in the near future. However, even if resolution were possible there are problems of principle that make using these techniques inappropriate. Therefore, we should proceed to respond to climate change whether or not we believe that the technical problems of execution can be solved.

Discounting

The first set of conceptual problems that I wish to identify is associated with discounting. There are three very serious problems here.

1. There are some things that are not and should not be discounted. No one asks: "What is the optimal rate of using up the national parks like Yellowstone and Yosemite? At what rate should we mine the topsoil at Gettysburg where our ancestors gave their "...last full measure of devotion"? On the contrary, we take it as our duty that we should preserve these things for posterity. This is precisely analogous to what many people think we should do with respect to the earth's ecosystems which are seriously threatened by rapid climate change. Using C/B

techniques as comprehensive instruments of planetary management assumes the whole issue away.

2. Another problem is that it is assumed that there should be a single discount rate.[8] But an all-encompassing discount rate actually covers up a number of quite different factors. For example, one of these is opportunity costs; another is uncertainty. Merging these two under one discount rate simply obscures what is at issue. We would reveal what is at stake much more clearly by talking about factors like opportunity costs and uncertainty separately.[9]

3. In any situation where there is a long-term temporal asymmetry between costs and benefits, such as global warming, discounting *imperils the future by undervaluing it.* The costs of averting the greenhouse effect occur in the present, but the benefits accrue in the distant future. The discounted value of harms that occur in 100 years are insignificant when compared with the present costs of avoiding them. As D'Arge, Schulze and Brookshire argue in *Carbon Dioxide and Intergenerational Choice*, "...a complete loss of the world's GNP 100 years from now would be worth about one million dollars today if discounted by the present prime rate".[10]

Differences In Kind

To apply for C/B techniques everything has to be subject to the same measure, preferably money. This is a perfectly reasonable requirement over relatively narrow ranges and with certain kinds of goods – e.g. a tractor contrasted with a new silo. But it is foolish to think that literally everything under the sun should be subject to the measure of money. Things are different in kind. Nor is there anything especially mysterious about this. For instance, we don't buy and sell public office. (To the degree that our institutions seem to be legitimating such sales they are appropriately criticized.) To see the folly, take the idea that public office should be bought and sold to its logical conclusion. Suppose that the Office of the President could be sold at any time. When any aspirant met the reservation price of the current office holder he/she would be sworn in. The obvious difficulty with this system is that there is no reason to believe that those who would pay the most for the office would discharge it responsibly.

The Congressional Metal of Honor cannot be bought, nor a chair in the Economics Department at the University of Chicago, at any price. New ivory cannot be legally purchased, nor can a new car without a seatbelt.[11] As I will show when I discuss the planetary trust, like many other things the earth's ecosystems are not simply another production factor to be subject to optimization procedures.

Distributive Issues

C/B techniques have no way to deal with distributive issues such as the unequal ability of countries to adapt to climate change. As a rule, technologically advanced countries will be better able to avoid and/or mitigate the negative aspects of climate change. For example, nations whose GNP is derived largely from their agricultural sectors could be especially hard hit if climate change disrupts production patterns. These nations may have few technical options open to them and limited access to credit markets to help mitigate the damage. Yet because of issues of the declining marginal utility of money, the "utility loss" will be magnified drastically. The C/B framework itself has nothing to say about such inequalities, except indirectly through interdependent utility functions – an avenue not developed in the economically oriented climate change literature. Yet it seems inappropriate in the extreme to try to assess what to do about global warming without assessing the impact of what we do, or fail to do, for those who are already, or will be, at the margins of subsistence.

Rationality

It is often asserted that the knowledge revealed by the application of C/B techniques must in Nordhaus' words "...underpin any rational decision".[12] This rests on a confusion about what rationality is. Any instrumental definition of rationality has two parts: an end, specified somehow; and means that are to be conserved in reaching the end. Maximizing discounted present value is one end; but in the context of climate change, stewardship of the earth and its ecosystems would seem to be a reasonable end to have in view. The literature of cost/benefit analysis confuses *a* definition of rationality with *the* definition.

This rationality paradigm in question is currently under attack from two directions. First, "ecological economics" and "sustainable development" protagonists argue that none of the models of the prevailing economic paradigm "...used by Washington, D.C., policymakers is consistent with the basic principles of physics, chemistry or biology".[13] In addition, these thinkers question whether the human species is reaching the limits of exploiting this planet.[14] Second, "socio-economics" argues that the neoclassical paradigm overlooks the role of values, culture and community in shaping human self-understanding and action.[15]

The Scope of Benefits

C/B techniques can shed no light on the question: costs and benefits *to whom*? It is usually assumed that the answer to this question is: presently existing persons taking into account future persons through the preferences of those now living. But aside from all the difficulties associated with discounting and with the intragenerational distributive issues among persons as discussed above, it is by no means clear that persons are the only relevant entities. The progenitors of modern C/B techniques, the classical utilitarians Bentham and Mill, held that the trait that makes something morally relevant is the ability to feel pleasure and pain, or happiness and unhappiness. On this basis, it is clear that the moral world is not bounded by concern with the human race. Animals obviously can experience these emotions, and if this is the touchstone of moral concern then they very clearly count. In *Animal Liberation*, Australian philosopher Peter Singer has drawn precisely this conclusion.[16]

Modern economists have tried to distance themselves from some of the difficulties associated with classical utilitarianism in part by substituting the notion of revealed preferences for the words "pain/pleasure" and "happiness/unhappiness." But this makes it even easier to grant animals moral standing. The ape reveals his preference for bananas by eating them, paramecium for warmth by moving to the other end of the petri dish. Along this dimension the revealed preference move really opens up the floodgates for broadening our concerns beyond humans. Perhaps this seemingly absurd conclusion has been avoided by an *ad hoc* marriage between theological notions about human dignity resting on man being created in God's image with the pleasure/pain principle – thus limiting the scope of the principle to humans.[17] As we will see the "created in God's image" way of bounding the moral community is undercut, though not conclusively disproved, by the findings of Charles Darwin.

The Planetary Trust

It is paradoxical that neoclassical economists would be interested in the climate change issue at all. The more one takes the axioms of the model concerning self-interest and discounting to heart, the more it is unclear why there should be any interest in a temporally distant problem. Indeed, the fact that the problem is studied at all shows that another model concerning the well-being of future generations has been invoked at a deeper and not explicitly recognized level. Those who argue for an insurance approach to climate change have already embraced a fiduciary or trustee conception of our responsibilities for the future. The argument is that we have an obligation; not contingent upon any discounted present value, to protect those who will come after us from the risks of rapid climate change. The implicit argument is that a person's position in time is irrelevant to moral obligation; we should protect future persons from risks we ourselves do not wish to bear. As the trustee conception is normally interpreted we have assumed a responsibility for passing on a world at least as good as we found. In what follows I will explicate this model, address in a very preliminary way the issue of moral standing raised above, argue that on the basis of an expanded conception of community we have an obligation to stabilize the climate, and show that such a move requires neither a substantial revision of our moral and religious beliefs (though it may imply substantial revision); nor is it likely to be especially economically onerous.

Edith Brown Weiss has worked out the features of the fiduciary model in her article "The Planetary Trust: Conservation and Intergenerational Equity."[18] In this model each generation has a responsibility to those who follow to preserve the earth's natural and human heritage at a level at least as good as what is received. The responsibility includes the obligation to preserve certain definite things, for example, intact ecosystems and national monuments. Our obligations are not discharged simply by achieving the highest discounted present value of consumption.[19] Of course, a certain level of wealth may be necessary to enable us to conserve certain resources, but wealth alone is not sufficient in discharging our obligations. Carefully constructed models that predict the economic consequences of various policies can help us determine the effects on aggregate wealth; but while these results could be illuminating they should not be taken to be conclusive.

As presented by Weiss the fiduciary trust framework imposes two duties on each living generation. One is the conservation of options so that future generations can survive and pursue their own visions of the good life. To ensure this we are obligated to preserve biological diversity so that the benefits of a diverse gene pool will be available to our descendants. We should conserve natural resources such as groundwater and/or invest in substitutes so as to leave the future as well off with respect to the resource as we are. Lastly, under this duty we should preserve our cultural heritage so that future persons can have enriched lives and learn from our mistakes.

The second duty concerns the conservation of quality and requires that we leave the world in at least as good a condition as we have found it. We discharge this duty in two ways. One is by conserving natural resources and/or investing in substitutes so that the real prices of those resources necessary for life do not rise in the future. The other duty is to conserve the unique natural resources of the world, such as places of unusual beauty, so that they can be enjoyed by our successors in the same state as they are enjoyed by us.

In her *In Fairness to Future Generations: International Law, Common Patrimony, and Intergenerational Equity,*[20] Dr. Weiss traces the implications of this concept of climate change. She argues that it imposes three specific duties on those who live in the present. First, it calls for measures to prevent rapid climate change such as reduction in practices which emit heat trapping gases into the atmosphere. "We need to evaluate these strategies against the normative goals of ensuring that our descendants have access to a planet with diversity and quality comparable to prior generations."[21]

Second, to the degree that we fail to prevent rapid climate change we need to take steps to minimize the damage that results from it. This category focuses on issues of irrevocable loss. Examples of steps we should take "...include gathering and conserving the germplasm for additional crops now neglected, and conserving the knowledge of traditional peoples of the utility of certain plants and animals, of ecosystems, and of practices adapted to harsh climate conditions."[22] She also advocates coastal zone management, particularly with respect to hazardous wastes and nuclear power plants. Third, we have an obligation to develop strategies to assist future generations in adapting to climate change. This category emphasizes the tools and planning we should provide our descendants. These would include monitoring changes in climate, long-term planning of water supplies, attempts to influence population settlement patterns and the like.

Let us now turn to the issue of the larger community, nascent but undeveloped in the writings of the classical utilitarians. It raises the question: who are the beneficiaries of the trust? Are the beneficiaries humans only, or humans and the rest of the biosphere?

Once we look at the world the way Darwin suggests, the logic for unique moral standing for humans collapses. The burden of Darwin's argument in *The Descent of Man*[23] is that humans differ from the other animals only in degrees. Whatever traits you can find in persons you can also find somewhere else in the animal kingdom. There is nothing that belongs to man alone by virtue of which humans can claim to be the only entities to whom morality applies. Darwin explicitly notes that "there is no evidence of a separate creation." If we are not divinely created and we can find no naturalistic way of distinquishing ourselves then there is no reason to think that we have exclusive moral standing.[24]

Some moral concern would seem to apply to all animals and, at least indirectly, to plants upon which animals depend. To the extent that rapid climate change is likely to undermine the stability and diversity of ecological systems, this framework imposes a strong duty on us to stabilize climate. "Stabilize climate" can be defined in this context as: that rate of change at which climate change is not a factor in the extinction of species. Many believe that climate change is a drastic threat to biological diversity. In *World Resources 1990-91*, for example, it says:

> Simply stated, if the world climate changes as fast as some scientists predict, many species would not be able to adapt quickly enough, and extinctions would be widespread because habitats would have shifted, shrunk or disappeared.[25]

Rapid climate change is likely to cause and/or accelerate extinctions in four ways: 1) the poleward migration of climate zones at rates faster than species can move; 2) changes in the wet/dry cycles in the tropics; 3) the loss of wetlands due to rapid sea level rise with re-establishment blocked by human diking; and 4) increases in ocean surface water temperatures disrupting the life cycles of species associated with coral reefs and ocean currents. These climate/species mechanisms suggest that all the arguments used to support the protection of endangered species can be marshalled to urge the protection of a stable climate.

If moral concern is not limited to persons one can reasonably ask: what do we owe the nonhuman members of the community? I do not propose a full-blown answer to this question here. But if we owe them anything, as I think my argument so far indicates we do, then we owe them existence. If they do not exist then we cannot discharge any of our other obligations to them. Accordingly, we have an obligation not to destroy or abruptly alter the ecosystems that the nonhuman members of the community inhabit and depend on. We should move to stabilize the climate both to preserve biological diversity for our descendants, and for what we owe plant and animal communities as such. As noted above, by "stabilize" I mean not cause the climate to change faster than the world's ecosystems can adapt to the change. Of course, this is not the only reason we should do this – all the insurance arguments associated with human well-being also apply.

But what if there is a trade-off between the expansion of human well-being and the preservation of species? As I have already suggested above human life is not a consumption, income or even utility optimization endeavor. We accept a great variety of constraints on such maximizing. Concern with stabilizing climate and protecting endangered species is simply added to a long and somewhat fluid list.

It might be argued that the goal of stabilizing climate is unrealistic for three reasons – one theological, one moral, and one economic. First, it might appear to require that persons give up one of the central elements of Judeo-Christian faith – human dominion over nature – and that this is just plain unlikely. Second, it might be argued that it requires too great a revision of our moral schemes, that our morality is too completely anthropocentric. Third, it might be argued that it costs too much, that we cannot afford to do it.

However, the theological tradition is rich and diverse – recall the Yahwist aspects of the creation story in the second book of Genesis, and other elements of the Judeo-Christian tradition that emphasize man's stewardship of nature. Second, our morality is already moving in the direction of a broader understanding of community. Indeed, it is possible to trace a long secular movement in the direction of giving moral standing to animals as the establishment of the first humane society in Great Britian in 1824 attests.[26] The Endangered Species Act and the Marine Mammals Protection Act are simply extensions of a longstanding historical trend.

The concern that stabilizing climate is too expensive can be called into question by appealing to the very findings of the C/B analysts. We can use their figures in a small numbers game – comparing the costs involved to those spent in cigarettes, or the arms trade, or beer, or any other set of expenditure figures generally negatively correlated with human well-being.

In "CO_2 Emission Limits: An Economic Cost Analysis for the USA," Manne and Richels "...investigate the costs of restricting carbon emissions to...their 1990 rate...through 2000, reducing them gradually to 80% of this level by 2020, and stabilizing them thereafter."[27] Here is what they say about their findings:

> The effects of a carbon constraint do not begin to have measurable macroeconomic consequences until 2010. At that point the rise in energy prices begins to have a significant effect upon the share of gross output available for current consumption. By 2030, roughly 5% of total annual macroeconomic consumption is lost as a result of the carbon constraint.[28]

But Manne and Richels are quick to point out that these very high cost figures are uncertain and can themselves be altered through policy:

> If emission controls are required there will be significant costs, but it is clear that the nation can reduce the size of the ultimate bill though R&D on both the supply and demand sides of the energy sector. According to our calculations, the combination of potential improvements could reduce the costs of a carbon constraint – perhaps by several trillion dollars.[29]

Of course, once we calculate the benefits of a stable climate the numbers become smaller still. A reasonable inference to draw from this is that we simply do not know the net costs of stabilizing climate in the intermediate term – between now and the year 2100.

Conclusion

What we do know is that taking the first steps in this direction is not likely to have any net costs at all. And this is all that is now being asked of us as we move toward a climate convention. Indeed, it might be cheaper to avert climate change than to let it occur. Wherever the numbers come out in the range we are discussing, securing the well-being of our descendants on this planet is worth at least this much.

Indeed, the inordinate attention given to the issue of the costs of stabilizing climate is diverting much needed attention from a far more presssing matter: stabilizing the human population. Human population growth is a key factor in destabilizing the biosphere and the ecosystems that make it up. The time is long past due when we should reorder our economies from growth to sustainability and our fertility patterns toward replacement or below.

Acknowledgments

This paper grows out of a working group on Equity and the Greenhouse Effect supported by the Ethics and Values Studies Program of the National Science Foundation. Other members of the group are William Moomaw, Peyton Young, Irving Mintzer and Lance Antrim. In addition, John Cumberland and Amanda Wolf provided especially useful comments. Defects remain my own.

References

1. See, for instance, Nordhaus' "How Fast Should We Graze the Global Commons?" *AEA Papers and Proceedings*, Vol. 72, No. 2, pp. 242-46.

2. Alan S. Manne and Richard G. Richels, "CO_2 Emission Limits: An Economic Cost Analysis for the USA" in the *Energy Journal*, April 1990, Vol. 11, Number 2.

3. Joshua M. Epstein and Raj Gupta, *Controlling the Greenhouse Effect: Five Global Regimes Compared* (Washington, D.C.: The Brookings Institution, 1990), p. 3.

4. David Pearce, Anil Markandya and Edward B. Barbier, *Blueprint for a Green Economy* (London: Earthscan Publications Ltd, 1989).

5. Stephen Piper, a graduate of the School of Public Affairs at the University of Maryland has reviewed the literature on national income accounts and concluded in a memorandum that: "a consensus is developing that the limitations of GNP as a measure of national income may cause distortions in national economic planning. In particular, the failure to account for the depletion of natural resources and reductions in environmental quality may give the impression of prosperity when, in fact, such prosperity cannot be sustained." His unpublished paper: "Revising National Income Accounting Measures to Reflect Natural Capital Depletion: A Review of Proposed Methodologies" considers six alternative accounting systems.

6. William R. Cline, "Economic Stakes of Global Warming in the Very Long Term." Unpublished paper.

7. Peter Passell, "Staggering Cost Is Foreseen To Curb Warming of the Earth," *The New York Times*, November 19, 1989, p. 1.

8. See Resources for the Future Position Paper *Intertemporal Discounting For the Fuel Cycle Social Cost Study* by Dallas Burtraw and Alan Krupnick, November 1990, for a discussion of the use of multiple discount rates.

9. Derek Parfit, "Energy Policy and the Further Future: The Social Discount Rate," in Douglas MacLean and Peter G. Brown, eds., *Energy and The Future* (Totowa, N.J.: Rowman and Littlefield, 1983), pp. 31-37.

10. Ralph C. D'Arge, William D. Schulze and David S. Brookshire, "Carbon Dioxide and Intergenerational Choice," in *AEA Papers and Proceedings*, Vol. 72, No. 2, p. 251.

11. See Walzer's *Spheres of Justice* (New York: Basic Books, 1983) for a discussion of this point.

12. Nordhaus, "To Slow or Not To Slow," paper presented to the February 1990 meeting of the American Association for the Advancement of Science.

13. Ralph C. d'Arge, "The Promise of Ecological Economics," *Journal of Soil and Water Conservation*, p. 430.

14. See Robert Costanza, "What is Ecological Economics" in *Ecological Economics*, 1, 1989, pp. 1-7.

15. See Amitai Etzioni, *The Moral Dimension* (New York: The Free Press, 1988).

16. Peter Singer, *Animal Liberation* (New York: New York Review Books, 1975).

17. This may not be the only theological underpinning brought in to shore up the neoclassical model. The notion of "consumer sovereignty" merges the Protestant idea of the unreviewability of individual belief with the purchasing behavior of suburban shoppers at K-Mart.

18. Edith Brown Weiss, "The Planetary Trust: Conservation and Intergenerational Equity," *Ecology Law Quarterly*, Vol II, 1984, No. 4. pp. 495-581.

19. Rawls' just savings principle has the implication that we have discharged our obligations to future generations if we have accumulated enough capital. See *A Theory of Justice*, pp. 284-293.

20. Edith Brown Weiss, *In Fairness to Future Generations: International Law, Common Patrimony, and Intergenerational Equity* (Dobbs Ferry, New York: Transnational Publishers, Inc., 1988).

21. *Ibid.*

22. *Ibid.*

23. Charles Darwin, *The Descent of Man, and Selection in Relation to Sex* (Princeton: Princeton University Press, 1981).

24. See James Rachels, *Created From Animals: The Moral Significance of Darwinism* (New York: Oxford University Press, 1990) for a brilliant discussion of this point.

25. *World Resources 1990-91*, p. 130.

26. Keith Thomas, *Man and the Natural World* (New York: Pantheon Books, 1983), p. 181. This work catalogues that steady decline in the anthropocentric morality of the West.

27. Manne and Richels, p. 65.

28. *Ibid.*, p. 68.

29. *Ibid.*, p. 73. Emphasis in original. See also Alan Miller's, Irving Mintzer's and my own "Rethinking the Economics of Global Warming" in *Issues in Science and Technology*, Vol. VII, No. 1, Fall 1990, pp. 70-73.

CO_2 AND SO_2: CONSISTENT POLICY MAKING IN A GREENHOUSE

Daniel J. Dudek, Alice M. LeBlanc, and Peter Miller

Environmental Defense Fund
1616 P Street, NW
Washington, DC 20036

Abstract

This past year has witnessed a resurgence of interest in problems of atmospheric pollution. Two broad-scale atmospheric problems have dominated the news: acid rain and global warming. Each has engendered significant national and international policy debate, but they have not been addressed as linked problems. Instead, the constituencies supporting or opposing various policy initiatives have pursued very narrowly focused objectives. This narrowed vision is a primary source of the complex and conflicting mandates of the current Clean Air Act. The reauthorized Act should be viewed as a major opportunity for reform which should begin with the recognition of trade-offs among environmental objectives. Chief among these opportunities are SO_2 control strategies which would also reduce CO_2, the primary greenhouse gas.

This report identifies and estimates the CO_2 implications of alternative acid rain control strategies. The primary competing control options under consideration are specific technologies such as flue gas desulfurization devices versus allowing each source to determine its least-cost solution including over-control at a smaller number of plants. While each option has its CO_2 effects, the market incentives provided by emissions trading would produce a pattern of pollution control investments which maximizes the reduction of CO_2 discharges. If the options assessed in this report are exploited and if the integrity of the market and its incentives are maintained, we expect a reduction of between 5.4-6.0% in utility CO_2 emissions otherwise expected to occur over the next 15 years. Each of the options analyzed, interfuel substitution, cogeneration, demand management, and energy efficiency investments, are economically and environmentally advantageous under emissions trading for acid rain control. The CO_2 reductions are a bonus.

As the United States enters the decade of the 90s, will policy makers rise to the challenge of developing tools for a new era of pollution reductions or will regional factionalism continue to shackle the future? The rest of the world is looking to the U.S. for its response to global environmental problems. The Clean Air Act is the first best chance

Copyright 1991 by Elsevier Science Publishing Company, Inc.
Global Climate Change: The Economic Costs of Mitigation and Adaptation
James C. White, Editor

to demonstrate the resolve to reduce our share of greenhouse pollution. We do not need to wait for the development of grand international agreements before progress on reducing greenhouse gases can be made. The market-based approach to controlling acid rain leads the way.

Introduction

In the absence of a comprehensive U.S. approach, energy policy is actually made by the Environmental Protection Agency under its mandates from the Clean Air Act (CAA). Because the emission of many atmospheric pollutants are the result of energy production decisions, the regulation of sulfur dioxide (SO_2), nitrogen oxides (NO_x), particulates, etc. directly affects those decisions. In the simplest terms, controlling these pollutants changes the relative costs between alternative energy sources, increasing the cost of those that produce greater emissions. These differential environmental control costs alter energy purchase and investment patterns changing both quantities of energy consumed and the method of production. Consequently, environmental policy influences energy production and use decisions more than any other factor except relative fuel prices.

The last twenty years has witnessed a continual evolution of the CAA both in terms of the pollutants that it controls and in the policies used to effect that control. In general, however, management has proceeded on a pollutant-by-pollutant basis without regard for any cross-pollutant effects. Ignoring these cross effects can have serious consequences for both the cost and performance of any environmental policy. This short paper will attempt to explore one particular set of pollutant trade-offs.

In the debate over reauthorization of the CAA, one of the most hotly contested titles concerned acid rain. The Bush Administration has proposed an innovative emissions trading program to reduce SO_2 and NO_x, the precursors of acid deposition.(1) Many previous proposals relied upon the use of mandated controls such as flue gas desulfurization (FGD) to achieve reductions. The Bush proposal allows affected sources to achieve reductions in any manner they choose including investment in over-control at other locations. Plants that can achieve reductions at lower cost are encouraged to produce excess reductions for transfer or sale to plants with higher reduction costs. Since the objective is to reduce emissions discharge as expeditiously and inexpensively as possible, it is critical to take advantage of the cost differences among plants and utilities to achieve these aims. Requiring one specific control technology, e.g. scrubbers, would obliterate the cost savings inherent in these plant differences.

The relevant cross pollutant of concern is carbon dioxide (CO_2) emissions. As hotly debated as the acid rain policy is the issue of U.S. commitment to control the emissions of greenhouse gases causing global climate changes. On a global basis, CO_2 emissions are thought to account for roughly 50% of the expected warming. In the U.S., which accounts for 23% of global CO_2 contributions, roughly 30% of total CO_2 emissions originate from electric utilities. Figure 1 shows the growth in utility CO_2 emissions under

EPA's base-case forecast for high SO_2 emissions. This forecast assumes no national acid rain control program and continued low energy prices.

The conventional wisdom is that action to reduce CO_2 emissions or even to damp their growth will be expensive. The main topic explored in this paper is the nature of the relationship between SO_2 and NO_x control on the one hand and CO_2 emissions on the other. Are there serendipitous gains in CO_2 reductions that could be achieved as a result of choosing one policy over another for acid rain control? If we are to expend resources controlling acid deposition precursors, why not increase the benefits of those expenditures by maximizing the opportunities for CO_2 reductions as well? Many of the options for reducing SO_2 emissions should also produce changes in CO_2 emissions since both gases are the by-products of fossil fuel combustion.

Reductions in CO_2

Acid rain control expenditures can both increase and decrease CO_2 emissions depending upon both the type of policy and the mix of control technologies. The policy choice matters because it may constrain control options available to affected sources, for example, by limiting interfuel substitution which could produce CO_2 savings. Over time these constraints would operate to stifle any incentive for innovation in pollution control technology as markets would have been foreclosed by regulators. Policy also matters because it is a significant determinant of cost which influences electricity demand and therefore atmospheric emissions including CO_2. Also, least-cost options tend to be CO_2 reducing. For example, when control technologies are mandated, there is no incentive for conservation by utilities. Not only is the overall acid rain program more expensive, but any CO_2 reductions are solely driven by electricity consumers' responses. Conservation is not tapped as an emissions reductions tool, even when least-cost.

Since the impacts of SO_2 control have both positive and negative implications for CO_2, it is important to assess cross-pollutant effects with a systematic accounting framework. The remainder of this paper develops this framework.

Interfuel Substitution

Some of the most direct CO_2 reductions under a flexible, market-based acid rain control program could be produced as a consequence of interfuel substitution. Since fossil fuels differ in their carbon content, they differ in the amount of CO_2 produced during their combustion. Table 1 presents average carbon content for the primary fossil fuels and illustrates the potential for CO_2 reduction. For example, substituting 1,000 Btu from coal combustion with 1,000 Btu from natural gas reduces CO_2 emissions by over 42%. Similarly, the substitution of natural gas for oil would reduce CO_2 production by over 28%.

While the calculations presented in Table 1 are straightforward, there is some uncertainty that these would be the exact reductions achieved when viewed from the vantage of creating a net reduction in greenhouse gas emissions. For example, as researchers have attempted to refine our understanding of the sources and sinks for methane (natural gas), attention has been focused on potential losses from pipelines and gas fields. If pipelines and gas fields leak, then substituting natural gas for coal may increase fugitive methane discharges to the atmosphere. Since methane is a more potent greenhouse gas, the benefits of reducing CO_2 from interfuel substitution might be offset. Research into the sources of methane increases in the atmosphere is continuing. Theoretical considerations indicate that naturally occurring leakage from a gas field is lessened once the field is developed and tapped (Gold, 1988). Gas that is stolen or erroneously metered could also be confused with leaks in the distribution system (see below).

Other factors which might alter the actual change in CO_2 produced as a result of interfuel substitution include any additional transportation of substituted fuels. For example, to the extent that nearby coals are substituted with more distant fuels, any CO_2 produced as a result of this added transport should be netted out. In practice, such changes may be very difficult to estimate as any increased transport would occur either by rail or water each with very different CO_2 implications.(2)

Residual Fuel Oil to Natural Gas

The amount of substitution of natural gas for residual fuel oil was estimated based on the American Gas Association's (AGA) and Gas Research Institute's (GRI) forecasts of fuel prices and the estimated value of SO_2 emissions permits assuming the Administration's proposed acid rain legislation is passed. The difference in the forecast prices of residual fuel oil (resid) and natural gas was used to calculate the cost per ton of SO_2 reduction achieved by fuel switching.(3) This price difference was compared to the value of a permit. No consideration was given to the capital costs involved in switching or to the value of NO_x reductions that would also be produced from the substitution. The assumption is that most, if not all, of the resid-fired plants are capable of firing natural gas and that they will have access to gas supplies.

Table 2 presents the results of these calculations. The cost-effective SO_2 emissions rate is the emissions rate above which the switch from resid to natural gas would yield a cost per ton of SO_2 reduced that is less than the expected market value of a permit. This analysis uses the equilibrium market values predicted by the EPA and ICF in their analysis of the Administration's acid rain emissions trading proposal which ignores any economic incentive to bank emissions over time. Given this opportunity and the time cost of capital, we would expect the equilibrium values of SO_2 permits to be monotonically increasing over the near term unless a radically new control technology shifts the supply curve and reduces the marginal cost of control. In the long run, as the existing fossil-fueled generating capacity is replaced, the supply of SO_2 credits should increase and prices stabilize. The total mmBtus of heat generated by resid-burning utilities with SO_2 emissions rates above the cost-effective rate were obtained from the NURF85 data base.(4) Total mmBtus generated from oil were obtained from the EPA high emissions base-case forecast.

If the relative price changes forecast by AGA materialize, then resid will be more expensive than natural gas by 1995. On that basis alone, one would expect natural gas to replace resid. GRI, however, expects gas prices to remain above residual fuel oil prices. Nonetheless, incentives to switch to natural gas are provided under emissions trading. To the extent that excess SO_2 reductions are more valuable than estimated by the EPA analyses of H.R. 3030 (S.1490), cost-effective substitutions of natural gas for resid would be relatively greater not only to meet SO_2 control obligations, but also to produce revenue from allowance sales. Another uncertainty in this analysis is natural gas prices. To the extent that gas substitution becomes aggressively exploited as an emission reduction strategy, natural gas prices could rise as a result of increased demand if supplies do not expand. Natural gas supplies are currently tightest in the Northeast. However, gas storage could overcome capacity restrictions while pipeline capacity is being expanded.

Regardless of relative prices, provisions of the acid rain title of S.1630 support substitution to natural gas by accommodating any emergency interruptions of gas supplies without any emission penalties if resid is used temporarily. Under these provisions (Section 405(h)), owners of plants using primarily natural gas can receive additional allowances from EPA to cover any SO_2 generated from switching to residual fuel oil during emergency weather conditions. By anticipating the potential operating conditions that this class of plants would face, the new Clean Air Act encourages natural gas substitution and the resulting CO_2 reductions.

Natural Gas Penetration Against other Fuels

In addition to forecasting gas prices, the American Gas Association has estimated that 500-1,000 additional billion cubic feet of natural gas will be sold as a result of acid rain legislation (AGA, 1989). The midpoint of this range would imply some 750 billion cubic feet of new sales (roughly 774 million mmBtu). If natural gas were to substitute for coal, however, the CO_2 savings per Btu would be almost twice as high.

A number of SO_2 control options which use natural gas have been identified. For example, gas reburn alone and gas reburn combined with sorbent injection would achieve about a 9% CO_2 reduction for a 20% natural gas substitution in existing large coal-fired units (AGA, 1989). Most current studies of the relative economics of alternative SO_2 and NO_x control technologies conclude that natural gas is likely to be too expensive to expect these alternatives to be very widespread (ICF, 1989). Nonetheless, these expectations about natural gas substitution are based upon future relative fuel prices, notoriously difficult quantities to predict. The critical point is that a flexible emissions trading program would not preclude the use of such options if they became economic at a later date.

Another indication of the potential for natural gas substitution is presented in a DOE study of fuel substitution without any new capital investment (DOE, 1989). In that study DOE estimated that roughly 6 billion mmBtu of coal could be replaced with natural gas. If indeed no new capital investments would be required, and if gas supplies were not constraining, this substitution would reduce cumulative CO_2 emissions by 2.44 billion tons in the 1995-2005 period. Clearly, gas supplies and prices would be the critical factors in determining how much of this substitution occurred.

Another longer run possibility for natural gas substitution under the acid rain titles of proposed legislation is offered by the repowering provisions. Converting large coal-fired boilers to 100% natural gas is estimated to reduce CO_2 emissions by 43%. Combined cycle conversion would increase this reduction to 64% (AGA, 1989).

Cogeneration

There are two types of CO_2 reductions that result from the replacement of standard electric utility generation by cogeneration. The first is a result of fuel switching if the fuel used by the electric utility has a higher CO_2 emissions rate than the fuel used by the cogenerator. The second type of reduction is the result of greater efficiency in the cogeneration unit in the production of both electricity and steam.

Estimates of new cogenerated and small power electricity were obtained from a report entitled *Background Documentation for 1989 Base Case Forecasts, Final Report* prepared for the EPA by RCG/Hagler Bailly. The difference between the new cogenerated electric power forecasts in the high and low emissions scenarios was taken as an estimate of additional cogeneration that would occur along with the Administration's proposed acid rain legislation. Because the legislation would cause electricity rates to increase and would require SO_2 emissions permits to be obtained for new electric utility capacity, it would stimulate development of new cogeneration capacity by industrial users who face higher rates and would strengthen a preference in utilities to buy new capacity from outside.(5)

The differences in forecasted electricity obtained from new cogenerated sources in the two scenarios were multiplied by 90% on the assumption that 10% of these estimates represent new small power sources that are not cogeneration units. The results of this calculation are the numbers listed as "new kwh cogenerated" in Table 3.(6)

Total CO_2 reductions from cogeneration are calculated by first adding the CO_2 emissions from the production of electricity in coal-fired electric utility power plants and the CO_2 emissions from industrial processes using regular industrial boilers to obtain total CO_2 emissions from the energy production that the cogenerated energy is replacing. Subtracting the total CO_2 emissions of the cogeneration facilities produces the estimate of CO_2 reductions attributable to cogeneration.

The offset requirement for new SO_2 sources would increase the relative costs of coal-fired cogenerating plants. Coal-fired capacity is expected to be 30% of expected increases in 1995 and 50% thereafter. After 2000 when offsets would have to be secured to maintain the national SO_2 cap, allowances are expected to cost between $650-$771 per ton. Independent power producers possess a number of significant cost advantages which should allow them to easily absorb these added costs. Of course, some concern has been expressed that independent power producers (IPPs) would be shut out of the market by anti-competitive behavior on the part of utilities, the main allowance holders.

To combat these perceived problems auctions and special reserves have been introduced. In any event, the offset provisions will not pose a barrier to the continued penetration of cogeneration.

Electricity Demand Response

One of the most easily quantified CO_2 reductions would stem from the reduction in electricity demanded as utility rates rise in response to SO_2 control costs. The strength of this response depends both upon the cost of the control program (and therefore the size of the rate hikes) and the responsiveness of demand (the price elasticity of demand). Another issue in the estimation of this effect is any time lag between the rate hikes and demand response.

Estimates of the long-run price elasticity of demand are at best uncertain. A fairly recent survey of elasticity analyses (Bohi and Zimmerman, 1984) published a "consensus" estimate for the residential demand sector, but found that estimates for the industrial and commercial sectors were too few and varied to permit identification of a similar "consensus" estimate. For example, while the survey found twenty studies that estimated the price elasticity of demand for the residential sector, there were only two for the commercial sector, and six for the industrial sector. The "consensus" estimate for the residential sector was -0.7, i.e. a 1% price increase would produce a 0.7% reduction in the quantity demanded. Estimates for the "commercial" sector ranged from -1.05 to -4.56 while estimates for the industrial sector ranged from -0.12 to -3.55. Despite the uncertainty regarding elasticity estimates for the commercial and industrial sectors, the consensus opinion is that they are more elastic than those for residential demand. In part, the uncertainty in price elasticities is compounded by the need to produce an estimate for the nation as a whole as responsiveness also varies across regions as well as sectors. The long-run elasticity used in Table 4 was -0.85 in part to account for greater responsiveness in the commercial and industrial sectors.

The potential for lags in response time also complicate the analysis. Electricity consumption decisions are conditioned, in part, by the efficiency of various capital equipment whether appliances or machinery. In the short run, consumers may have very limited options in response to price increases. This is not to say that they are nonexistent, as simply devoting more attention to use decisions and patterns can reduce consumption. However, over the long run, adjustments can also include investment in complementary goods such as more energy efficient appliances.

Table 4 shows the effect of forecast electricity rate increases on demand for electricity and on CO_2 emissions resulting from electricity generation. Price increases were taken from ICF's analysis of the Administration's proposed acid rain legislation. Rate hikes from that report were adjusted upward to reflect the percent change in real terms from rates in 1988. The short-run elasticity of demand for electricity of -0.25 was used to estimate the demand response in order to avoid double-counting with the conservation investment analysis which follows. It was also assumed that the reduction in electricity consumption displaced electricity generated at coal-burning facilities. To the extent that reduced demand displaces natural gas fired peaking units first, these estimates are biased upwards by the difference in CO_2 between coal and natural gas.

Energy Conservation Investments

Emissions trading will allow utilities to invest in conservation programs as a cost-effective SO_2 emissions control measure. To the extent the programs are successful, they will result in reduced CO_2 emissions as well. Examples of cost-effective conservation measures include increasing the efficiency of residential appliances such as refrigerators and freezers, removing excess fluorescent lighting and reducing air flow rates in commercial spaces, and installing more efficient lighting and motors in the industrial sector.

Estimating the amount of conservation that can be induced as cost-effective SO_2 emissions control is not trivial. Two cost curves were developed to determine feasible conservation levels. The first is the expected marginal cost per ton of SO_2 emissions reductions under the Administration's proposal plotted against total amount of reduction below 1980 levels.(8) As would be expected, marginal cost increases as more tons of SO_2 emissions are eliminated.

The second curve represents the marginal cost of conservation measures per ton of SO_2 emissions reductions plotted against total amount of reductions below 1980 levels. This curve was derived as a weighted average of regional conservation cost curves from New York State, the Pacific Northwest, and the Midwest from a set of engineering-economic analyses. Each of these curves represents the maximum kwh saved for a given level of investment per kwh.(9) These curves were translated into total tons of SO_2 reduced and dollars per ton of SO_2 reduced using the assumption of a 2.0 pounds per mmBtu SO_2 emissions rate and a standard electric utility heat rate of 10,000 Btu/kwh. A 2.0 lb/mmBtu SO_2 emissions rate is the average emissions rate for all coal to be burned under the proposed legislation, as estimated by ICF.

The two curves were then compared for each of the three years (1995, 2000 and 2005) to find the point at which the following two conditions hold: (1) the marginal cost of conservation measures per ton SO_2 reduced equals the marginal cost of other SO_2 control measures – i.e. the marginal costs on the two curves are equal; and (2) total tons SO_2 reduced from both curves equals the mandated level of SO_2 reductions. This result is equivalent to determining the level at which the conservation programs are cost-effective SO_2 control measures.

Once this level was determined, the CO_2 reductions for the kwh saved by conservation were calculated using the standard CO_2 emissions rate for coal and the standard heat rate for the electric utility. It was assumed that the conservation measures displace the use of coal, the dominant fuel consumed by the nation's electric utilities. To the extent that more expensive sources of electricity such as natural gas peaking units are eliminated through conservation first, this analysis overstates the SO_2 and CO_2 reductions.

There are two conditions which must be met for these conservation benefits to be realized. The first is the adoption of the emissions trading program for acid rain. Without a system which credits conservation investments with the reductions they produce, conservation remains an opportunity without motivation or incentive. The second condition is the removal of any remaining institutional barriers against conservation investments. This latter condition is more difficult to evaluate. However, one potential remedy may be to increase the incentives created by the Clean Air Act.

Increases in CO_2

In order to properly estimate the effect of alternative acid rain control policies upon greenhouse gas emissions, any potential source of emission increases must also be assessed. Certainly, relative to no-control scenarios, investments to control acid rain precursors will affect CO_2 emissions as energy production and use decisions are altered.

Scrubbing SO_2

The use of flue gas desulfurization, or scrubbing, as an SO_2 emissions control measure increases CO_2 emissions primarily because extra energy is required to operate the scrubbers. The extra energy is obtained from additional coal burning which results in higher levels of CO_2 emissions for the same amount of electricity generation. In addition, some CO_2 is produced in the reaction of the limestone agent with SO_2. The range of the total increase in CO_2 emissions attributable to scrubbing is from 2% to 6%, with a probable average of 4%. (Memo from Bruce Braine, ICF, to Rob Brenner, EPA, December 1988.)

Assuming a 4% increase in CO_2 emissions as a result of scrubbing, Table 6 shows the extent of additional CO_2 emissions under acid rain emissions trading. Estimates of new scrubbing capacity come from the *Economic Analysis of Title V of the Administration's Proposed Clean Air Act* prepared by ICF for the EPA. The total increase in CO_2 over the period from 1995 to 2005 represents about one tenth of one percent of the total CO_2 emissions from electric utilities forecast in that time period under the high emissions base case.

In contrast to the estimates presented above, more new scrubbing capacity would come online if mandated by regulators. An examination of the effects of the Wise Bill during the period 1995 to 2005 indicates an addition of almost 125 million tons of CO_2 emissions resulting from increased scrubbing (ICF, 1989). The freedom of choice in SO_2 control strategy allowed under emissions trading would clearly lead to fewer additions to CO_2 emissions from new scrubbing capacity.

The Wise Bill (H.R. 3211) would require the top 20 SO_2-emitting powerplants in 1985 to reduce SO_2 emissions by 90% through the use of flue gas desulfurization devices. These Phase I reduction requirements would be augmented by a second phase during which all sources above 75 megawatts capacity would have to meet a 1.2 lb SO_2/mmBtu requirement. In each phase, subsidies would be provided to utilities employing scrubbers to meet their control obligations. As Figure 3 illustrates, these provisions would lead to a dramatic increase in scrubbing and CO_2. Furthermore, the actual increase in CO_2 is likely understated since the Wise Bill provisions would probably prevent the emergence of an allowance market.

Switching to Fuels with Lower Energy Content

With the enactment of acid rain legislation that allows flexibility in choice of control methods, it is expected that some utilities will switch to the use of coal with a lower sulfur content. Some of the switching will occur between high-sulfur and low-sulfur bituminous coals from Appalachia. This substitution will have little effect on CO_2 emissions as there is little variation among bituminous coals. However, some midwestern utilities will switch from high-sulfur bituminous coal mined in Appalachia to low-sulfur sub-bituminous coal from more distant western areas. Because sub-bituminous coal has on average a higher moisture content than bituminous coal, it can cause a utility's heat rate to rise and result in more coal burned per unit of electricity generated. The estimated increase in CO_2 emissions resulting from a switch from bituminous to sub-bituminous coal is 2%.

ICF's analysis of acid rain control technologies assumes that no sub-bituminous coal can be used in boilers designed for bituminous coal, although their assessment of the Administration's acid rain proposal notes that 40 to 50 million tons per year of sub-bituminous coal could penetrate the bituminous market. Data Resources Inc. (DRI), on the other hand, estimates that 29 million tons per year in Phase 1 and 25 million tons per year in Phase 2 will represent a switch from high-sulfur bituminous coal to low-sulfur sub-bituminous coal. Based on these forecasts Table 8 displays estimates of additional CO_2 emissions. The high end of the range of cumulative CO_2 increases assumes the high end of ICF's estimates of sub-bituminous market penetration for this entire decade.(10)

Potential Effects on Other Greenhouse Gas Emissions

Leakage of methane (natural gas) in the production, transmission and distribution processes is an issue that should be addressed because methane is a far more potent greenhouse gas than CO_2. Leakage occurs naturally in gas fields as a result of pressure exerted by the gas. Production, by removing some of the gas, reduces the pressure and the naturally occurring leakage. However, production equipment is another source of leakage. Unfortunately, little research has been conducted to measure leakage before and after a field is tapped.

Leakage also occurs in the transmission of gas through pipelines and its distribution in individual customer grids. The magnitude of this leakage appears to be related to the age and material composition of the transmission and distribution systems. While the leakage is difficult to measure, it is clear that the amount of "unaccounted for " gas in the United States also includes theft, measurement inaccuracies caused in part by meter sensitivity to temperature, and variations in gas purchase and sale billing dates.

Several studies and surveys have recently been conducted to estimate the extent of methane leakage worldwide and in the U.S. An *Engineering Technical Note* by the American Gas Association dated November 1989 estimates, based on a survey of members, that 0.06% is emitted in transmission and 0.28% in distribution. In addition, the AGA estimates 0.2% is lost in production. The Interstate Natural Gas Association of the U.S. estimates that transmission pipelines account for emissions of 0.13%.

Methane Leakage from Natural Gas Operations reports the results of an investigation by the Alphatania Group of Sweden in August 1989. The report presents information on gas leakage in the U.S., Canada, Japan and 10 European countries. It concludes that there is no justification for estimating methane leakages from the gas industry worldwide at more than 1% of supply and that even this figure is regarded by many as too high. It also states that there is strong evidence that the only significant methane leakage occurs in older distribution systems found in some countries. Based on the countries surveyed, the following estimates are made for methane leakage: production 0 to 0.2%; transmission 0 to 0.13%; new distribution systems 0.03 to 0.3%; and old distribution systems up to 1%.

The Netherlands Energy Research Foundation presented a study in Paris in June 1989 entitled *CH_4/CO Emissions from Fossil Fuels Global Warming Potential.* The study addresses whether fuel switching strategies to mitigate the greenhouse effect would be affected by methane leakages. It concludes that global warming differences in the burning of oil and coal as opposed to gas would be offset by methane leakages if they occurred at the high levels of 9 to 13%.

It is difficult to compare the potency of methane and CO_2 as greenhouse gases because of complexities involved in modeling and forecasting their effective lifetimes in the atmosphere. However, it is possible to compare methane leakage from additional natural gas use brought about by acid rain legislation to coal bed methane leaked in the coal mining process. If methane leakage from the production, transmission, and distribution of natural gas is 0.5%, and on average 19 cubic meters of methane is leaked per ton of mined coal, the methane leakage from coal mining is approximately seven times that from natural gas use on an energy equivalent basis. Of course, nothing in this comparison accounts for the fact that emissions from both sources could be controlled.

Conclusions

This report has identified major interrelationships between SO_2 management decisions and CO_2 consequences. Table 9 summarizes the estimates for CO_2 reductions and increases. The cumulative totals indicate that *acid rain control via emissions trading provides a serendipitous opportunity to eliminate between 5.4-6.0% of all utility CO_2 emissions over the decade 1995-2005.* Only the market-based approach embodied in emissions trading will give affected sources the flexibility to choose among the emission control options assessed in this report. Programs which force the adoption of one technology over another, particularly those proposals which would require the use of scrubbers, would forego major CO_2 reduction opportunities and cause increased CO_2 emissions. *The major unspoken policy question surrounding acid rain is whether the U.S. should capture these CO_2 benefits.* When set against the backdrop of ongoing international discussions concerning climate change policy and the criticism that the U.S. has received on this issue, it is difficult to understand why the relationship between SO_2 and CO_2 has been ignored. The Cooper-Synar Bill (H.R. 5966) would remedy this deficiency by requiring new stationary sources to offset all CO_2 emissions. The U.S. should strive to adopt an acid rain program that incorporates incentives to reduce SO_2 emissions most efficiently. Further, given increasing concerns about global climate change, that program should be designed to maximize opportunities for CO_2 reductions as well. Is the U.S. so wealthy that we can choose not only the most expensive approach to solving the acid rain problem, and at the same time increase the cost of any ultimate policy action on CO_2?

Another way to understand the potential magnitude of these CO_2 effects is to compare them with the CO_2 implications of the energy and emission futures that EPA has developed as baselines from which alternative policies can be evaluated. The base case analysis produced by EPA this year differs sharply from prior practice in that two alternatives are presented; one scenario with high emissions and an alternative with lower emissions. Table 10 presents the differences in critical assumptions underlying the alternative SO_2 emissions forecasts.

The standard approach in analyzing SO_2 policies has been to compare the policy results against alternative future emissions scenarios as depicted in Figure 4. Each of these alternative futures represents the status quo since neither presumes any acid rain control. Unfortunately, there is no way to know which alternative future will materialize. Consequently, as in the case with SO_2, potential CO_2 emission changes should be gauged against both the high and low emissions alternatives. To the extent that the low emissions future occurs independent of any acid rain program, some of CO_2 reduction benefits outlined in this paper would be eliminated. For example, one of the major differences between the high and low tracks in CO_2 terms is the amount of new cogeneration. Both base cases forecast large additions to cogeneration capacity. Will the larger amount of new cogeneration in the low emissions base case occur independent of acid rain legislation? Perhaps, but factors related to that legislation, such as higher electricity rates and SO_2 control costs, should spur more new investments in independently produced cogenerated power.

Figure 4 presents the high and low SO_2 base case forecasts translated into CO_2 terms. The figure presents CO_2 emissions forecasts only to the year 2005 and concentrates on the decade 1995-2005 when acid rain policies are expected to have the most impact. Cumulative CO_2 emissions from utilities over the decade are expected to be 28.43 billion tons under the high scenario and 23.76 billion tons for the low track. The 4.48-billion-ton difference between the two scenarios represents roughly a 16% CO_2 reduction from the high track. Clearly, this is not a trivial difference. Nonetheless, it begs the question of whether in fact the U.S. economy will actually be on the high or low emissions track. To the extent that the high emissions track represents business as usual, it is worthwhile to ask how acid rain control would affect future CO_2 emissions.

Throughout this report the potential for CO_2 savings from acid rain legislation have been emphasized. Table 10 presents the key factors presumed to account for the differences in emissions between the high and low scenarios. After investigating a number of CO_2 reduction opportunities and utility reactions under acid rain policies, one can examine the assumptions of Table 10 and inquire which scenario is more likely to occur under emissions trading. Certainly, to the extent that natural gas substitution is encouraged and natural gas supplies do not expand, gas prices will be higher. Electricity prices are likely to rise under any acid rain control program. In response, the quantity of electricity demanded is likely to fall. Cogeneration and unplanned nonfossil fueled capacity are also likely to be greater with any acid rain legislation that emphasizes emissions trading. In sum, the assumptions of the low emissions scenario match closely the expected effects from acid rain emissions trading.

Regardless of any other changes influencing future emissions, emissions trading for SO_2 and NO_x will preserve the least-cost path for CO_2 management. Emissions trading allows the use of whichever current or future pollution control techniques will reduce discharges most efficiently. As with coal switching versus scrubbing, natural gas substitution for residual fuel oil, and energy efficiency measures, there are ample opportunities to reduce both acid rain precursors and CO_2 simultaneously. Because emissions trading involves a market, it is extremely difficult to forecast actual outcomes. The best that can be done is to indicate what could happen. What new technologies, opportunities, or constraints will emerge over the 15 years assessed in this paper is uncertain at best. Whatever changes those years bring, we can be assured that an emissions market will adjust to the maximum benefit of both the economy and the environment.

TABLE 1. Carbon Dioxide Differences Among Fuels

Fuel	Carbon Content (gC/kBtu)	% CO_2 Reduced	% CO_2 Reduced
Coal	25.109		
Oil	20.256	19.33	
Natural Gas	14.454	42.44	28.65

SOURCE: Edmonds and Reilly, 1985.

TABLE 2. Substitution of Natural Gas for Residual Fuel Oil

	AGA PRICE FORECASTS			GRI PRICE FORECASTS		
	1995	2000	2005	1995	2000	2005
Total Oil Use (mmBtus)	2070	2190	1720	2070	2190	1720
Cost-Effective SO_2 Emissions Rate (lbs/mmBtu)	0.0	0.0	0.0	0.68	0.026	0.46
Affected mmBtus (above cost effective emissions rate)	2070	2190	1720	1556.64	2183.43	1419.00
CO_2 Reductions (millions of tons/year)	46.58	49.28	38.70	35.02	49.13	31.93

TABLE 3. CO_2 Reductions due to Cogeneration(7)

	1995	2000	2005
New KWH Cogenerated (billions)	18.0	74.7	114.3
CO_2 Generated by Electric Utilities from Coal, displaced by Cogeneration (million tons)	18.18	75.45	115.44
CO_2 Generated from Industrial Processes, displaced by Cogeneration (million tons)	21.78	136.01	208.11
Total CO_2 Displaced by Cogeneration (million tons)	39.96	211.46	323.55
CO_2 Generated by Cogeneration from Coal and Gas (million tons)	27.68	166.90	255.37
Annual Net CO_2 Reduction Attributable to Cogeneration (million tons)	12.28	44.56	68.18

Cumulative CO_2 Reduction 1995-2005 -- 423.92 million tons

TABLE 4. Effect of Higher Electricity Prices

	Short	Run	Elasticity	Long	Run	Elasticity
	1995	2000	2005	1995	2000	2005
% Price Increase	0.23	1.70	1.95	0.23	1.70	1.95
Electricity Demand (billions kwh)	3080.7	3524.0	3957.6	3080.7	3524.0	3957.6
Demand Change (billions kwh)	-1.771	-14.977	-19.293	-6.023	-50.922	-65.597
Demand Change (billion mmBtu)	-0.018	-0.150	-0.193	-0.060	-0.509	-0.656
CO_2 Change (million tons/year)	-1.79	-15.13	-19.49	-6.08	-51.43	-66.25

Cumulative CO_2 Reductions 1995-2005 --
128.85 - 438.00 Million Tons

TABLE 5. Effect of Conservation Investments in Reducing CO_2

	1995	2000	2005	2010
Marginal Cost of SO_2 Emission Reductions ($/ton)	$120	$500	$500	$500
Potential Electricity Savings from Conservation (GWh/year)	32,214	86,283	96,672	108,313
Potential SO_2 Emissions Reductions from Conservation (million tons)	0.35	0.91	1.02	1.15
Potential CO_2 Emissions Reduced from Conservation (million tons)	32.54	87.15	97.64	109.40

Cumulative CO_2 Reductions 1995-2005 -- 728.62 MILLION TONS

TABLE 6. Effect of New Scrubber Capacity on CO_2 Emissions under the Administration's Proposal

	1995	2000	2005
Change in Scrubber Capacity (GW)	.5	22.0	22.0
Total New Gwh Scrubbed	2,269	99,829	99,829
Increase in CO_2 Emissions (Million tons)	.09	4.03	4.03

Cumulative CO_2 Increase 1995-2005 -- 30.48 Million tons

TABLE 7. Effect of Scrubbing on CO_2 Emissions Wise Bill

	1995	2000	2005
Change in Scrubber Capacity (GW)	51.2	73.0	74.9
Total new Gwh Scrubbed	232,319	331,251	339,872
Increase in CO_2 Emissions (Million tons)	9.39	13.38	13.73

Cumulative CO_2 Increase 1995-2005 -- 124.7 Million Tons

TABLE 8. Effect of Coal Switching on CO2 Emissions

	DRI FORECAST			ICF FORECAST
	1995	2000	2005	2005
Switch to Sub-bituminous Coal (Million tons)	29	25	25	40-50
Efficiency Loss (Million tons of Coal)	0.58	0.50	0.50	0.8-1.0
Increase in CO_2 Emissions (Million tons)	1.12	0.96	0.96	1.54-1.92

Cumulative CO_2 Increase 1995-2005 -- 10.01 - 19.20 Million Tons

TABLE 9. The CO_2 Balance Sheet

SOURCE of REDUCTIONS	CUMULATIVE AMOUNT (mm tons)	SOURCE of INCREASES	CUMULATIVE AMOUNT (mm tons)
Resid to Natural Gas	411.440-459.600	Added Scrubbing	30.480-124.700
Cogeneration	423.920	Sub-Bituminous Coal	10.010-19.200
Demand Response	128.850		
Conservation	728.620		
TOTAL	1,692.78-1,740.99		40.580-143.900

Cumulative CO_2 Reductions 1995-2005 -- 1,548.88 - 1,700.41 Million Tons

Table 10. Differences Between High and Low Emissions Scenarios

ASSUMPTION	HIGH CASE	LOW CASE
oil and gas prices	lower	higher
electricity demand	higher	lower
repowered capacity	lower	higher
unplanned coal capacity	sooner	later
unplanned non-fossil fuel capacity	lower	higher
cogeneration	lower	higher
fossil fuel plant lifetime	longer	shorter
nuclear power plant lifetime	shorter	longer

Source: ICF Resources, EPA, April, 1989.

Table 11. Fuel Price Forecasts

Fuel	1995	2000	2010
AGA	1989 $/mmBtu		
natural gas	$2.69	$3.43	$5.63
resid (No. 6)	$3.40	$4.20	$7.76
GRI	1988 $/mmBtu		
natural gas	$3.20	$3.59	$5.64
resid (No. 6)	$3.05	$3.58	$5.23

Figure 1. Utility CO_2 Emissions Under the High Base Case

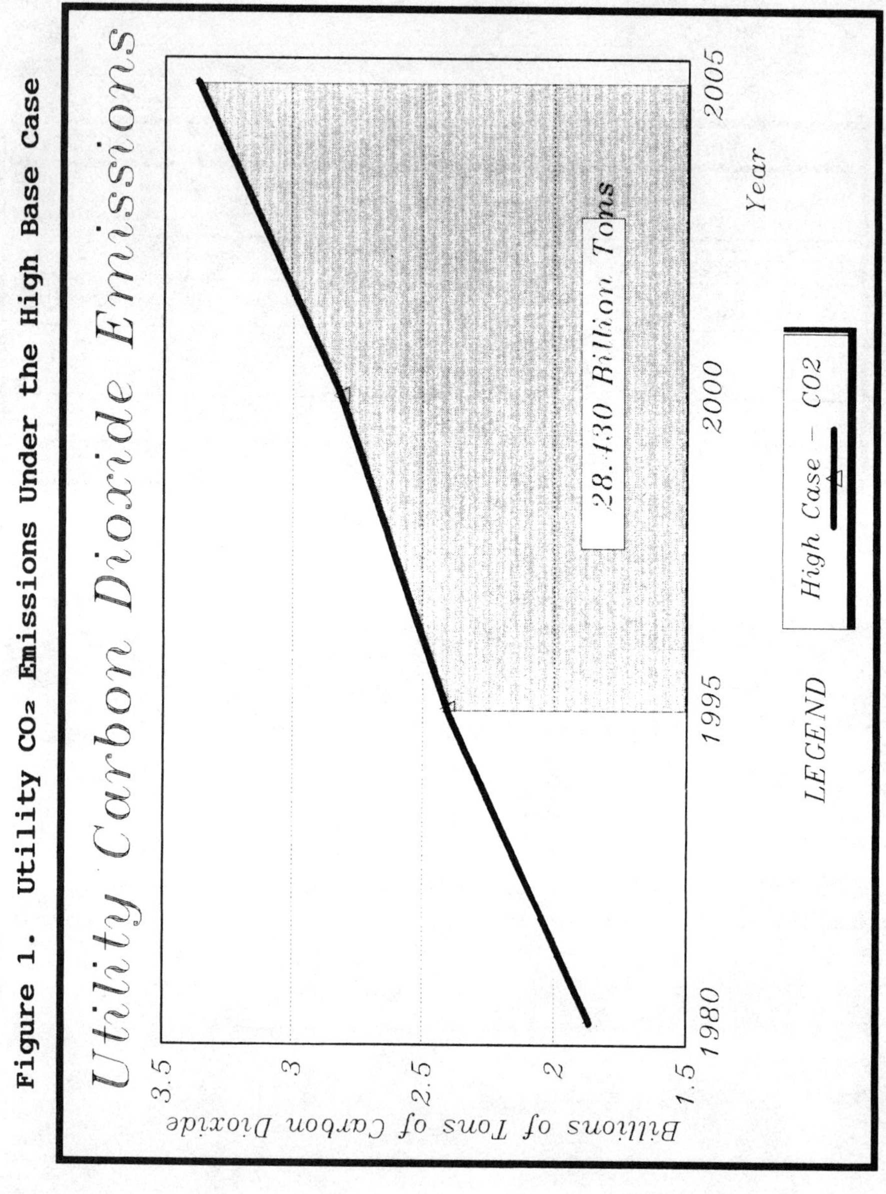

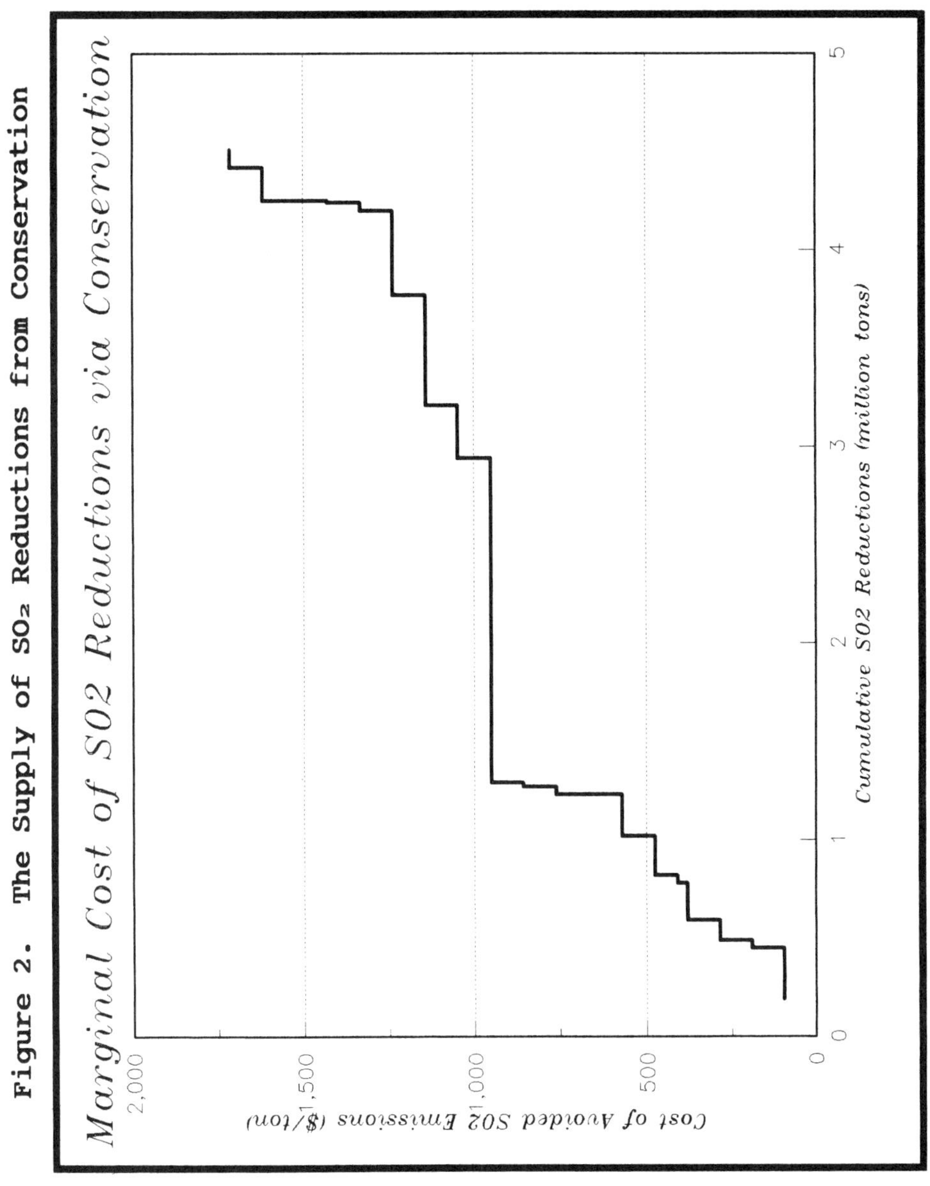

Figure 2. The Supply of SO_2 Reductions from Conservation

Figure 3. CO_2 from Scrubbing Under Alternative Acid Rain Proposals

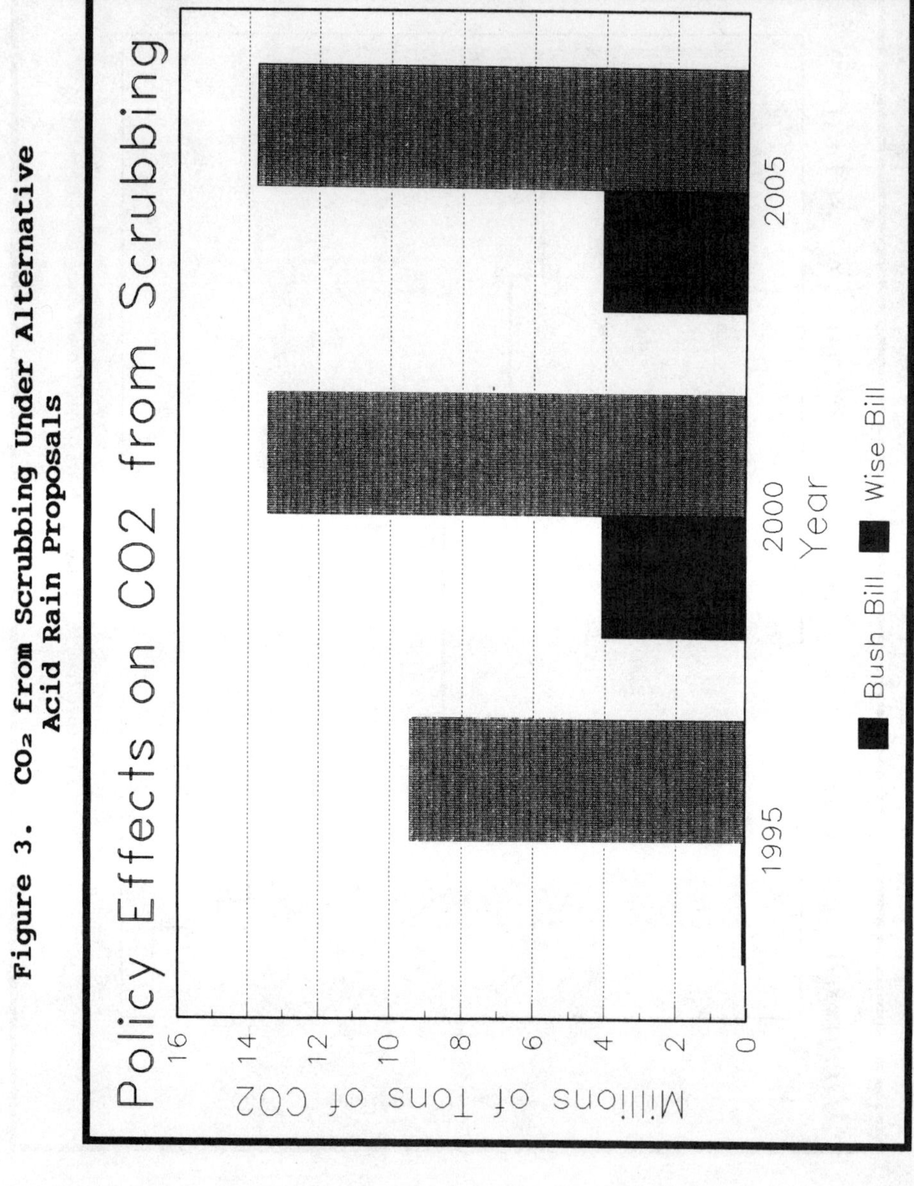

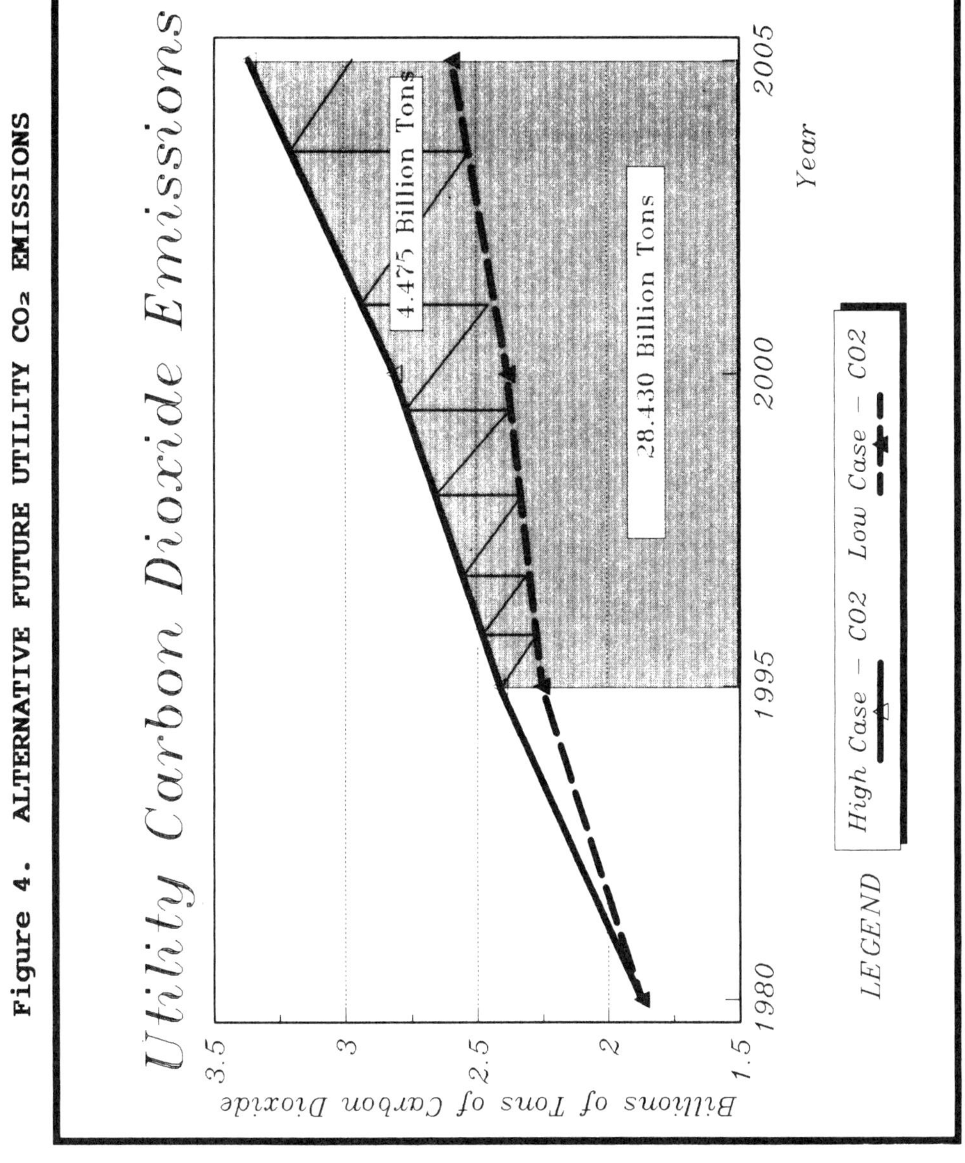

Figure 4. ALTERNATIVE FUTURE UTILITY CO_2 EMISSIONS

Acknowledgments

The authors wish to express their appreciation for the financial support of the Joyce Foundation. This paper has benefited from the comments of Bob Hahn, Elizabeth Nixon, Michael Oppenheimer, Tim Searchinger and Jim Tripp. Others including Lori Frayweek, Michael Gibbs, Thad Huetteman, John Palmisano, Paul Wilkinson, and Jim Williams have assisted this research by contributing valuable information and suggestions. The contents of this report are the responsibility of the authors however.

Notes

1. For a technical analysis of the implications of the Bush Administration's acid rain legislation see ICF Resources (1989). For a more complete discussion of the emissions trading policy see Dudek (1989).

2. A recent report from Shearson Lehman concerning the implications of acid rain legislation for coal markets concludes that barge transport of coal should increase if legislation is passed (Shearson Lehman, p. 37-39).

3. The natural gas prices used in this analysis are for interruptible supplies and represent the prices normally charged to utilities. Both the AGA and the Gas Research Institute (GRI), among others, forecast natural gas and residual fuel oil prices. Their forecasts for interruptible supplies to utilities are shown below in Table 11.

4. The NURF database identifies the primary fuels used by individual units. To the extent that resid-fired units are already using some natural gas, the estimates of CO_2 savings presented are overstated. The extent of this bias is unknown.

5. The status of cogeneration units under the several acid rain proposals is not unambiguous. Cogenerating units are likely to be subject to the offset requirements for new sources. However, due to the large proportion of cogeneration units that will be gas-fired and the greater efficiency of the cogeneration process, offsets required by cogenerators should be substantially less than those required for added utility capacity.

6. Assumptions made to calculate the CO_2 savings include the following. New cogenerated electric power replaces power previously produced from coal-fired facilities. In 1995, the new cogeneration sources are 30% coal-fired, 47% gas-fired combined cycle and 23% gas-fired combustion turbine. In 2000 and beyond, the mix is 50% coal-fired, and 25% each combined cycle and combustion turbine. Boiler efficiency of industrial boilers is 80%. Efficiency for coal-fired cogeneration is 80%, for gas combined cycle is 78% and for combustion turbine is 70%. The average heat rate for a coal-fired electric utility is 10,000 Btu/kwh. The ratio of electric to thermal energy (e/t) is 0.13 for coal-fired cogeneration facilities, 0.9 for combined cycle and 0.65 for combustion turbine. It was also assumed that new coal-fired cogenerated thermal energy would replace energy produced from coal in industrial processes and likewise for natural gas.

7. The CO_2 generated by electric utilities from coal is calculated by converting the kwh in column 1 (new kwh cogenerated) to mmBtus using the average electric utility heat rate and multiplying by the CO_2 emission rate for coal (202 lbs/mmBtu). The CO_2 generated by cogeneration is calculated separately for coal- and gas-fired cogeneration units and summed in row 3. To do this, an estimate of the fuel input to cogeneration is obtained and translated into CO_2 emissions. For example, for coal-fired units, the percent cogeneration that is coal-fired according to the assumptions is multiplied

by the kwh in row 1 and converted to mmBtus using the conversion 3413 Btus/kwh. An estimate of mmBtus of thermal energy produced is obtained by solving for t in the appropriate equation for coal-fired steam turbines, $e/t = 0.13$. The total mmBtus is divided by the efficiency of the cogeneration system, in this case 80% to obtain fuel input which is converted to tons of CO_2 emissions based on the emissions rate for coal. A similar process is carried out for gas-fired combined cycle and combustion turbine cogeneration units. The CO_2 generated by industrial processes is estimated by taking the thermal energy calculated as the value of t above and dividing by the efficiency of the industrial boiler and estimating the level of CO_2 emissions for the appropriate fuel type, coal or gas.

8. The marginal costs of SO_2 control were taken from ICF's analysis of the economic implications of the Administration's acid rain proposal (ICF, 1989).

9. Each of these curves expresses the relationship between the cost of conservation measures in $/kwh saved and the total number of kwh saved. Total kwh saved in the three curves were converted to percent kwh saved. A weighted average was taken to produce a "national" curve. The percentages for kwh saved in the national curve were applied to total projected U.S. electricity use forecast by ICF for the EPA's base case emissions. This procedure was repeated for 1995, 2000 and 2005 to obtain a nationwide conservation cost curve in terms of kwh saved and $/kwh.

10. DRI believes that most plants would comply with Phase I requirements by blending sub-bituminous coals with other higher sulfur coals. In Phase II, however, DRI believes that utilities will switch to compliance coals to avoid de-rating plants as required if all sub-bituminous coals were used. Consequently, DRI estimates sub-bituminous coal use to fall over time.

References

American Gas Association, *An Evaluation of Alternative Control Strategies to Remove Sulfur Dioxide, Nitrogen Oxides and Carbon Dioxide at Existing Large Coal-Fired Facilities*, EA-1989-1, January 13, 1989, 30 pp.

American Gas Association, natural gas substitution forecasts, telephone conversation with Nelson Hay.

American Gas Association, "Natural Gas Price Forecasts", *TERA 1989 MidYear Base Case Forecast*.

Bohi, Douglas, and Mary Beth Zimmerman, "An Update on Econometric Studies of Energy Demand Behavior," *Annual Review of Energy*, 1984, pp. 127-149.

Data Resources Inc., *Coal Market Forecasts*, 1989.

Department of Energy (DOE), *A Preliminary Analysis of U.S. CO_2 Emissions and the Substitution of Natural Gas for Coal in the Period to 2010*, DOE/NBB-0085, February 1989, pp. 14-15.

Dudek, Daniel J., "Emissions Trading: Environmental Perestroika or Flimflam?" *Electricity Journal*, November 1989, pp. 32-43.

Energy Information Administration, *Electric Power Annual 1987*, U.S. Government Printing Office, 1988.

Geller, H., E. Miller, M. Ledbetter, and P. Miller, *Acid Rain and Electricity Conservation*, American Council for an Energy-Efficient Economy, Washington, D.C., 1987.

Gold, T., Letter to *Nature*, 1988.

ICF Resources, *CO_2 Implications of Scrubbers*, Memorandum from Bruce to Rob Brenner, EPA, December 1988.

ICF Base Case Report, April 1989.

ICF Memorandum, Analysis of EDF Proposal, May 1989 (source of CO_2 estimates).

ICF Resources, *Economic Analysis of Title V (Acid Rain Provisions) of the Administration's Proposed Clean Air Act Amendments (H.R.3030/S.1490)*, September 1989, 31 pp.

ICF Resources, *Forecasts from Analysis of Wise Bill (H.R. 3211)*, Memorandum from B. Braine, R. Stuebi to Tricia Pitton, EPA, November 1, 1989, 6 pp.

Miller, P., H. Geller, and J. Eto, *Their Potential for Electricity Conservation in New York State*, American Council for an Energy-Efficient Economy, Washington, D.C., 1989.

Northwest Power Planning Council, *1989 Supplement to the 1986 Northwest Conservation and Electric Power Plan*, Portland, OR, 1989.

Shearson Lehman, *Acid Rain Legislation: Coal Market Effects*, 1989.

HOW SHOULD WE ADDRESS ECONOMIC COSTS OF CLIMATE CHANGE?

Richard Schmalensee

Council of Economic Advisers
Executive Office of the President
17th and Pennsylvania Avenues, NW
Room 315
Washington, D.C. 20500

The Council of Economic Advisers has been concerned with the issue of global change since well before we discussed it in the 1989 *Economic Report of the President*. We were heavily involved in the White House Conference on Science and Economics Research Related to Global Change last spring; Chairman Boskin was one of the three co-chairs, and I was a member of the U.S. delegation.

One of the outputs of that Conference was a recognition, in the United States and abroad, that this area of policy, like most areas of policy, involves not only questions of science; it involves science, economics, and other disciplines. In particular, one of the things that has come out of that Conference has been the recognition within the U.S. Government that we need to increase support for research on the economics of global change. This reorganization will be reflected in next year's budget proposal.

I want to focus today on the economic uncertainties this research must seek to reduce or resolve. You have heard about some of them from earlier speakers; I will discuss those and a few others as well. Let me begin by asserting that, even if we had only one global circulation model, and we knew it could correctly predict regional impacts, rainfall and other factors, we would still not know enough to make coherent policy. I will concentrate on explaining why.

The Issues

The relevant issues in this area concern the input side (where do emissions come from); the impacts side (what do emissions do and what would mitigation of emissions do); and the policy side. On the policy side, it is important to distinguish between the questions of how much to do and how to do it.

Copyright 1991 by Elsevier Science Publishing Company, Inc.
Global Climate Change: The Economic Costs of Mitigation and Adaptation
James C. White, Editor

The how-to-do it question involves considerations of cost-effectiveness, flexibility, and cost minimization. There is a fair bit of economics research to be done on that question, once you come to grips with the fact that the issue really is global. What is under consideration in the negotiations beginning here early in 1991 is an international agreement involving nations with very different positions in terms of costs, benefits, stage of development and so forth. The negotiators must try to design an agreement that passes the equity test necessary to get it adopted and passes the efficiency test necessary to minimize its global burden.

The second question is how much to do; that is, what total cost should the world incur? In this connection I would admit that economists do get carried away occasionally with attempting to value the consequences of alternative policies. But I would submit that there is really no alternative to a systematic exploration of the consequences, economic and otherwise, of proposed policies and of no-action baselines. There is a good deal more to be done on both cost-benefit analysis and, if you will, cost-consequence analysis, in which these consequences that cannot sensibly be valued must instead be clearly described.

Economics Research

Stepping back from issues of analytical tools and methods, I think it really is extraordinary, in view of the volume of active discussion and the number of calls for strong actions, how little is known about the economic and other consequences of changing rates of emission of greenhouse gases and, if the climate models are right, thus changing the rate at which the climate changes. That is what we are talking about in these policy debates, after all. Stabilizing emissions in the standard models does not stabilize climate; it slows change. But the consequences, economic and otherwise, of slowing the rate at which the climate changes, have received very little systematic analysis.

This is an absolutely key economics research area. I agree with earlier speakers that we will not know all the answers within the relevant time horizon. But that is no reason to downgrade the importance of research; it does seem to me there is some high value to be had fairly quickly. In particular, much of what has been done has been concerned entirely with the United States. In international fora, however, small island nations point out that sea level rise has a rather different meaning to them than it has to us. It is generally important to think about global consequences of different rates of climate change.

Baselines

Another important area for research is on the emissions side. It does not seem to be much appreciated by noneconomists just how uncertain are very long-term emissions forecasts. They depend on population forecasts, GNP growth forecasts, technology forecasts, and sectoral change forecasts involving, for instance, shifts from manufacturing into services – all over extraordinary time horizons. All of those things are hard to predict individually, and errors generally add. It is, therefore, not too surprising that long-term emissions forecasts vary a good deal. We are never going to get this right, in the sense of removing all uncertainty from 50-year forecasts. But my sense is, given the level of effort so far, a bit of incremental effort can reduce the uncertainty substantially.

Emissions forecasts not only affect climate change projections; they have very serious implications for cost estimates. If you start out with a model that assumes CO_2 equivalent emissions are going to grow at 10 percent a year in the absence of any action, the costs of stabilizing are going to be much higher than if you assume that emissions will be almost stable absent any action. Figuring out the relevant range of baselines thus has enormous implications for the cost debate.

Abatement Costs

There has been a good deal of debate on the costs of reducing greenhouse gas emissions, but there still is a wide range of opinion. There seems to be consensus that the CFCs and related gases are relatively cheap to abate, and nobody seems to have much of a clue about the costs of reducing net emissions of methane or nitrous oxide. The debate is almost entirely about CO_2, which is disturbing since these other gases matter. Here too, I think there is a fair amount that can be done fairly quickly to reduce some of the uncertainty.

It is important first of all – and this is going on now at the Stanford Energy Modeling Forum and under the auspices of the OECD in Paris – to try to understand why different analyses reach different conclusions. These efforts seek the anatomy of the differences. The first stage of the debate – and certainly this was the case at the White House Conference – consisted of people shouting numbers and objectives at each other. That is not terribly productive, and I am pleased that people are now trying to agree on why they disagree, to pursue what some have called second-order agreement.

Most of the work on CO_2 abatement that has involved explicit modeling has been national in scope. But as we have learned in recent months in connection with the Persian Gulf, we have world markets for oil. Even for coal and natural gas, single-country analyses can be misleading because of trade. To the extent that we are dealing with a global issue and contemplating global action, it is important to think about world market impacts, how they would affect costs, and where the costs would be borne, north or south.

There is some work on this now at the OECD, in part as a consequence of a strong U.S. push for it. The OECD work will be the first attempt to do a global-scale general equilibrium model of these consequences of CO_2 abatement policies. It will inevitably have major defects, and there will be more to be done; it will also be an important first step.

Two Approaches to Policy

Finally, let me note that a large issue in the abatement cost controversy comes from a difference that I will characterize – or, better, caricature – as a difference between an economist's and an engineer's approach to what should be done when markets do not work the way we like. The usual economist's approach is to say that, if a market is not working well, there must be some failure, some information problem, some structural problem, and to consider policies which aim at that failure. If we have an environmental problem, it is because there is an externality, and perhaps we should impose that external cost on the polluter through a charge, through marketable permits, or through other means.

Instead of saying, "Let's look at the market and see if we can identify a failure," the engineering approach has involved saying, "Let's look at outcomes and see if we can find shortcomings." In terms of the engineering analysis of energy technologies in particular, this approach has led to the conclusion that market participants are making fairly widespread, systematic mistakes.

One version of this analysis says that motorists would be better off if Corporate Average Fuel Economy (CAFE) limits were raised because they would save money. This, of course, neglects the other reasons why people buy cars and often opt for large engines. Other analyses suggest that business has systematically spent large quantities of money on lighting systems that are clearly inefficient. As far as I can tell, there is a fair amount of support for that conclusion.

However, even if this conclusion is correct, it does not answer the policy question it poses. Do you deal with this apparent market failure by saying, "Here are lighting standards you must comply with"? Or do you deal with it by saying, "What's wrong with that market? Let's try to figure it out. Let's get the information out there; maybe we should jump-start it." That is, one can reduce choices and overrule the market or design a policy that makes the market work better.

One of the reasons why there is a wide difference in estimates of the costs of reducing CO_2 emissions is the tension between these two approaches. The economist, in stereotype, says, "The market must be doing it perfectly." The engineer, in stereotype, says, "No. My analysis says the market has it all wrong." Now, most markets, in fact, perform well but imperfectly. How energy markets actually perform, how we ought to deal with them, and how we can effectively deal with them with available policy instruments will, I think, have an important bearing on the outcome of this debate.

DEALING WITH THE ECONOMIC COSTS OF CLIMATE CHANGE MITIGATION: A PERSPECTIVE FROM THE AUTOMOTIVE INDUSTRY

John W. Shiller and Paul D. McCarthy

Ford Motor Company
The American Road
P.O. Box 1899
Dearborn, Michigan 48121

Mark A. Shiller

The University of Michigan
Ann Arbor, Michigan 48109

Abstract

There is growing international awareness that common everyday human activities may be inadvertently changing the climate of the world through the intensification of the natural greenhouse effect. This has prompted a number of nations as well as the international community to explore what can be done to minimize or avoid climate change before it might occur and at what cost. The challenge of minimizing the economic costs of climate change mitigation both nationally and globally is the subject of this paper, particularly as it relates to the automotive industry and its customers.

Many of the necessities and comforts of our daily life are made possible by motorized vehicles. They contribute not only to the time-efficient creation and delivery of many essential human needs but also to the privilege of expanded personal mobility. From many points of view, our current motor vehicle transportation fleet represents a time-tested balance of values, including cost efficiency, safety, environmental needs, and the direct requirements and desires of vehicle owners. That balance has never been static as product improvements are made in response to consumer "votes" in the market place. Any strategy involving highway vehicles devised to address climate change must be carefully designed to minimize costly disruptions to this market process.

The challenge will be to find a cost-effective method that allows for the continued development of the various transportation systems required by consumers worldwide

Copyright 1991 by Elsevier Science Publishing Company, Inc.
Global Climate Change: The Economic Costs of Mitigation and Adaptation
James C. White, Editor

while simultaneously addressing other important global, national, and public goals. Command and control type fuel economy standards are the least effective way to address that challenge. They typically obstruct the flexibility needed to obtain lowest cost solutions and create unnecessary market distortions. There are significantly better ways to promote environmental protection that minimize disruptions to economic growth around the world, based on the concept of the carbon fee. A carbon fee approach appears to be superior to worldwide "bubble" trading schemes that use marketable emission permits. Some of the basic economic elements necessary to minimize cost are discussed along with some examples.

Introduction

Concern that human activities may be inadvertently changing the climate of the world through the intensification of the natural greenhouse effect recently has been expressed by the Intergovernmental Panel on Climate Change (IPCC, August 1990). The IPCC is the international group assigned with the task of defining the scientific understanding of the greenhouse effect and of identifying international policy options to address that issue.

The IPCC was established in 1988 by the United Nations Environment Programme (UNEP) and the World Meteorological Organization (WMO). This was done in response to a 1987 resolution by the United Nations General Assembly in reaction to the Brundtland Report titled *Our Common Future* on global environment and development (Brundtland, 1987).

The theme of the Brundtland Report, presented to the UN General Assembly October 19, 1987, covers more than just the greenhouse issue. Its objective is to move the global community toward achieving the goal of sustainable development defined in terms of environmental protection, promotion of economic development, and security enhancement. Sustainable development, as portrayed in the Brundtland Report, demands changes in the domestic and international policies of every nation. It calls for the voluntary partial limitation of national sovereignty in favor of increasing interdependence among nations in economics, environment, and security. It would mean a stronger role for the United Nations and the international community in world affairs because sustainable development must be managed jointly by all nations. (An example of this type of cooperation is the 1991 allied war with Iraq to end the occupation of Kuwait.)

The Brundtland Report proposes new forms of international cooperation designed to break out of existing national behavior patterns and to foster social change linked to sustainable global economic development and environmental protection. Embodied in the global climate change issue are most of the social concerns that the United Nations has struggled with ever since its inception 46 years ago. Thus, the greenhouse issue is a much broader matter than just an environmental concern.

The IPCC completed its First Assessment Report on the greenhouse effect in 1990. That study concluded that man-made emissions are increasing the atmospheric concentrations of certain greenhouse gases. If left unchecked, it could lead to an additional predicted warming of the earth's surface between 2 and 5 degrees C sometime during the next 100 years (IPCC, August 1990). A temperature range was used by the IPCC to give some indication of scientific uncertainty.

Some studies of available surface temperature data suggest that an increase of 0.3 to 0.6 degrees C already has occurred over the last 100 years (IPCC, July 1990). The size of this warming is loosely consistent with the predictions of climate models but it is also of the same magnitude as natural climate variability. Therefore, more time will be required to gather the data necessary to substantiate the actual trend, cause, and validity of temperature projections.

The details of future climate change stemming from temperature fluctuations and subsequent effects also are obscured by substantial scientific uncertainty. Some climate changes may be beneficial while others detrimental. Agricultural potential could increase or decrease depending on location. The availability of water and biomass might be significantly affected both positively and negatively in different regions (IPCC, July 1990). Arid and semi-arid regions may be more sensitive to climate change as existing climate problems might be compounded. Mean sea level may rise by as much as 2 feet over the next century threatening some low-lying islands and coastal zones. On average, those changes may be more of a drawback than an advantage according to the IPCC.

In view of the scientific uncertainty and possible global consequences, it would be prudent at this time to improve our thinking about identifying responsible policy alternatives and to examine how best to minimize the probable economic costs of climate change mitigation. Some background material is presented first, followed by consideration of how much control is needed and potential control approaches.

What Gases Are Important?

Carbon dioxide was identified by the IPCC with slightly over half (55%) of the current greenhouse enhancement that may lead to climate change (IPCC, June 1990). Other greenhouse gases which make up the balance are methane 15%, chlorofluorocarbons (CFCs) 11 & 12 17%, other CFCs 7%, and nitrous oxide 6% (Figure 1).

Water vapor is known to be the strongest greenhouse gas but is not included in Figure 1 because its concentration is determined essentially by climate system mechanisms and appears not to be affected by human water vapor sources and sinks. However, the impact of other greenhouse gases on the hydrological cycle is extremely important to research because water can both enhance and inhibit the greenhouse effect, thus influencing its ultimate magnitude. Unfortunately, the details of this influence are not adequately understood.

An index has been developed to help quantify the relative importance of the various greenhouse gases. The index is called the Global Warming Potential (GWP) and was used to develop Figure 1 (IPCC, April 1990, Section 2). The GWP index is the best available, but is still preliminary because much of the information needed to generate unquestionably correct values is not available. Nevertheless, the index will be very useful for planning purposes until better information becomes available.

The GWP index developed by the IPCC varies over a wide range for different substances. For example, over the next century, a given tonnage reduction in methane emissions for the current calendar year is worth 21 times more than reducing CO_2 emissions in the same year by an equivalent amount. Examples of the index for other gases are: nitrous oxide 290, CFC-11 3,500, and CFC-12 7,300, all relative to CO_2 on a mass basis. The index is computed from the position and strength of the absorption bands of the gas, its lifetime in the atmosphere relative to CO_2, its molecular weight, and the time period over which climate effects are of concern.

How Are Highway Vehicles Related to Greenhouse Gas Emissions?

Highway vehicles emit both CO_2 and CFCs along with small, generally negligible, amounts of CH_4 and N_2O. A discussion of each of these gases, as related to highway vehicles, follows:

Carbon Dioxide

The combustion of gasoline and diesel fuel resulting in CO_2 and water is a necessary process for releasing the stored fuel energy. Of all the greenhouse gases, CO_2 will be the most difficult and costly to control because, unlike the emissions of other substances, its formation is necessary to provide for efficient energy release. For a specific carbon-containing fuel, the mass of CO_2 emitted from an automobile is directly proportional to fuel consumption or inversely proportional to fuel economy assuming mileage doesn't change with efficiency improvements.

In order to estimate the magnitude of the contribution of highway vehicles to climate change, the IPCC greenhouse enhancement distribution of Figure 1 can be further allocated among the various sources of CO_2 based on available emissions data (IEA, 1989a and 1989b); the result is illustrated in Figure 2 for calendar year 1987. The white pie slices represent CO_2 emissions from six sources: oil, coal, natural gas, waste combustion, cement calcination, and deforestation (Figure 2).

CO_2 resulting from oil combustion, of which highway vehicles play a part, represents about 17.2% of the total potential for global greenhouse enhancement. Based on IEA and IPCC data and the fact that almost 100% of current highway vehicle operation depends on oil, CO_2 from highway vehicles worldwide represents about 5.9% of the potential for greenhouse enhancement (IEA, 1989a & 1989b, IPCC, June 1990). CO_2

from all US highway vehicles (cars & trucks) represents about 2.3% of potential greenhouse enhancement while CO_2 from US cars alone represents 1.3%.

Chlorofluorocarbons

The Montreal Protocol, as renegotiated in June 1990, provides for the complete phaseout of all CFCs by the year 2000. This phaseout will significantly reduce potential greenhouse enhancement even though the Protocol was undertaken for a different purpose (preservation of stratospheric ozone).

The automotive industry, in response to the renegotiated Protocol, will phase out CFCs including air conditioning, foam blowing, and cleaning applications. Substitute materials with low greenhouse potential are being developed as replacements. For example, CFC-12 is being replaced with chlorine-free HFC-134a. In addition, programs to recycle existing CFCs are being accelerated.

The scheduled phasing out of CFC-12 will reduce the global warming impact of a car's air conditioning refrigerant 84% on average over the next century based on the IPCC GWP index. This change is significant. Once complete in the United States, it would be roughly equivalent, in term of greenhouse impact, to reducing CO_2 emissions from all US highway vehicles 35%. In addition, the refrigerant's potential to destroy stratospheric ozone will be completely eliminated with the changeover (IPCC, April 1990, Section 2).

Methane

Most emissions of methane come from agricultural activities and natural sources (enteric fermentation, rice production, wetlands, biomass burning, oceans, and freshwater lakes). The contribution to greenhouse enhancement by methane emissions from highway vehicles is small (less than 0.1 percent) based on IPCC data and emission inventory estimates. Clearly, the impact of vehicle methane emissions is negligible compared to vehicle CO_2 or CFC emissions.

Nitrous Oxide

Nitrous oxide is a trace gas that is produced mostly from a wide variety of biological sources in soils and water. Until recently, the combustion of fossil fuels was thought to be an important source of atmospheric N_2O. Now it has been found that earlier estimates are incorrect because artifact N_2O was being produced in the sampling flasks being used to collect N_2O from combustion sources (IPCC, April 1990). It is not now considered significant by the IPCC for any combustion source including highway vehicles.

How Much Control Is Needed?

No one really knows just how much control is desirable for climate change mitigation. CO_2 proposals have ranged from reducing only the growth in emissions, to stabilizing emissions by 2000, to reducing emissions 20% by 2005, and/or 50% by 2025. The most widely expressed CO_2 control proposal is emission stabilization. Such a step is usually seen as the most that could be negotiated given existing uncertainties, provided that sustainable development of the world's economy would not be threatened by such action.

How much control each nation would be responsible for also needs to be negotiated considering current emissions, past control efforts, future plans, and national needs. Table 1 contains CO_2 emissions data by region or group of nations (IEA 1989a & 1989b). The table also includes population and per capita emissions statistics (uncorrected for non-energy petroleum use, deforestation, and cement calcination CO_2 emissions). Per capita CO_2 emissions vary by region as does the amount of CO_2 per dollar Gross National Product (GNP, measured in 1987 US dollars, WRI, 1990). Examination of data such as this will help identify opportunities for emission reduction and aid in the negotiation process.

CO_2 per \$GNP is a measure of energy efficiency for the production of goods and services. The pounds of CO_2 emitted per dollar GNP varies from a low of 2.0 for OECD countries to a high of 7.5 for Asia. The US by itself emits 2.5 pounds CO_2 per \$GNP, 11% below the 2.8 average for all nations. If deforestation (as well as non-energy and cement production corrections) were to be included in the table, the spread in pounds CO_2 per dollar GNP between nations would become larger.

Largely due to planned CFC reductions, total national US greenhouse emissions (in terms of CO_2 equivalents) will be essentially stable out to the year 2010 without additional controls (Cristofaro, 1990). The European Community has formally adopted a CO_2 emission stabilization program by the year 2000 at 1990 levels, with some member states, in particular Germany and the Netherlands, promising to achieve net reductions by that time. The government of Japan adopted a target of stabilized CO_2 emissions on a per capita basis by 2000. These countries also will phase out CFCs. In contrast, increased emissions are expected for a number of developing nations even with additional controls as they try to improve their economic situation.

Environmental ministers and others from 137 countries met in Geneva November 6 to 7, 1990, at the Second World Climate Conference (SWCC) to discuss policy. A Declaration was issued to outline the agreement reached (SWCC, November 1990). The conferees found that, while there exists great uncertainty regarding the magnitude, timing, and regional effects of climate change due to human activity, they stressed, "as a first step, the need to stabilize, while ensuring sustainable development of the world economy, emissions of greenhouse gases not controlled by the Montreal Protocol."

Official international negotiations will start February 1991 to formulate a framework for a climate change convention. A "convention" by definition is a general international policy agreement used to help guide how to address specific issues. The November 1990 Second World Climate Conference was used as the launching mechanism for generating support for such a convention. Completion is anticipated June 1992 leading into the UN Conference on Environment and Development to be held in Brazil at that time.

What Approaches Should Be Favored to Manage Mitigation Costs?

There are a number of essential elements for the effective management of the economics of climate change mitigation once a decision has been reached as to how much control would be needed and by what timetable. Because selecting a course of action will involve considerable uncertainty, policies must be flexible enough to incorporate future scientific findings. Policies that can be easily reversed or expanded should be favored. Approaches that can be justified for reasons in addition to climate change (such as CFC control) would prove to be of highest value.

International Coordination Is Necessary

International coordination will be necessary to reduce emissions. No one nation can single-handedly do enough to significantly offset the greenhouse effect. Even with the full cooperation of all 24 member countries of the Organization for Economic Cooperation and Development (see OECD, List), net reductions in global emissions could not be accomplished if emissions from the developing world increase at an unlimited rate.

Disproportionate efforts among the nations to control emissions inevitably would lead to economic dislocations. Because of the global nature of the greenhouse effect, each nation is subject to unique and distinctive climate changes (positive and/or negative, depending on location) that are a result of the collective actions of all. Although the disadvantages and benefits of climate change are dispersed among all nations, the economic rewards of not curbing emissions are not distributed, but accrue to the nation that doesn't participate.

If climate change must be addressed, nations will have to learn how to share costs and develop cooperative policies. Any comparative economic advantages must be shifted from those nations who do not participate to those who do through negotiated strategies fashioned by the international community. Without such a coordinated approach there exists a strong incentive for numerous nations not to participate.

The international cooperation of Third World and Eastern Bloc nations is crucial. These nations want to expand economically. Unfortunately, they generally do not have the capabilities, business climate, or the funds needed to quickly build their capital stocks with state-of-the-art, environmentally sound technology. Older technology with greater CO_2 emissions usually provides a faster way to improve economic conditions over the short term. Therefore, poorer nations have the greatest incentive not to cooperate unless their concerns are addressed.

Without full international participation, the positive steps taken by developed countries are likely to be nullified. By trading technologies appropriate to the situations in most Third World and Eastern Bloc nations, by coordinating allowable emissions, and by conducting joint research into improving energy efficiency, both the economic and environmental challenges may be addressed. All nations will need to carry their "fair share" of the burden if the policy negotiated is to be sustainable. The definition of "fair share" needs to be developed through negotiation and experience in order to discover what will actually work.

Command and Control Mandates Should Be Minimized

Command and control regulatory approaches that target the average fuel economy of new vehicles have potential to reduce emissions in only one way: through increasing the efficiency of new vehicles. Market-based approaches, however, recruit the initiative and creativity for saving fuel from all consumers. In the transportation sector, these policies conserve fuel and reduce emissions for both old and new vehicles by motivating travelers to conserve, not just new vehicle buyers at the point of purchase. Because of such effects, the cost of reducing emissions is far less than under command and control tactics.

Typically, command and control mandates remove the flexibility needed to obtain solutions with the lowest cost to society and with the fewest market distortions. By obstructing the reallocation of costs across industries or across different greenhouse gases, such nonmarket policies raise the social cost of any required reduction in greenhouse gas emissions.

Command and control measures also can inhibit the range of products available to meet diverse consumer needs. They can even be counterproductive. For example, higher fuel efficiency standards combined with relatively inexpensive fuel encourages increased vehicle usage. In addition, when fuel prices are low, command and control can encourage people to hold on to their older and less fuel-efficient vehicles rather than purchasing a new, more expensive, high-efficiency car.

The demand for energy is usually less responsive to market price in the short run than it is in the long run. This occurs because it takes some time for energy saving investments to be put in place and for people to change their habits. Such a delay does not automatically justify command and control regulations because the market ultimately will respond in a more efficient way. Mandating that changes take place in very short periods of time can multiply the costs to achieve the same outcome as would be eventually provided by the market. In general command and control efficiency standards will always have significant disadvantages when compared to methods that incorporate social costs into the price, such as charges and user fees.

The Efficiency Of The Market Should Be Harnessed

Simply left to its own workings, the market will fail to deal with climate change unless higher costs are associated with the decision to consume fossil fuels and emit other greenhouse gases. In order to take advantage of the efficiency of the market, the question becomes how best to compel economic decision makers to internalize these higher costs. Two approaches that employ a market mechanism to control greenhouse gas emissions are the carbon fee and tradeable emission permits.

Tradeable emission permits are receiving attention as a possible approach to capture the world's emissions in one giant "bubble." In this approach, permits to emit greenhouse gases would trade on an open market. Because the total supply of permits is fixed there is little uncertainty about the emission level that would result as long as enforcement is sufficient. Because the permits would be tradeable, reductions would take place where they are least costly. If permits can be traded internationally and between gases (properly indexed) then the cost of total reductions in greenhouse warming potential could be minimized. With the proper supply of permits, the market price will reflect the social costs associated with the emissions. While theoretically sound, there are a number of practical reasons why this may not be the preferred approach.

Determining the proper amount of emissions for which to issue permits is one of the major problems. Given the large degree of uncertainty about the extent of global warming, and the relationship between future emissions and future temperature increases, there is little basis for confidence that the optimal level of emissions can be identified. Carbon fees appear to be a more reasonable domestic policy instrument when there are likely to be changes in our knowledge about what we should be doing to delay or prevent global warming.

Emission permits would be expensive to administer and difficult to enforce. These problems would be particularly acute if permits were traded on world markets. Enforcement would create many burdensome and costly reporting requirements that would be unnecessary with a fee-based approach. Without adequate enforcement, meaningful reductions in greenhouse gas emissions are unlikely, due to the large incentives to emit beyond owned permits.

Participation in the permit market is likely to be particularly expensive for small emitters. In contrast, carbon fees applied at the point of production ensure the participation of all economic actors regardless of size. Because a carbon fee is simpler to administer and enforce than a market for emission permits, carbon fees imply a lower cost for a given reduction in CO_2 emissions. Moderate carbon fees are a particularly appropriate instrument to initiate reduced growth in CO_2 emissions. Although proposals to increase cost are usually unpopular, a carbon fee (and/or its equivalent for other gases) appears to be the most direct method to deal with the contribution of CO_2 to the greenhouse effect with minimal economic disruption.

Carbon Fee

A carbon fee would assign an associated cost of climate change with the consumption of all carbon-containing fossil fuels. Fuels vary in the amount of carbon they contain (Figure 3) as well as in their heating value (Figure 4). Most carbon fee proposals would not apply the charge directly to agriculturally based fuels because they are renewable (i.e., plants pull carbon from the atmosphere). The carbon fee, however, would apply indirectly, to the extent that fossil fuels are used in the production of agriculturally based fuels.

A broad-based carbon fee will ensure that reductions in CO_2 emissions will occur for uses that are of lowest value to individuals and society. A carbon fee helps ensure that, if there are social costs to CO_2 emissions, they are borne by the emitter. This would help avoid excessive production and consumption. A carbon fee provides the greatest opportunity to achieve greenhouse emission reduction goals. These reductions would be achieved at a lower cost than can be provided by command and control efficiency standards alone.

A carbon fee also raises revenue. This revenue could be used to offset other taxes, compensate groups adversely impacted by higher energy prices, and spent on other environmentally sensitive projects such as replanting tropical rain forests. The Congressional Budget Office has estimated that carbon fees of $100 per metric ton could raise from 110 to 120 billion dollars of additional revenue in the year 2000 (1988 dollars). Lower carbon charges would raise less revenue and save less energy, but also would have a smaller impact on economic growth. Command and control measures burden the economy like a tax without any offsetting increase in revenue. Because command and control measures are inherently less direct and comprehensive than an appropriate fee, the burden for any given reduction in greenhouse gas emissions is likely to be much higher as well.

Historical energy unit prices charged at the point of production (at the mine opening or wellhead) are plotted in Figure 5 based on 1987 dollars. On an equal energy basis, oil has been consistently the most expensive fuel over the period 1970 to 1988 because it has advantages that justify a premium over other energy alternatives. It also has had the greatest price fluctuations over that period. High oil prices have stimulated fuel switching, conservation, efficient technology, and higher production and exploration. The peak price for oil reached in 1981 was not sustainable because of such activities.

When investing in energy saving technologies, the return comes from the energy saved multiplied by the price of the energy. Investment decisions by individuals and organizations are based on the trade-off between the up-front cost and the discounted rate of return from the investment. If energy is, in fact, worth more than its price, there will be little investment. Energy saving investment can be encouraged by increasing the price of energy, lowering the interest rate, or subsidizing the price of the investment. The most direct and comprehensive approach is to change the price of energy.

A carbon fee generates greater reductions in CO_2 emissions than would an equivalent increase in the price of any single fuel. By raising the price of all carbon-based fuels, carbon fees moderate the motivation for fuel switching. To the extent that fuel switching takes place, it will result in further reducing CO_2 emissions beyond the greatest reductions caused by higher fuel prices alone. Fee revenues also impact producers. Because the increase in price is due to the fee, the additional revenue does not flow to producers so that further stimulation of additional fuel production and exploration does not occur.

Figure 6 shows the effect of a carbon content fee on the price of coal, petroleum, and natural gas (assuming equal energy content). The slope of the coal line is the largest because it has more carbon per energy unit than the other alternatives. The slope of the natural gas line is the smallest. The intercepts are based on average 1987 prices (SA, 1990).

The advantage of a carbon fee is that it encourages both industry and consumers to carefully choose the fuels they will use while internalizing costs. It also would increase the present value of developing energy efficient technology. However, the carbon fees probably needed to achieve stabilization in the near term, followed by reductions in the long term, may be very disruptive to the economy as a whole, particularly if the fee is rapidly imposed.

Some analyses, recently released, suggest that a broad-based carbon fee of the order of $100 per metric ton (which would raise gasoline prices by about 25 to 30 cents per gallon) would be needed to sustain bringing about emission stabilization followed by long-term reductions in annual CO_2 emissions from present levels (DOE, September 1990 and CBO, August 1990). However, large fees could be disastrous to some industries and geographic areas.

The Department of Energy (DOE) estimates that mine opening coal price increases of 180%, oil wellhead price increases of 70%, and commercial natural gas price increases of 49% would be needed to accomplish reductions. The US Congressional Budget Office (CBO) suggests that over the longer term, carbon charges of $100 per ton could hold the level of GNP at least 1 percent lower than without the charges (assuming a gradual phase-in). Because of this concern, they suggest that charges lower than $100 per ton should be considered initially. Once in place, the rate could be adjusted as necessary to reflect resolution of scientific and economic uncertainties.

Other Considerations

Coal is the nation's most abundant fossil fuel. It supplied 24% of the energy needs of the US in 1987, much of it used for electricity generation. But being the fuel with the largest carbon content, it would be the most affected by a carbon fee. At a fee of about $65 per metric ton of carbon, coal would cease to be America's cheapest energy source, becoming equal to natural gas (Figure 6, based on average 1987 prices, SA, 1990).

The uncertainties in relying on foreign sources of fuel should be considered. The 1990-91 Middle East crisis/war highlights the possible costs associated with foreign source uncertainties. The peak price for oil reached shortly after Iraq invaded Kuwait was $35 per barrel up from a pre-invasion level of about $18. This peak price is illustrated by the "Post-Kuwait invasion" line in Figure 6. This temporary increase resulted, not because there was any oil shortage, but because people were worried that there would be.

Certainly the size and composition of US energy demand would be affected by price. Higher prices would have a significant impact on the economy. In the latter half of the 1980s, the United States spent about 3% of Gross Domestic Product (GDP) on fossil fuels (1% of GDP on fuels for motor vehicle operation) based on prices at the wellhead or mine opening, not retail prices (SA, 1990).

All policies that generate fuel savings will reduce our productive potential even with improved fuel efficiency technology in all energy sectors. It also will take time for improved technology to be developed, made available, and put in place. Therefore the short-term effects of a carbon fee (or any other less desirable energy saving alternative) should be considered before implementation. Slow phase-in will be required to minimize the impact on economic growth. Low initial carbon fees can be adjusted over time to reflect resolution of scientific and economic uncertainties. Commitments to increasing the cost of energy in the future will increase investments in energy savings today as long as people know the increases will indeed be put in place.

International trade flows can quickly undo greenhouse gas reductions that are undertaken unilaterally. As an example, raising the cost of coal in the US alone would raise the domestic cost of products like steel, encouraging production abroad. This would increase imports of steel to the US as the market shifts to lower cost producers. Such a policy would hurt local industries and help foreign producers. The costs of unilateral action can be quite damaging to specific industries. Any costs to be increased need to be spread across national borders through negotiation.

Additional Approaches

Governments also could help reduce emissions and save energy by encouraging carpooling through advertising or the granting of special right-of-way privileges to ridesharers. Strategies designed to keep the flow of traffic moving, such as staggered working hours, will be particularly helpful. Mass transit systems, if designed properly and used, also could have beneficial effects.

Any new approach being considered, however, should focus on satisfying the genuine transportation needs of people while reducing emissions rather than just concentrating on reducing the quality, value, and convenience of transportation in order to discourage use. In addition, it is important to consider the total net impact to greenhouse gas emissions implied by all phases of suggested policies. As an example, mass transit proposals should be evaluated in terms of the amount of greenhouse gases generated during construction, maintenance, and operation.

The rationing of gasoline, mandating sale quotas for alternative fueled vehicles, or restricting travel (such as no-drive days) could all be given the force of law. But the political and economic costs of such measures always have been understood to be far too high to implement except in extreme/temporary circumstances. Clearly, these measures remove the freedom of consumers to decide the best way to save energy given their particular transportation needs and requirements.

From society's viewpoint, a combination of an emission reduction program along with some planned adaptation to global warming may be the most reasonable strategy. It may turn out that the costs of some climate change mitigation tactics are significantly greater than the costs of adapting to the impacts of greenhouse enhancement. Thus, some adaptation measures may be appropriate.

Support for scientific climate research should be continued. Research will generate the information needed so that climate mitigation programs can be periodically adjusted to be commensurate with recognizable environmental benefits. Because of significant uncertainties in knowledge, any response to mitigate climate change must be iterative, with periodic feedback from the scientific community so that control programs can be readjusted as necessary.

What Is the Cost of Potential Climate Control Targets?

Available economic cost estimates for controlling global climate change tend to be as controversial as benefit estimates. The reasons for controversy are based on questions regarding the future validity of underlying economic assumptions, the ability of selected models to reflect reality, and the assumed impact of future technical innovation. Despite the controversy, economic projections are of value in decision making provided that all of the limitations of the necessary simplifying assumptions are recognized.

CO_2 Control Costs

As discussed earlier, the cost of stabilizing or reducing CO_2 emissions was estimated by the US Congressional Budget Office (CBO) and the Department of Energy (DOE) to be a reduction in GNP of about 1 percent (DOE, 1990; CBO, 1990). For the US, emission stabilization/reduction would cost about $50 billion per year based on this estimate. For the world, emission stabilization would cost more than $150 billion per year. If world GNP grows at an average annual rate of 3 percent, the total cost out to 2050 (discounted 5%/year) of a 1 percent loss in GNP every year would be over $5 trillion (DOE, 1990).

On the other hand, some have suggested no-cost, or even negative-cost in addressing global warming. That is, equal or greater than historic GNP growth was projected out into the future based on optimistic technological innovation and other suppositions (Miller, 1990). Technological innovations, however, are never guaranteed even if incentives are used to promote research. Considering the amount of energy research that already has been conducted, what is currently known about available options, and the difficulties experienced in making major advances in fuel efficiency, the no-cost projections appear unlikely, especially over the next several decades.

The DOE points out that, between 1973 and 1985, the average price of energy rose 47% relative to non-energy products at the consumer level and rose 80% at the industrial level (DOE, 1990). During that same period, US and OECD CO_2 emissions were essentially constant. This historical occurrence suggests that any attempt to stabilize energy use today would be at least as economically disruptive as the oil shocks of the 1970s.

Of course, changes in GNP are not a complete measure of all the costs that may result from climate change mitigation options. Unavailable product options, job dislocations, lifestyle changes, product utility changes, and functional changes (such as tradeoffs between fuel efficiency and safety) usually are not easily quantifiable in dollars and are often overlooked.

CFC Control Costs

It is clear that the cost of eliminating (not just stabilizing) the use of all CFCs in the United States will be much less than the cost of just stabilizing CO_2. The cost of CFC elimination was estimated by the DOE to be about $3 billion (present value) over the next decade (DOE, 1990). This assumes that the timing of the Montreal Protocol is not accelerated. Costs would be considerably higher under a faster phase-out schedule. In general, greenhouse gases that are the least costly to control should be the ones first dealt with.

Based on an analysis of a Canadian CFC phase-out acceleration proposal, it has been estimated that a ban of all new car and light truck air conditioners that use CFCs, beginning September 1991, would result in costs for Canada of at least $1.3 billion. By comparison, the cost of the Montreal Protocol imposed CFC stepped phase-out (100% by the year 2000) to be sustained by Canadians was estimated to be between $22 million and $54 million over the next decade (Abt, 1990). Thus, CFC control costs might swell up to 60 times if faster phase-out than specified in the Protocol is required.

What Has Been Done To Improve Automotive Fuel Efficiency?

Corporate Average Fuel Economy (CAFE) standards were enacted in 1975 and first took effect in 1978. The goal was to double average 1974 US new car fuel economy by 1985. The average fuel economy of import and domestic fleets was kept separated (Figure 7). Average new domestic car fuel economy increased 104% while new imported car fuel economy increased 38% over 1974 levels. However, not all of this increase can be attributed to CAFE requirements.

The domestic US car market was moving towards higher fuel economy before CAFE took effect (see Figure 8). For model years 1978 to 1983, gasoline price increases were sufficient to keep market demand above CAFE requirements (Leone, 1990). Manufacturers were not constrained by CAFE during this period in the sense that consumers, on average, were buying more fuel-efficient vehicles than required by CAFE (Figure 8). Technology decisions during this period were influenced by market conditions and not just by CAFE requirements.

However, from the peak reached in 1981, actual gasoline prices started to fall. In response, the market demand for fuel economy softened. During this period of declining prices, the historical impact of CAFE regulation was estimated to have increased the average fuel economy of new vehicles 1 to 1.5 mpg over what the market would have selected in its absence. A gasoline fee, if it would have been imposed in 1984, was estimated to have had the same impact as the CAFE regulation, but at an 85 percent lower cost to society (Leone, 1990).

The real burden or benefit to society of CAFE rules depends critically on timing. CAFE regulation is least burdensome to society when future fuel price increases are higher than anticipated. Because the US CAFE regulation was imposed prior to an unanticipated oil shock, the costs of the regulation were less than they otherwise would have been in the early 1980s. The precision and foresight required in formulating appropriate future standards is beyond the scope of any government or other organization in our rapidly changing world. Additional increases in CAFE will be particularly costly and environmental goals thwarted if future oil prices turn out to be relatively low.

Currently there are many design changes under development for future fuel efficiency, but the technologies are new and the expenditures for the benefit of the consumer are often marginal. The easiest ways to significantly increase fuel economy have been exploited. Future increases in efficiency will be relatively more expensive. Because CAFE is a law, the industry is obligated to apply such technologies when required, at times even when the implementation cost exceeds projected lifetime consumer benefit.

The Greenhouse Potential of a 40% Fuel Economy Increase

The United States Congress is reviewing various proposals to increase CAFE standards for cars and light trucks (up to 8,500 pounds GVW) by as much as 20 to 40% (eg., 20% by 1996 and 40% by 2001). However, an increase by as much as 40% likely would force consumer sacrifices such as significantly lowering vehicle weight, size, load capacity, automatic transmission usage and air conditioning availability, as well as other changes, possibly including increasing North American diesel engine usage, conceivably to European levels, if other problems (future US Clean Air Act and market requirements) can be solved.

The effect of a 40% fuel economy increase for US cars and light trucks on greenhouse enhancement is developed in Figure 9 and explained below. To estimate maximum impact, it is assumed that passenger cars and light trucks are able to increase their fuel economy instantly, beginning from the 27.5 mpg 1990 US base standard for cars and 20.2 mpg for light trucks. Because it takes many years for the in-use car and truck fleet to turnover (about 7 years to replace 50% of the cars and about 15 years to replace 90% of the cars) any CO_2 impact during the vehicle turnover period would be smaller than the instant change assumed here.

As shown earlier in Figure 2, CO_2 emissions from US highway vehicles (all cars and trucks) contribute 2.3% to potential greenhouse enhancement when all greenhouse gases are considered. The relative amount of the 2.3% CO_2 emitted by passenger cars, light trucks, and heavy trucks, as groups, is illustrated in the first bar of Figure 9. If heavy trucks are deleted from the total, the contribution for the remaining cars and trucks is 1.82% (second bar).

Instantly moving all US in-use cars and light trucks to 1990 CAFE levels results in a 15% decrease to 1.55% (third bar in Figure 9). Finally, instantly increasing fuel economy for both domestic and imported vehicles by 40% would have less than a half percent maximum impact (a 0.44% difference between the third and fourth bars) on greenhouse enhancement. Any actual impact would be smaller during the new vehicle phase-in period. If such a change is really necessary, there are more cost-effective ways (e.g., a carbon or gasoline fee) than relying on CAFE.

What Is the Role of Alternative "Clean" Fuels?

Alternative fuels have some potential to reduce CO_2 emissions relative to gasoline as indicated in Figure 10 (Acurex, 1986). The figure includes the effects of fuel type/source, combustion emissions during vehicle operation, the CO_2 generated during fuel production, and emissions due to process efficiency variations in fuel production. The values in Figure 10 are approximate as they may change with technological improvements.

As for renewable fuels (such as ethanol), only the fossil fuels used in their production need be included (net versus gross CO_2 in Figure 10). Market-value based allocation of fossil CO_2 to all by-products that are created during the production of a renewable fuel also is important to evaluate. Such considerations cause the estimated greenhouse impact of renewable fuels to vary widely in the open literature. In the final analysis, what will be found to be practical in the market place is the important consideration for estimated impact. The greenhouse benefit of renewable fuels depends upon farm yield, fossil energy inputs, and the usage, value, and allocation of CO_2 emissions among by-products.

Methanol-fueled vehicles have potential for higher engine efficiency (brake thermal efficiency) in the 3% to 15% range. Because of this, methanol vehicles range from somewhat better to twice as poor as conventional fuels in terms of their greenhouse warming potential, depending on the primary source used to produce the fuel. Methanol made from natural gas would provide the least amount of additional CO_2 and would be directionally helpful to climate change mitigation.

The greenhouse impact of electric vehicles, hydrogen fuel, and other alternatives depends upon the energy source used and the efficiency achieved. To be practical, reasonable consumer requirements need to be achieved before massive adoption of any new energy alternative should be made. To date, the relatively low cost of gasoline and diesel fuel and their very desirable inherent advantages have naturally restricted the usage of other energy sources. This represents the challenge that all potential clean-fuel alternatives must face.

The Clean Air Act Amendments of 1990 require a pilot clean-fuel program to be undertaken in California, starting model year 1996. Minimum volume requirements are 150,000 vehicles per year for model years 1996, 1997 and 1998, and 300,000 vehicles per year thereafter. Pilot programs help evaluate the wisdom of alternatives under consideration before massive commitment. Federal reformulated-fuel and clean-fuel programs also were included in the 1990 Amendments.

Alternative fuels are not a new concept. Consideration of a wide variety of energy strategies started at the very beginning of the development of motor vehicles (Shiller, 1990). The very early days of the motorized vehicle era were characterized by no firm ground rules or guidelines upon which to base vehicle design and associated energy strategy. The fathers of the automobile nudged knowledge forward in tentative bits and pieces (many of their early ideas are recorded in the automotive magazine *The Horseless Age*, a monthly journal that was devoted to motor vehicle interests during their early development {ca 1895-1918}).

It is interesting to note the rich variety of energy strategies that were reported including motorized vehicles built to operate on gunpowder, calcium carbide/acetylene, compressed air, ether, compressed springs, carbonic acid, electric battery, alcohols, coal gas, crude petroleum, gasoline and other energy sources (Ingersoll, 1895 & 1897a, b, c, d, e; Staner, 1905; Hagen, 1977). During the development of these ideas and with advancements in both engine technology and fuels, the advantages of liquid fuels became apparent. Gasoline emerged as the fuel of choice because of important practical considerations such as driving range, convenience, safety, and cost.

As an illustration of one of the major advantages of liquid fuels, Figure 11 shows the relative amount of useful mechanical energy that can be derived per unit fuel system mass (energy per unit mass or specific energy). Liquid fuels also have relatively high power densities (specific power or power per unit mass).

A low value for specific energy means a more limited driving range. This can be offset to some extent if one is willing to use a more massive on-board energy storage system (limited by vehicle design, function, and size) or if less conservative safety factors are used (such as operating closer to the rated bursting pressure of compressed gas tanks). As illustrated in the figure, the amount of useful mechanical energy available to operate current-technology motorized vehicles, at a combined fixed mass of fuel and its on-board storage system (relative to gasoline), varies widely. Of the various energy strategies listed, liquid fuels provide the greatest useful energy.

Over 97% of the energy used by the transportation sector in the United States comes from oil because of that fuel's inherent advantages. The challenge for alternative fuels will be to provide as much of the desirable benefits of petroleum-based fuels as possible. To the extent that alternative fuels cannot match gasoline in terms of cost, performance, vehicle utility, and fuel availability, consumer resistance will hamper acceptance. In the end, the consumer will be the final judge of how well that challenge is addressed.

Conclusions

Any response to an "enhanced" greenhouse effect, if it is found to be necessary, needs to be based on international cooperation in order to work. Although flexibility is needed so that each country will be able to take into account its own unique situation, the need for international cooperation and coordination remains. Disproportionate efforts among the nations inevitably would lead to economic dislocations. International trade flows can quickly undo greenhouse gas reductions that are undertaken unilaterally. Without full or major international participation, the positive steps taken by participating countries likely would be nullified.

The challenge in addressing the greenhouse issue will be to find a cost-effective method that allows for the continued development of the various transportation systems required by consumers. Command and control type fuel economy standards are the least effective way to address that challenge. They typically obstruct the flexibility needed for finding lowest-cost solutions and create unnecessary market distortions. In general command and control efficiency standards will always have significant disadvantages when compared to methods that incorporate social costs into the price for energy.

The concept of a carbon fee on all fossil fuels is a method that ensures that the social costs of greenhouse gas emissions are borne by emitters. The method also would accomplish objectives at lower cost than command and control regulations. In addition, the carbon fee approach is simpler to administer and adjust than policies that rely on the trading of marketable emission permits.

Despite its advantages, the short-term effects of a carbon fee should be considered before implementation. Slow phase-in very likely will be required to mitigate the main economic costs of reducing the growth in CO_2 emissions. The cost of stabilizing and ultimately reducing CO_2 emissions might be quite large. Some studies suggest that CO_2 emissions could be stabilized at a cost of a reduction in GNP of about 1%. The magnitude of the optimal carbon fee and the appropriate emission reduction objectives still need to be established.

CO_2 from highway vehicles worldwide represents less than 6% of the current potential for greenhouse enhancement. With the scheduled phasing out of air conditioning refrigerant CFC-12, its contribution to global warming impact per vehicle will be reduced 84% over the next century. This change is significant. It is roughly equivalent, in terms of greenhouse impact, to reducing CO_2 emissions from US highway vehicles 35%. In addition, the refrigerant's potential to destroy stratospheric ozone will be completely eliminated with the change-over.

Support for scientific climate research should continue. Periodic feedback from the scientific community is needed so that response programs can be readjusted as necessary. It is very important to reduce the uncertainty regarding what needs to be done to limit costs.

Alternative fuels have some potential to reduce CO_2 emissions relative to gasoline. However, the main challenge facing alternative fuels will be to provide as much of the desirable benefits of petroleum-based fuels as possible.

Table 1
1987 CO_2 Emissions by World Region

Emissions by Continent or Group of Nations	Billion* Tons CO2 Combustion	Percent by Region	1987 Population In Million	Metric Tons Per capita	lb CO_2 Per $GNP
Africa	0.6226	2.929	590.96	1.05	4.0
Latin America	0.9493	4.466	414.64	2.29	2.9
Asia	3.3437	15.730	2586.47	1.29	7.5
Eastern Europe	5.2623	24.755	420.15	12.52	3.8
Middle East	0.5900	2.776	177.68	3.32	3.9
Mediterranean	0.0059	0.028	1.05	5.57	3.0
OECD#	10.1972	47.971	817.16	12.48	2.0
Ocean Navigation	0.2862	1.346	0.00	-	-
Totals	21.26*	100.0	5008.1	4.24	2.8
Deforestation	8.00	-	5008.1	1.59	-
Cement Calcination	.51	-	5008.1	0.10	-

* Does not include deforestation and cement calcination listed on the last two lines in the table above.

Includes 24 countries (see OECD, List).

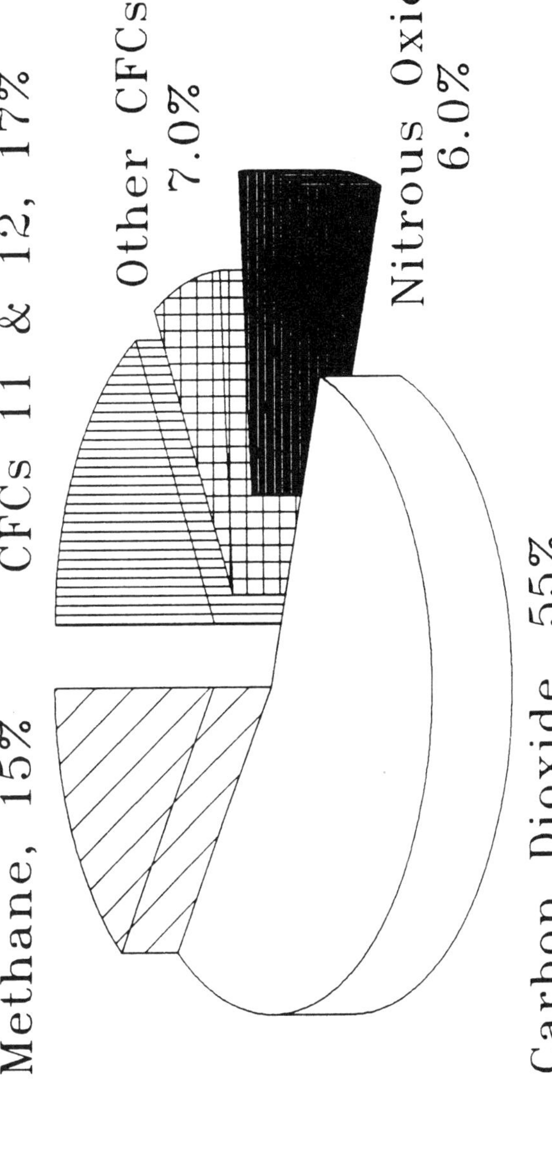

Figure 1

The Relative Importance Of Greenhouse Gases

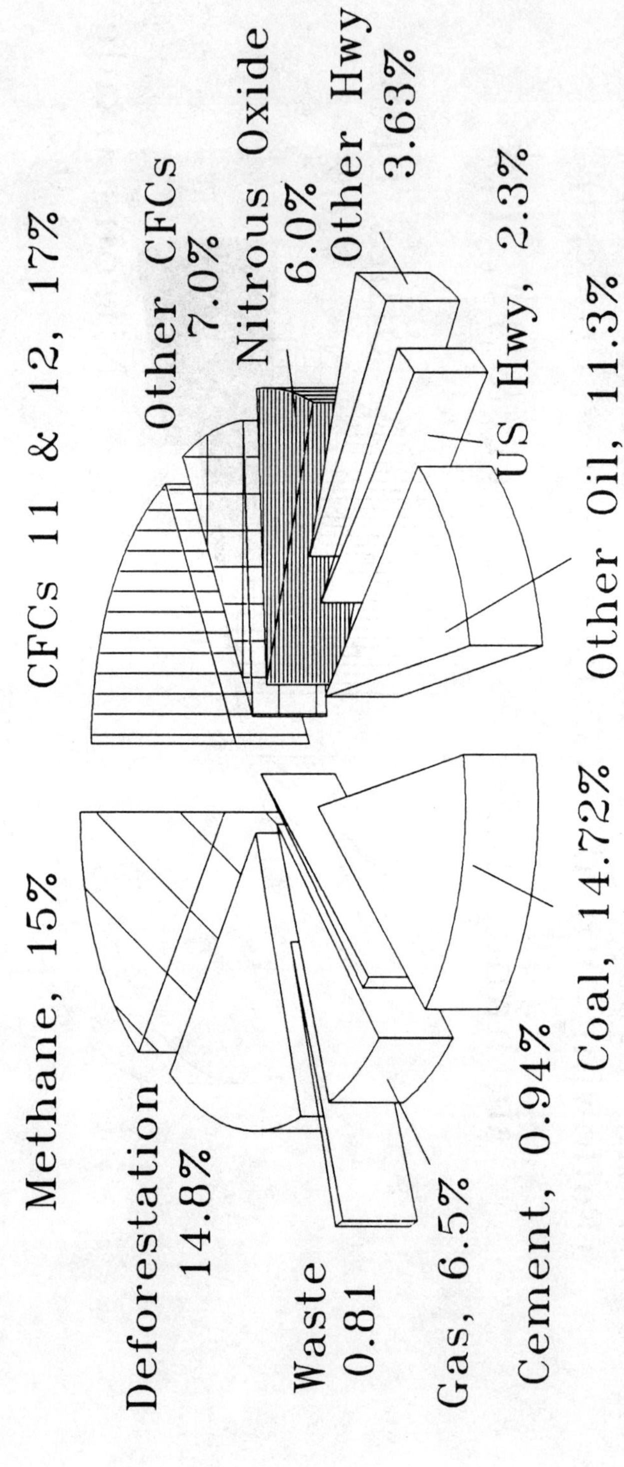
Contribution To Greenhouse Enhancement
CO2 (White Pie Slices) Allocated By Fuel Type/Usage
Methane, 15%
CFCs 11 & 12, 17%
Deforestation
14.8%
Other CFCs
7.0%
Nitrous Oxide
6.0%
Other Hwy
3.63%
Waste
0.81
Gas, 6.5%
Cement, 0.94%
Coal, 14.72%
Other Oil, 11.3%
US Hwy, 2.3%

Figure 2

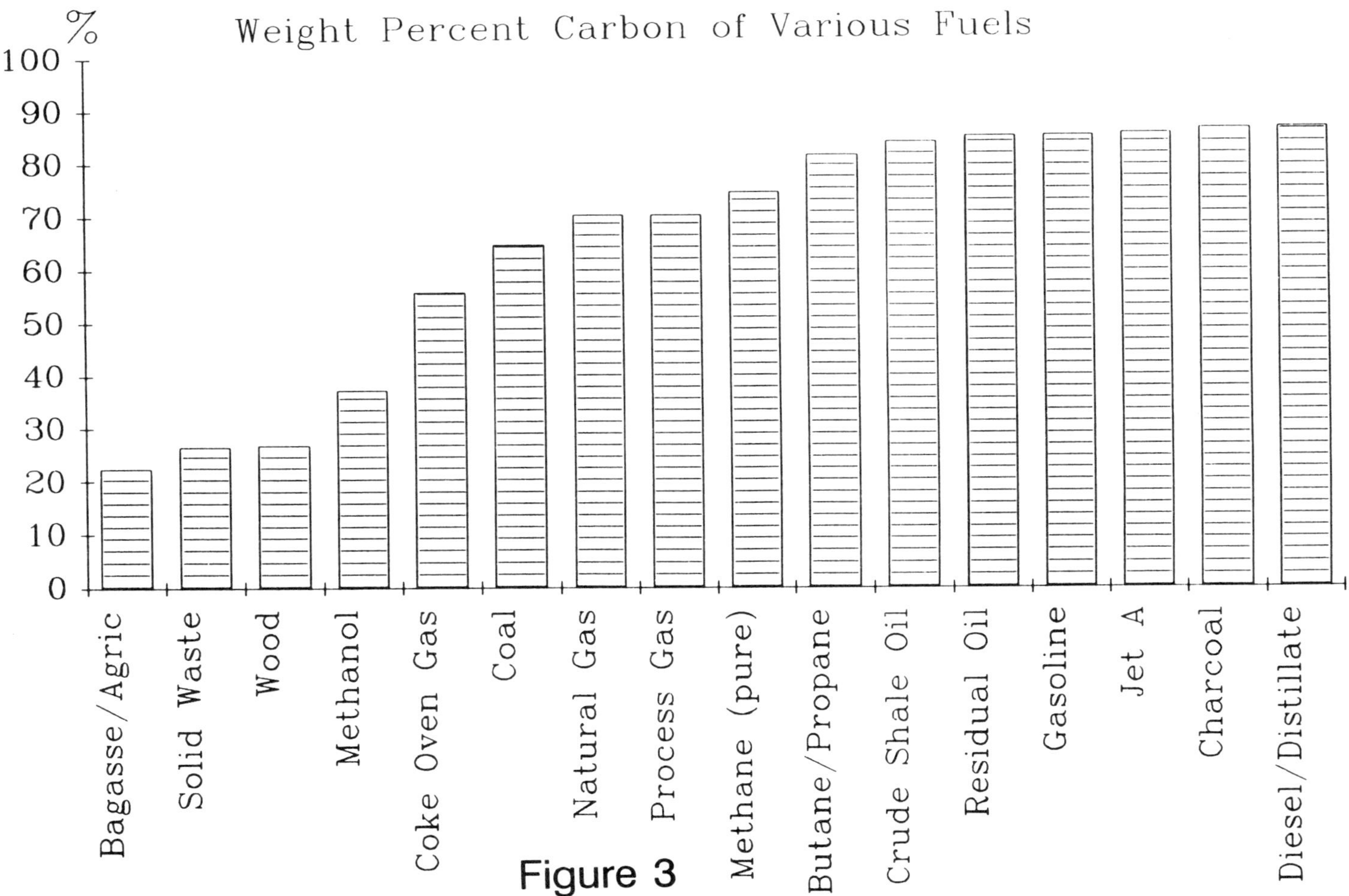
Weight Percent Carbon of Various Fuels
%
100
90
80
70
60
50
40
30
20
10
0
Bagasse/Agric
Solid Waste
Wood
Methanol
Coke Oven Gas
Coal
Natural Gas
Process Gas
Methane (pure)
Butane/Propane
Crude Shale Oil
Residual Oil
Gasoline
Jet A
Charcoal
Diesel/Distillate

Figure 3

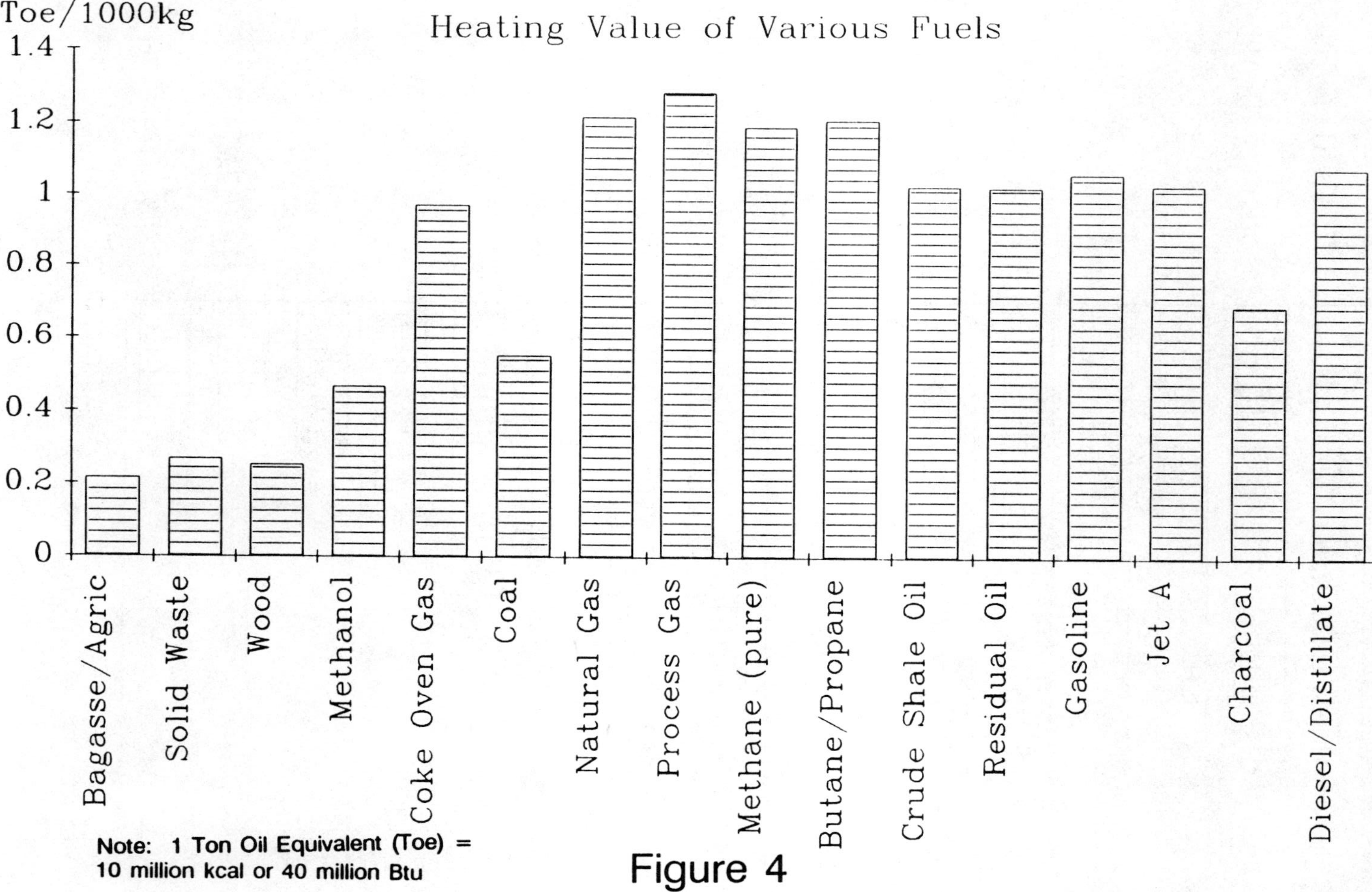

Note: 1 Ton Oil Equivalent (Toe) = 10 million kcal or 40 million Btu

Figure 4

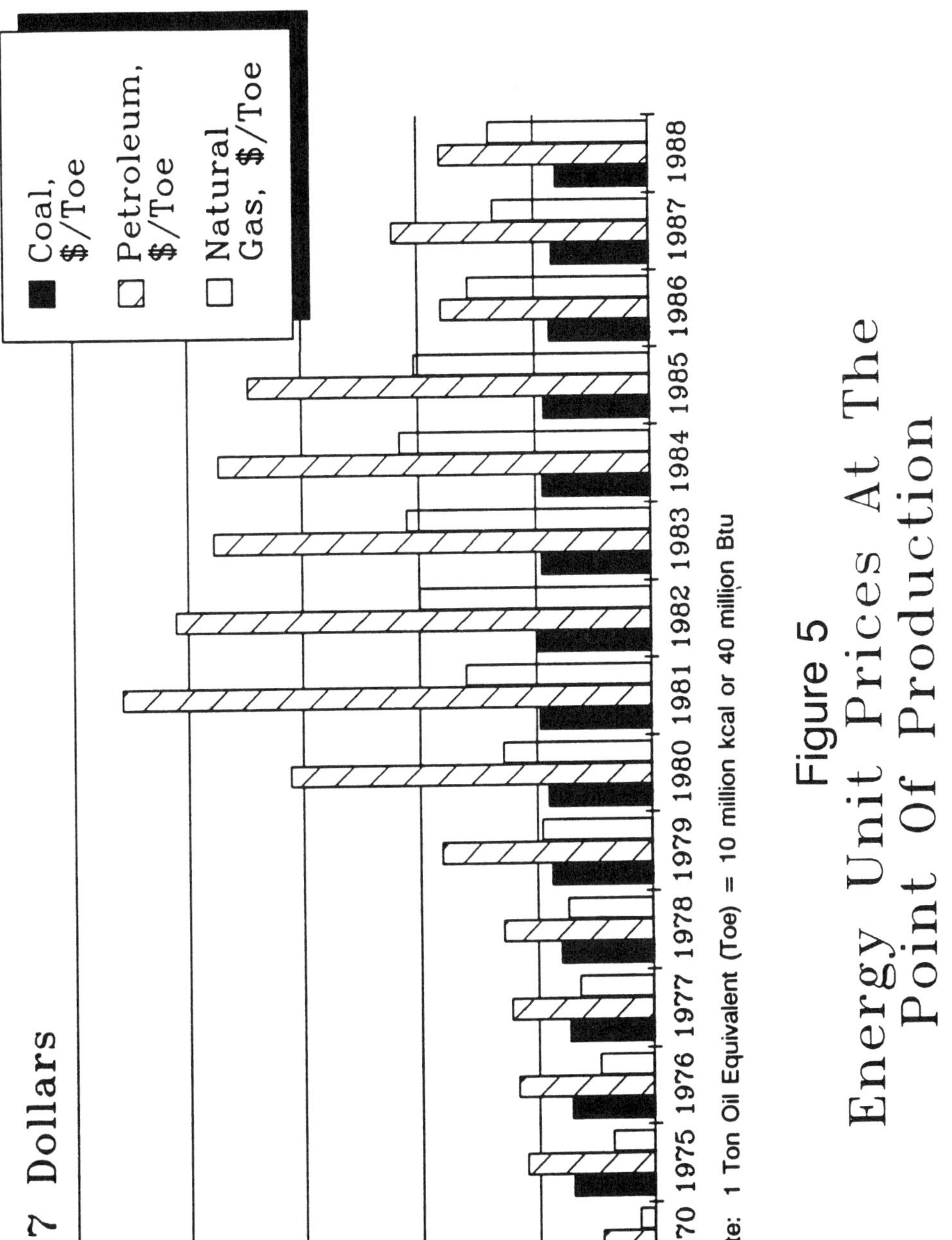

Figure 5
Energy Unit Prices At The Point Of Production

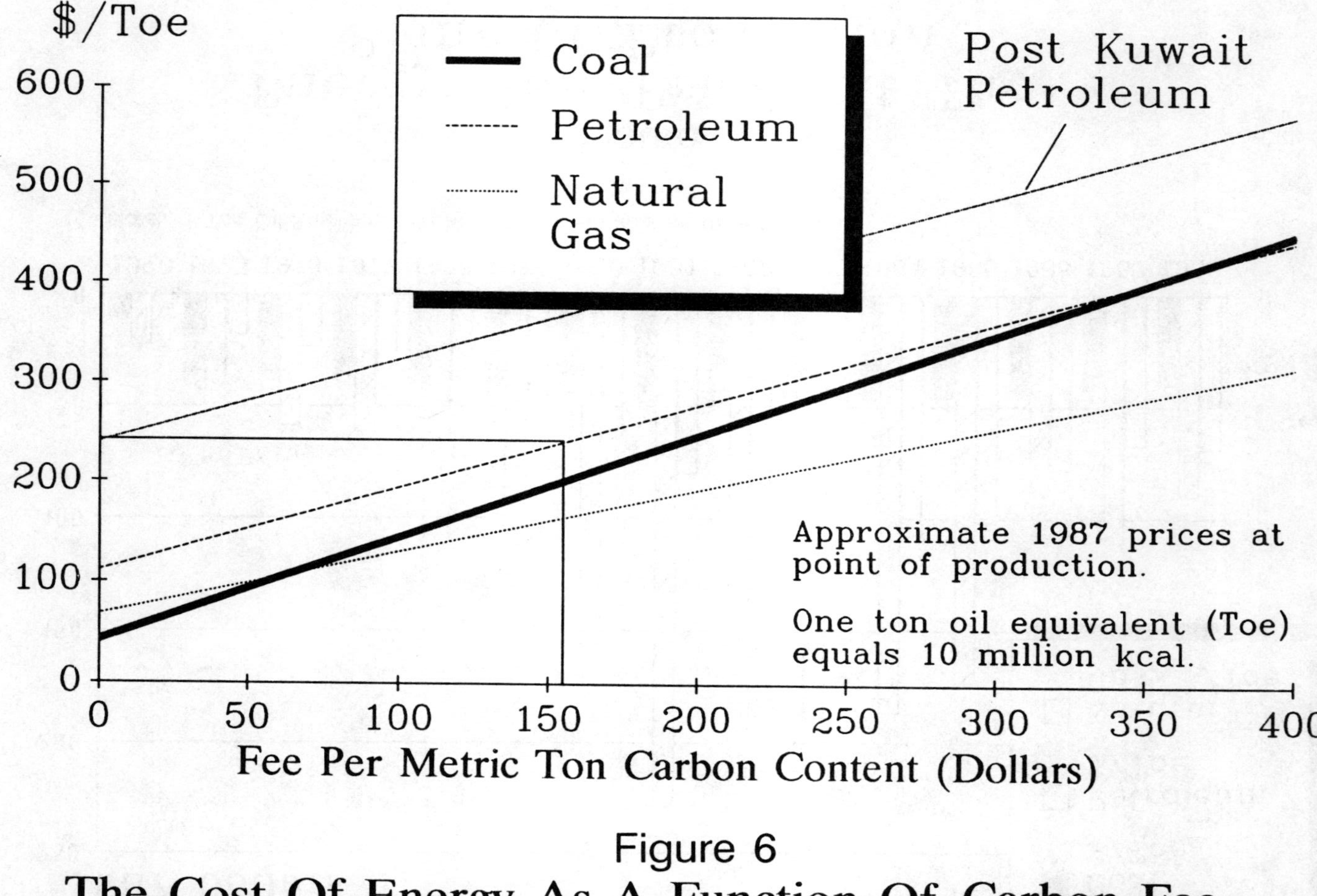

Figure 6
The Cost Of Energy As A Function Of Carbon Fee

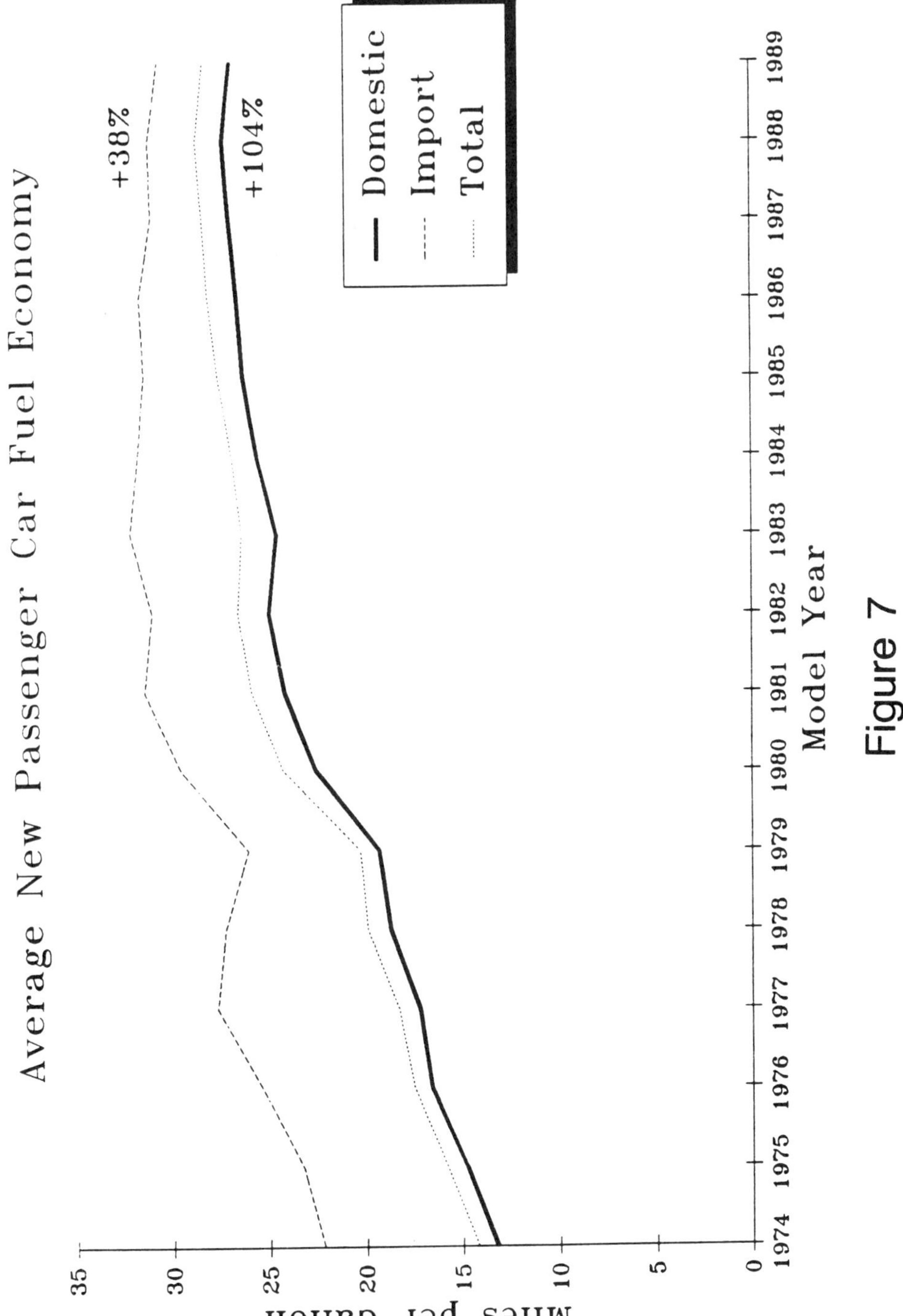
Average New Passenger Car Fuel Economy
+38%
+104%
Domestic
Import
Total
Miles per Gallon
0
5
10
15
20
25
30
35
1974
1975
1976
1977
1978
1979
1980
1981
1982
1983
1984
1985
1986
1987
1988
1989
Model Year

Figure 7

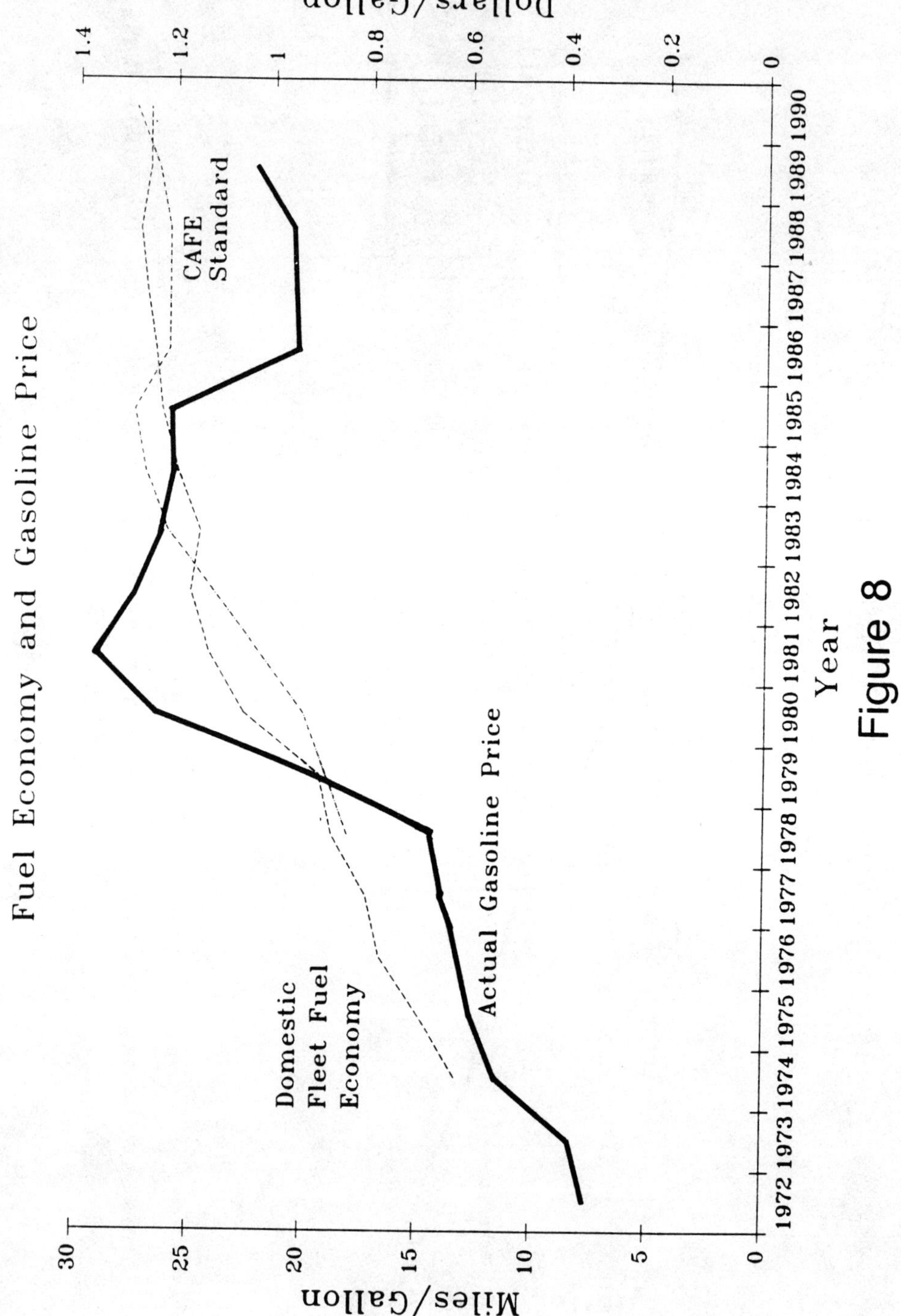

Fuel Economy and Gasoline Price
Miles/Gallon
0
5
10
15
20
25
30
Dollars/Gallon
0
0.2
0.4
0.6
0.8
1
1.2
1.4
1972 1973 1974 1975 1976 1977 1978 1979 1980 1981 1982 1983 1984 1985 1986 1987 1988 1989 1990
Year
Domestic Fleet Fuel Economy
Actual Gasoline Price
CAFE Standard

Figure 8

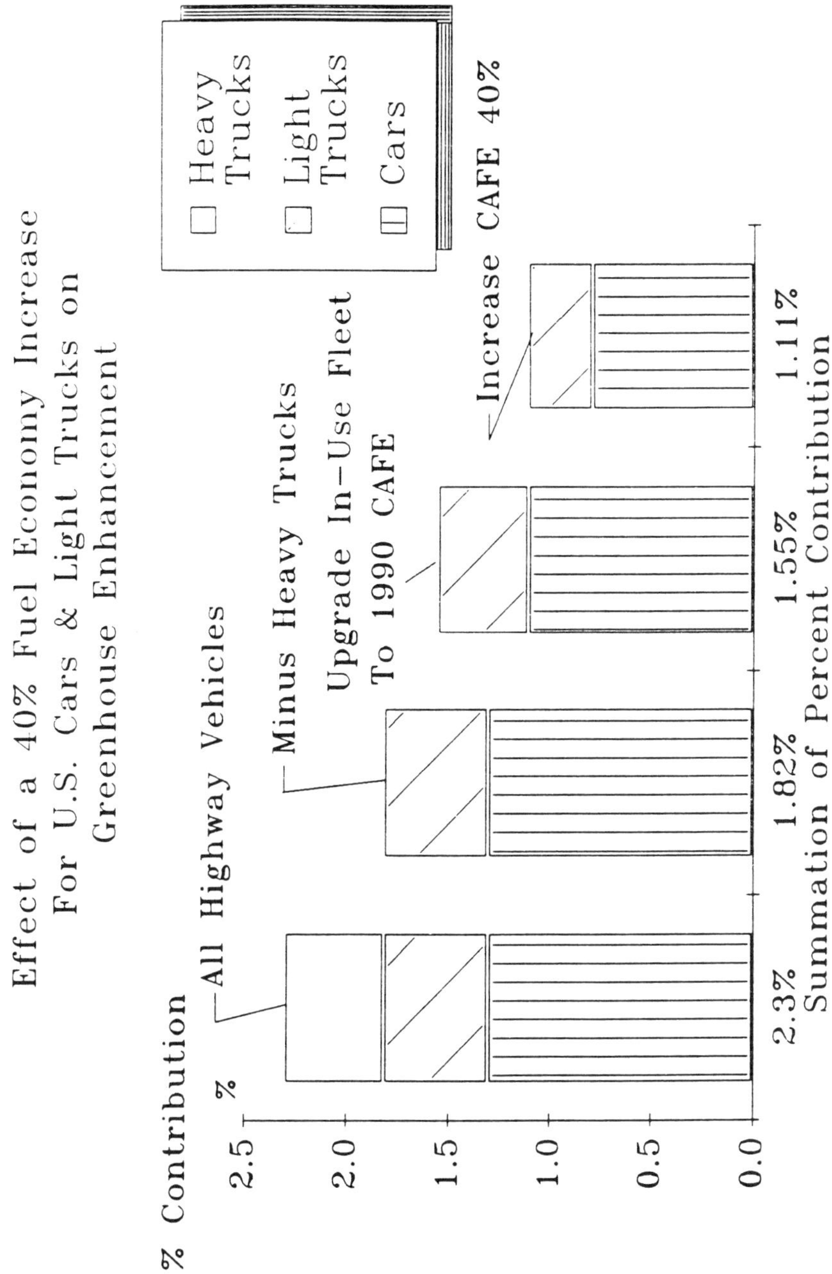

Effect of a 40% Fuel Economy Increase
For U.S. Cars & Light Trucks on
Greenhouse Enhancement

Heavy Trucks
Light Trucks
Cars

% Contribution

2.5
2.0
1.5
1.0
0.5
0.0

All Highway Vehicles
Minus Heavy Trucks
Upgrade In-Use Fleet
To 1990 CAFE
Increase CAFE 40%

2.3% 1.82% 1.55% 1.11%
Summation of Percent Contribution

Figure 9

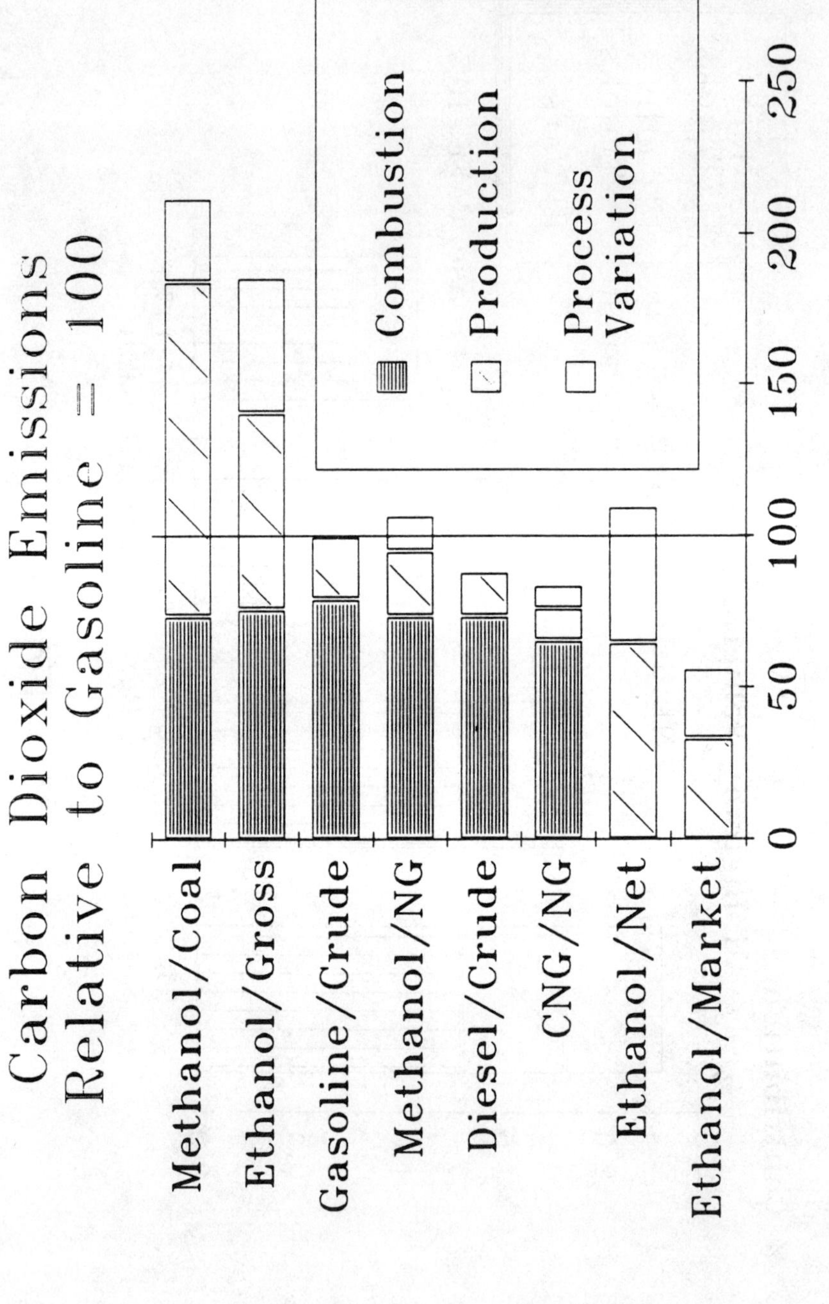
Carbon Dioxide Emissions
Relative to Gasoline = 100
Methanol/Coal
Ethanol/Gross
Gasoline/Crude
Methanol/NG
Diesel/Crude
CNG/NG
Ethanol/Net
Ethanol/Market
Combustion
Production
Process Variation
0
50
100
150
200
250

Figure 10

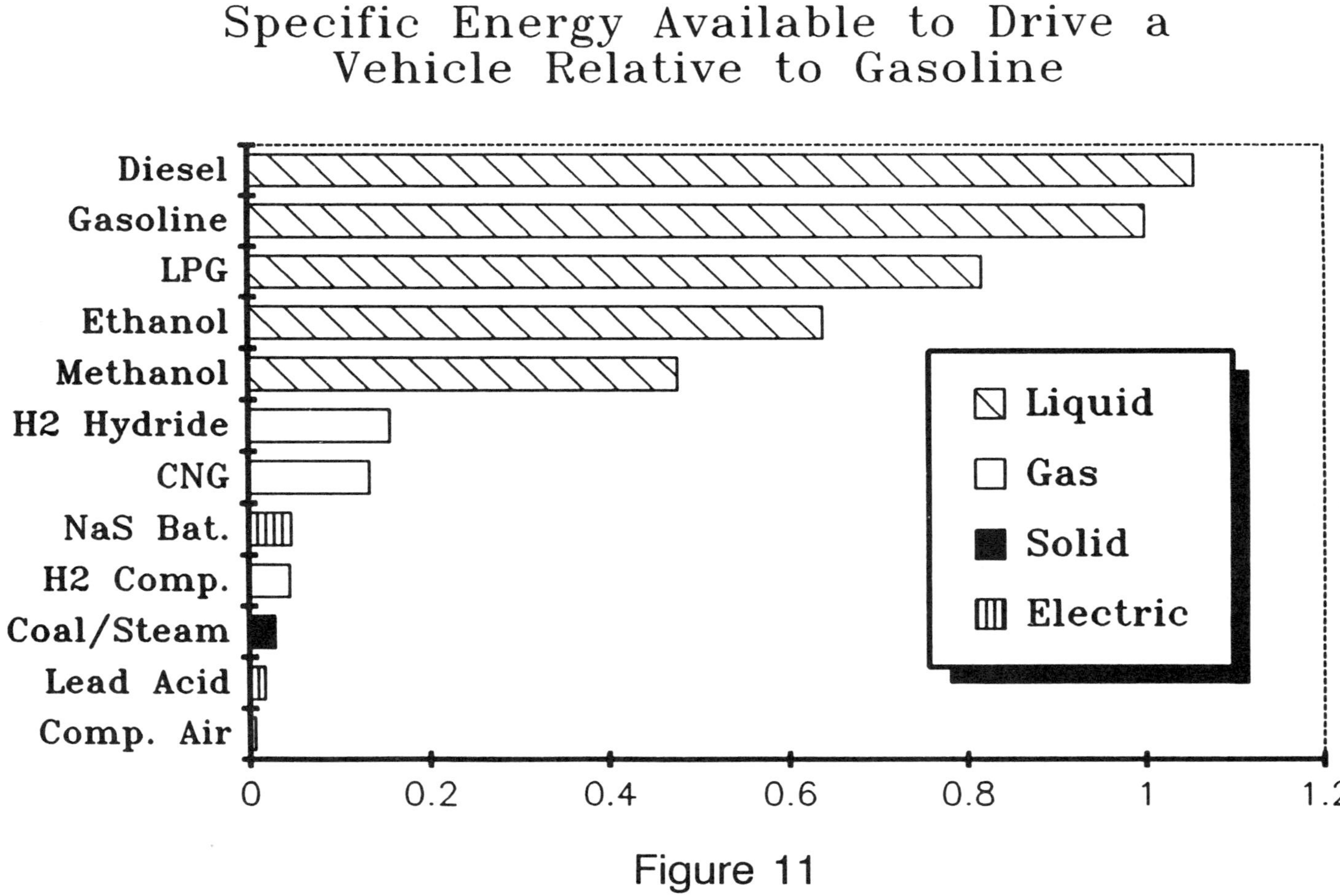
Specific Energy Available to Drive a
Vehicle Relative to Gasoline
Diesel
Gasoline
LPG
Ethanol
Methanol
H2 Hydride
CNG
NaS Bat.
H2 Comp.
Coal/Steam
Lead Acid
Comp. Air
0
0.2
0.4
0.6
0.8
1
1.2
Liquid
Gas
Solid
Electric

Figure 11

References

Abt, 1990, *An Assessment of the Economic Impacts of a Ban on the Use of Chlorofluorocarbons in Car and Light Truck Air Conditioners*, Project Report to Environment Canada, Abt Associates of Canada.

Acurex Corporation, November 1986, *California's Methanol Program, Evaluation Report, Volume I, Executive Summary*, for the California Energy Commission at the request of Governor George Deukmejian.

Brundtland, April, 1987, *Our Common Future* (or The Brundtland Report), after commission chairman Gro Harlem Brundtland (Prime Minister of Norway). Final report of the World Commission on Environment and Development, published in more than twenty languages; English version from Oxford University Press.

CBO, August 1990, *Carbon Charges as a Response to Global Warming: The Effects of Taxing Fossil Fuels*, Congress of the United States, Congressional Budget Office, Washington, D.C.

Cristofaro, Alex, December 1990, *The Cost of Reducing Greenhouse Gas Emissions in the United States*, Air and Energy Policy Division, EPA.

DOE, September 1990, *The Economics of Long-Term Global Climate Change, A Preliminary Assessment*. Report of an Interagency Task Force, United States Department of Energy, Office of Policy, Planning and Analysis.

Hagen, D. L., 1977, *Methanol as a Fuel: A Review with Bibliography*, SAE 770792. Also in SAE/PT-80/19, 1980.

IEA, 1989a, *Energy Balances of OECD Countries 1986/1987*, International Energy Agency, Paris.

IEA, 1989b, *World Energy Statistics and Balances 1971-1987*, International Energy Agency, Paris.

Ingersoll, E. P., Editor, November, 1895, *The Horseless Age*, New York.

Ingersoll, E. P., Editor, January, 1897a, "The Oil Famine Bugaboo," "A Gunpowder Motor," *The Horseless Age*, Vol. II, No. 3, New York.

Ingersoll, E. P., Editor, February, 1897b, "Acetylene as a Motive Agent," "Motor Cabs (Electric) in New York," *The Horseless Age*, New York.

Ingersol, E. P., Editor, March 1897c, "Acetylene Motors," *The Horseless Age*, New York.

Ingersoll, E. P., Editor, September, 1897d, "Riker Electric Victoria," "Carbonic Acid Carriage Motor," "The Worthley Steam Carriage," "New Motor (Electric) Fire (Fighting) Apparatus," "Alcohol as a Fuel for Motors," *The Horseless Age*, New York.

Ingersoll, E. P., Editor, October, 1897e, "Compressed Air Vehicles of the Pneumatic Carriage Company," *The Horseless Age*, New York.

IPCC, April 1990, *Greenhouse Gases and Aerosols*. Peer Reviewed Assessment for Intergovernmental Panel on Climate Change, WG1 Plenary.

IPCC, April 1990, *Section 2, Radiative Forcing of Climate*. Peer Reviewed Assessment for WG1 Plenary.

IPCC, June 1990, *Policymakers Summary of the Scientific Assessment of Climate Change*. Report to Intergovernmental Panel on Climate Change from Working Group I, Final draft.

IPCC, July 1990, *Overview and Conclusions, Climate Change; A Key Global Issue*. Report of the Intergovernmental Panel on Climate Change.

IPCC, August 1990, *Overview, IPCC First Assessment Report*, WMO, UNEP.

Leone, Robert A.; Parkinson, Thomas W.; May 1990, *Conserving Energy: Is There a Better Way?* A Study of Corporate Average Fuel Economy Regulation, Boston University and Putnam, Hayes & Bartlett, Inc.

Miller, Alan; Mintzer, Irving; and Brown, Peter; 1990, "Rethinking the Economics of Global Warming," *Issues In Science and Technology*, Volume VII, No. 1, National Academy of Sciences, National Academy of Engineering, Institute of Medicine.

MVMA, 1989, *World Motor Vehicle Data 1989 Edition*, Motor Vehicle Manufacturers Association of the United States, Inc., Economic & International Affairs Department.

MVMA, 1990, *Motor Vehicle Facts & Figures '90*, Motor Vehicle Manufacturers Association of the United States, Inc., Public Affairs Department.

OECD, List. The 24 member countries of the Organization for Economic Cooperation and Development are: Australia, Austria, Belgium, Canada, Denmark, Finland, France, Germany, Greece, Iceland, Ireland, Italy, Japan, Luxembourg, the Netherlands, New Zealand, Norway, Portugal, Spain, Sweden, Switzerland, Turkey, the United Kingdom, and the United States.

SA, 1990, *Statistical Abstract of the United States, 110th Edition*, US Department of Commerce, Bureau of the Census, Washington, D.C.

Shiller, J. W., 1990, "The Automobile and the Atmosphere," *Energy Production, Consumption, and Consequences*, National Academy of Engineering, John L. Helm, Editor, National Academy Press, Washington, D.C.

Staner, H. Walter, Editor, January 14, 1905, "Alcohol as a Fuel for Motor Cars," *The Autocar*, Vol. XIV, No. 482.

SWCC, October/November 1990, *Final Statement of the Second World Climate Conference Scientific/Technical Sessions*, Annex II as reported to the United Nations General Assembly, Forty-fifth session, agenda item 81, A/45/696/Add.1.

SWCC, November 1990, *Ministerial Declaration of the Second World Climate Conference*, Annex III as reported to the United Nations General Assembly, Forty-fifth session, agenda item 81, A/45/696/Add.1.

WRI, 1990, *World Resources 1990-91, A Guide to the Global Environment*, World Resources Institute, Oxford University Press, p. 350, ISBN 0-19-506229-9.

WEIGHING THE COSTS AND BENEFITS OF CLIMATE CHANGE POLICIES

Mary Beth Zimmerman

The Alliance to Save Energy
1725 K Street, N.W.
Suite 914
Washington, D.C. 20006-1401

Introduction

Are the benefits of avoiding climate change worth the investments necessary to do so? Most economists agree that at least some efforts – especially in the area of energy efficiency improvements – are worthwhile. Others believe that the immediate economic and environmental benefits of mitigation (emission control) policies make substantial reductions in greenhouse gases (GHGs) worthwhile, ensuring both continued economic growth and avoidance of severe climate change.

To date, neither mitigation cost nor benefit estimates have been complete in their assessment of policy welfare implications. In general, attempts to measure policy benefits have been either fairly simplistic in their characterization of a climate-changed environment or incomplete in the types of benefits accounted for. Differences in models and assumptions mean that consistent and thorough comparison of policy cost and benefit estimates have not been possible.

This paper presents a framework for ensuring measurement of policy costs and benefits from a consistent baseline, as well as for identifying the various ways in which policies may generate costs and benefits. Use of this framework indicates that, in general, the costs of climate change – and therefore the potential benefits of climate change policies – have tended to be under-reported. On the other hand, the costs of mitigation policies have tended to be overestimated.

Economics is the science of using scarce resources efficiently and policy costs and benefits are discussed from this perspective. The focus is on the effects that both climate change and climate change policies might have on the way resources are used, and in turn the effect of resource use on welfare. The welfare implications of nonmarket conditions in particular are problematic, but necessarily addressed in any evaluation of major environmental public policies.

Copyright 1991 by Elsevier Science Publishing Company, Inc.
Global Climate Change: The Economic Costs of Mitigation and Adaptation
James C. White, Editor

Even in this broad framework, an economic approach to evaluating climate change is not the only, or even the broadest, approach possible. In its traditional form, economic analysis focuses on the utility or well-being of individuals. It is largely unsuccessful in evaluating the overall well-being of a society, or a number of societies over generations. It also assumes that the total value of natural resources is the utility those resources can provide to human beings. Despite these limitations, it is a useful tool for evaluating the resource implications of policy choices.

A Framework for Evaluating Climate Change Policies

A Basic Model of Welfare Growth

Much of the work done to estimate the welfare implications of climate change mitigation policies has been based on a relatively simple, traditional model of economic growth. In this framework, output is determined both by the quantity of labor, capital and natural resources available – the factors of production – and the efficiency with which they are used.

Two types of output are generated in production: consumer goods and capital investments. (Figure 1.) Consumer goods generate immediate welfare and capital goods add to the resources available for production in future years. Taking account of the changes in capital stock from year-to-year allows an assessment of how much output can be produced; the larger the portion of output set aside (through consumer savings) for investment in new capital, the greater output will be down the road.

Year-to-year changes in the amount of labor, capital and natural resources from which each year's output can be generated under this basic model are characterized in Figure 2. Changes in the availability of labor and natural resources tend to be given less attention in the basic growth model. Changes in the size of the labor force from year-to-year are usually taken as a given (unaffected by choices made in the marketplace).[1] Many energy models account for reductions of finite energy sources, but rarely deal with other changes in the natural resource base.

The change in welfare from year-to-year made possible by these changes in the factors of production are portrayed as a simple "growth path" in Figure 3. The y-intercept indicates the current level of welfare and the slope shows how welfare changes over time. When capital accumulation is the only truly endogenous factor contributing to growth in the resource base, model results will tend to be very sensitive to the rate of savings and anything which affects the rate of savings. A higher savings rate lowers the y-intercept (reflecting lower initial consumption), but increases the slope of the growth curve.

The growth paths presented in this paper are illustrative and include no scale of measure, setting aside for the moment difficulties in actually measuring welfare. Ideally, measures of well-being include all changes in the well-being or utility of each individual. In practice, welfare is neither directly measurable or comparable from person to person.

An Extended Model: Adding Environmental Goods and Investments

People may derive benefits directly from natural resources and environmental quality, such as the excitement of viewing the Grand Canyon or the enjoyment of breathing fresh air.[2] Economists make it clear that although such benefits (and costs) can be external to market purchases, these "externalities" nonetheless affect welfare. The simple growth model described above includes only consumption goods and does not acknowledge the fact that consumers may derive benefits directly from natural resources.

The possibility of producing environmental goods as well as market goods has important implications for the choices consumers can make in using resources to generate welfare. Cleaner air, for instance, can be produced by requirements to limit solvents in paints; the mix of output will include a different kind of paint and fewer volatile hydrocarbons in the atmosphere. As with all public goods, optimum purchase of environmental goods is made collectively through public policy.

Changes in environmental quality can also constitute investments in the natural resource stock available for future production. Environmental investments are entirely analogous to capital investments; trees planted today, for instance, increase the opportunity for future production. Likewise, failure to keep air or water clean can erode the natural resource stock, lessening its economic value, or even generate disease which reduces labor productivity. Many watersheds and air basins in the United States are substantially cleaner than they were 10 or 20 years ago, representing an addition to the resource base which would not have been possible without prior investment.

A more complete model of our basic economy would look like that portrayed in Figure 4. There are now four types of output: Two types of consumer goods (market and environmental) and two types of investment goods (capital and natural resource).[3]

Using this framework, changes in the resource base from year-to-year are portrayed in Figure 5. There are now several more endogenous choices to be made in addition to the basic intertemporal choice between consumption and savings. Consumers choose between both market and environmental goods and different types of investments.

Environmental consumption and investment goods can be purchased through public dollars spent planting new trees or on water sewage treatment. They can also be purchased through policies designed to change the ways resources are used in production (through regulations or taxes), the assumption usually made for climate change modeling. The changes in resource use result in an increase in production of environmental goods and a decrease in consumption of market goods.

Measuring Welfare

Once environmental goods and investments are added to the picture, difficulties in measuring welfare must be dealt with directly. Market prices are usually taken as a clue to the value of products bought in the market, on the presumption that no one will pay more for a product than it is worth to them.[4] There are also various techniques to estimate the value of environmental goods. Once they are taken into account, we might expect the welfare path described in Figure 3 to be higher than a GNP-based growth path would indicate.

A shift in consumption from marketed goods to environmental goods will result in a loss of measured GNP, but a gain in welfare (assuming the shift occurred because people preferred the environmental goods more). (Figure 6.) In this case, the loss of output, as measured by GNP is illusionary; part of the output has merely been redirected to nonmarket products.

Discounting is another issue which emerges when welfare is measured over time. Standard economic practice provides for discounting – or reducing – the value of future costs and benefits. The reasons for discounting vary, involving elements such as the productivity of capital investment and people's time value of money (they would rather have $100 now instead of next year). In general, the higher the discount rate, the less important future events become in policy analysis. There is substantial debate over the appropriate rate at which future welfare should be discounted. A discount rate of zero is suggested by those who believe it is inappropriate to give greater weight to the welfare of current generations than the future generations. Because the growth path framework outlined here compares changes in welfare on a year-to-year basis, the growth path can be shifted to reflect different chosen discount rates.

Implications for Policy Evaluation

In practice, most climate change modeling reflects some aspects of natural resource use, but generally overlooks the potential for producing environmental consumption and investment goods. By overlooking the additional choices that people have in the ways they use resources – and the connections between these choices and welfare – the basic growth framework has a tendency to underestimate the value of climate change or other environmental policies.

In the extended model, consumers are free to select from among both environmental and market goods. Because of the fact consumers can shift among the types of goods they consume, increased consumption of environmental goods need not subtract from savings. There is therefore no *a priori* reason to expect investment to fall as a result of public policies which trade off market goods for environmental goods. Without an effect on savings, the cost of the environmental good is simply the value of the foregone market goods, with no implications for future output.

In using the basic model, however, an implicit assumption is usually made that consumers perceive the reduction in market goods as a loss of income, despite the increased availability of environmental goods. As a result, consumers are perceived as both reducing their expenditures on consumption and their savings in order to adjust to the lower level of income. The reduction in savings reduces capital investment, reducing production in future years. When purchase of environmental goods is presumed to affect capital accumulation, the estimated "cost" of purchasing the environmental goods will rise significantly.

A similar problem occurs when models fail to account for deterioration of the natural resource base resulting from production. Pollution from industrial agricultural discharge, for instance, can reduce fishing and other water-based economic activities. When environmental resource loss is not taken into account, the potential for future growth is overestimated because more and higher quality resources are presumed to be available in future years than will actually be the case. (Models which exclude consideration of natural resource availability implicitly assume no deterioration of the base.) In addition, because most modeling does not allow for the possibility that environmental investments made today may increase output in future years, all environmental policies are viewed as simply reducing current output and future capital stock.

Given the limitations on consumer environmental choices built into the basic growth framework, it is not surprising that reliance on these models suggests an inevitable "tradeoff" between purchase of environmental quality and economic well-being. The extended framework, by endogenizing environmental investments, makes it clear that they are complementary objectives.

Assumptions About Resource Availability

The growth path generated by any model is based on what we believe is available in terms of both resources and productive capabilities. If available resources decline in some way—because of an oil embargo or destructive earthquake, for instance—output will decline as well and the anticipated growth path will have to be readjusted downward. Likewise, an unexpected improvement in technology can increase expected growth.

In some cases, of course, it is not the resource base itself that changes, but our knowledge of that base. The discovery of a new oil reserve, or the downward estimation of existing natural gas supplies, will change our view about future welfare potential even though the amount of oil or gas actually in the ground did not change. In short, estimated growth paths can be expected to change over time as we learn more about our environment.

Comparing Mitigation Policy Costs and Benefits

Climate as a Resource

Climate—or, more accurately, stable climate—has only recently been thought of as a natural resource in the same way that air or water quality are. Climate can actually best be thought of as a bundle of environmental qualities, each of which affects human welfare. Average temperature changes are important, but so are factors such as the number of extreme high temperature days and weather severity. Even predictability is an important attribute of climate: three inches a year of predicable rainfall affects irrigation needs differently than does five inches of unpredictable rainfall.

Like clean air or pure water, these aspects of a stable climate are enjoyed for their own sake; they have "standing value." The stability of climate also indirectly affects welfare through its effects on more traditional environmental resources, such as agricultural yields of food, fuel and other commodities.

Whereas we once thought of climate as an inexhaustible resource, we now know that climate stability is a depletable resource, vulnerable to increased concentrations of GHGs. Simply put, our activities today affect the degree to which we can expect to enjoy the benefits associated with a stable climate in the future.

Modeling the Welfare Effects of Climate Change

New information about the climatic effects of greenhouse gas emissions should change our estimation of the size of the natural resource base just as new information about oil supplies would. In all likelihood, our estimation of the future natural resource base should be adjusted downward. (Figure 7.)

We can label the new, lower growth path the "climate change" curve. The area between the top and bottom growth paths can be viewed as the "cost" of the climate effects of greenhouse gas (GHG) emissions over time; it is the total difference in welfare between what was thought to be possible[5] and what is now thought to be possible given the ways we use fossil resources and our knowledge about climate effects.

Using the extended framework, policies designed to restrict CO_2 or other GHGs can be viewed as investments in the natural resource base—in particular, the future capacity of the atmosphere to absorb GHG emissions without disrupting climate stability. By altering productive processes now—through a carbon tax, restrictions on fuel use, or efficiency standards—a portion of the atmosphere's carrying capacity is reserved for availability in the future in order to avoid the loss of productivity illustrated by the lower climate change growth curve.

This second growth path, the one built explicitly with information about climate effects on the resource base, is the appropriate baseline from which to assess climate change policies. It is what we should expect to happen given current resource use practices. Unfortunately, the higher curve is usually used as the basis from which climate change policy costs are measured. Although labeled as "reference" or "business-as-usual," the higher curve does not realistically reflect the potential for growth in a world of unchecked GHG emissions.[6]

The Benefits of Mitigation Policies

When viewed as an investment in climate stability, efforts undertaken now to reduce greenhouse gases have the effect of avoiding some of the anticipated downward shift in the natural resource base, and thus increase the level of welfare expected in the future. In effect, it provides the opportunity to "shift" the welfare curve back up towards the level we previously expected.

To the extent that it is impossible to undo warming effects already physically committed to, or even to eliminate GHG emissions overnight, mitigation policies cannot put the welfare curve back where we originally thought it would be, but rather the benefits of reduced GHG emissions place the welfare growth path somewhere in between. (Figure 8.)

Including Mitigation Policy Costs

Steps taken now to reduce carbon dioxide and other greenhouse gases can themselves have costs in terms of foregone welfare from other goods and services which would otherwise have been produced. These costs also limit the extent to which mitigation policies can raise the growth path.

Information about both the costs and benefits of climate change mitigation policies can be used to construct a third welfare curve, representing the net changes in welfare on a year-to-year basis from undertaking a mitigation strategy. In the early years the net welfare shift might be negative, given the fact that some policies will involve incurrence of costs before benefits are realized. In later years, we would expect net benefits, reflecting the accumulated improvements in welfare from avoiding climate change. The "mitigation policy welfare curve" might look something like that portrayed in Figure 9.

A consistent analysis of policy costs and benefits would compare the areas between the climate change and mitigation curves. The area in which the mitigation growth curve drops below the climate change curve should be regarded as a loss of welfare; the area in which the mitigation curve rises above the climate change curve should be regarded as an improvement to welfare. If the improvements exceed the losses, the mitigation policy represents a net improvement for society.

Estimating Climate Policy Costs and Benefits

Estimating the Welfare Effects of Climate Change

What does the climate change welfare path really look like? Opinions vary greatly as to how much climate change might lower long-term prospects for growth, ranging from negligible to potentially catastrophic. Much of the difference in view depends upon which of the models of welfare growth described above is employed.

To look at the entire picture of changes, it is helpful to go back to Figure 5 and look at the different ways in which potential growth in welfare can change.

Environmental consumer goods

- *Direct welfare effects* such as preferences for moderate and stable weather conditions.[7] As with all public goods, the direct "consumption" of climate by one person does not reduce its effect on another, so that the total welfare implications of environmental changes are additive across all people, over many generations and may be quite large. People can also receive utility from knowing that future generations will not suffer from unstable climate conditions, even though they may not experience climate changes themselves.

- *Effects on the standing value of other environmental resources.* Climate changes are also expected to change some of the other environmental goods we tend to consume directly (in their standing state). Loss of forests and wetlands reduces the recreational and other welfare we derive from them. Again, losses are additive.

Changes in the natural resource base

- *Effects on the productivity of environmental resources.* This is the area in which climate change costs have received the most attention. To the extent that climate change and instability affect the productivity of biological processes, they affect the productivity of agriculture, fishing and other activities which depend upon biological growth.

- *Land use changes.* The loss of productivity from changes in land use patterns facilitated by climate change has been less fully evaluated. Most obviously, some amount of land is lost from the resource base due to sea level rise. However, we can also expect some loss in productivity from shifts in the location of productive processes. Assuming productive activities are optimally located to take advantage of existing resources, transportation and other requirements, any change in land use patterns will reduce productivity. This effect is unlikely to be noticed, but will nonetheless reduce total output.

Changes in other factors of production

- *Labor productivity.* If viruses and other diseases increase in geographic range with warming, the result may be exposure of some populations to diseases for which they have little natural immunity and perhaps no effective treatment. The result, of course, is a loss of welfare from poorer health, but also a decline in output from a less productive labor force.

- *Capital resources.* Substantial capital may be necessary to accommodate rising sea levels, changing water supplies, increased temperatures, etc. In diverting capital from the uses to which it would otherwise be put, these expenditures permanently lower both GNP and welfare.

The changes described above are equilibrium changes – they reflect the change in productivity and resource base after adjustment to new climatic conditions has occurred. Unfortunately, climate will not change from one equilibrium level to another, but will remain volatile as long as GHG emission levels exceed the atmospheric carrying capacity. It is more realistic to think about costs under conditions of continued ecological and thus economic disequilibrium. Economic changes which involve large capital investments, human migration, etc., can be expected to have substantial transitional or dislocation costs associated with them. This is especially true when the rate of change is fairly unpredictable, as it is with climate change.

Most of the costs outlined above are not reflected in the traditional model of economic growth, and have not been included in many of the analyses of climate change costs (or, conversely, the benefits of climate change policies). Even in the area of capital accumulation, which many growth models focus on, the capital resource implications of adaptation to climate change have not been thoroughly explored.

Estimating Mitigation Policy Costs

Carbon taxes, energy efficiency standards, and many other policies designed to reduce GHG emissions shift output to less energy intensive goods and shift production to methods which require less energy. In general, the use of less energy will require the use of more labor, capital, and/or nonenergy resources. If the most economic means of production are already in use, the expected result will be a reduction in the amount of goods and services which can be produced.

If there are major inefficiencies in current energy markets, however, we cannot presume that energy is already used in an economically efficient manner. Just as public policy can discourage energy use, it can encourage the use of energy in ways which would not otherwise be cost-effective. In the United States the electric utility sector, accounting for 35 percent of energy use, is regulated in a way which traditionally discourages the use of the least costly means of producing energy services.[8] In many countries, energy prices are heavily subsidized. Both situations result in a failure to adopt cost-effective energy efficient technologies which would reduce the overall costs of production and increase output. Policies which improve market signals increase output at the same time they reduce emissions.

Engineering or "bottom-up" estimates of the amount of cost-effective efficiency improvements available to be tapped suggest that sizable reductions could be made in U.S. carbon emissions in ways which also increase the output of market goods and services.[9] The desirability of including the potential for additional cost-effective energy efficiency improvements in models of the type described above is well acknowledged, but has been difficult to implement in practice. As a result, climate change mitigation costs are undoubtedly overestimated. The opportunity to tap cost-effective efficiency improvements may well mean that the mitigation welfare curve is never lower than the climate change curve, even in the short run. In this case, the mitigation curve would look just as it does in Figure 8.

Models designed to assess the costs of climate change policies will also be incomplete unless they take account of ancillary benefits which policies can have in the form of nonclimate environmental (or national security) improvements. As is the case for assessing the climate-related benefits of mitigation policies, unless the model explicitly recognizes the opportunity to purchase environmental goods, it is unlikely to account for the ancillary benefits of reducing fossil fuel use. The estimated costs of climate change policies are reduced when noncarbon benefits of reducing fossil fuel use are taken into account.

Conclusions

Consistent comparison of the costs and benefits of climate change policies can be facilitated by the use of growth paths to reflect expected changes in welfare over time and under differing circumstances. Climate changes, like any other changes in the availability of labor, capital or natural resources, will alter the rate at which welfare can grow.

A consistent evaluation of policy costs and benefits must consider the effects that both climate instability and mitigation policies can be expected to have on capital accumulation. Changes in the capital stock must be evaluated carefully, however; policies which increase consumption of environmental goods do not necessarily reduce investment in capital stock. In addition, climate instability itself can indirectly reduce the productivity of the capital stock, reducing its availability for output in future years.

Finally, both climate change and mitigation policies have implications for the natural resource base which must be taken into account for an accurate assessment of welfare effects. Climate instability represents a depletion of one portion of the natural resource base and can reduce potential output in future years in several ways. On the other hand to the extent that mitigation policies have ancillary (nonclimate) environmental benefits, they enhance local and regional natural resource bases.

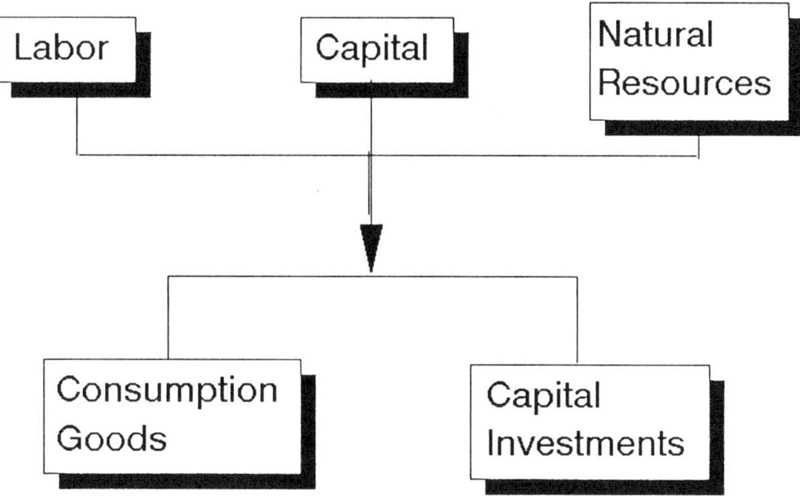

Figure 1: Production under the basic model

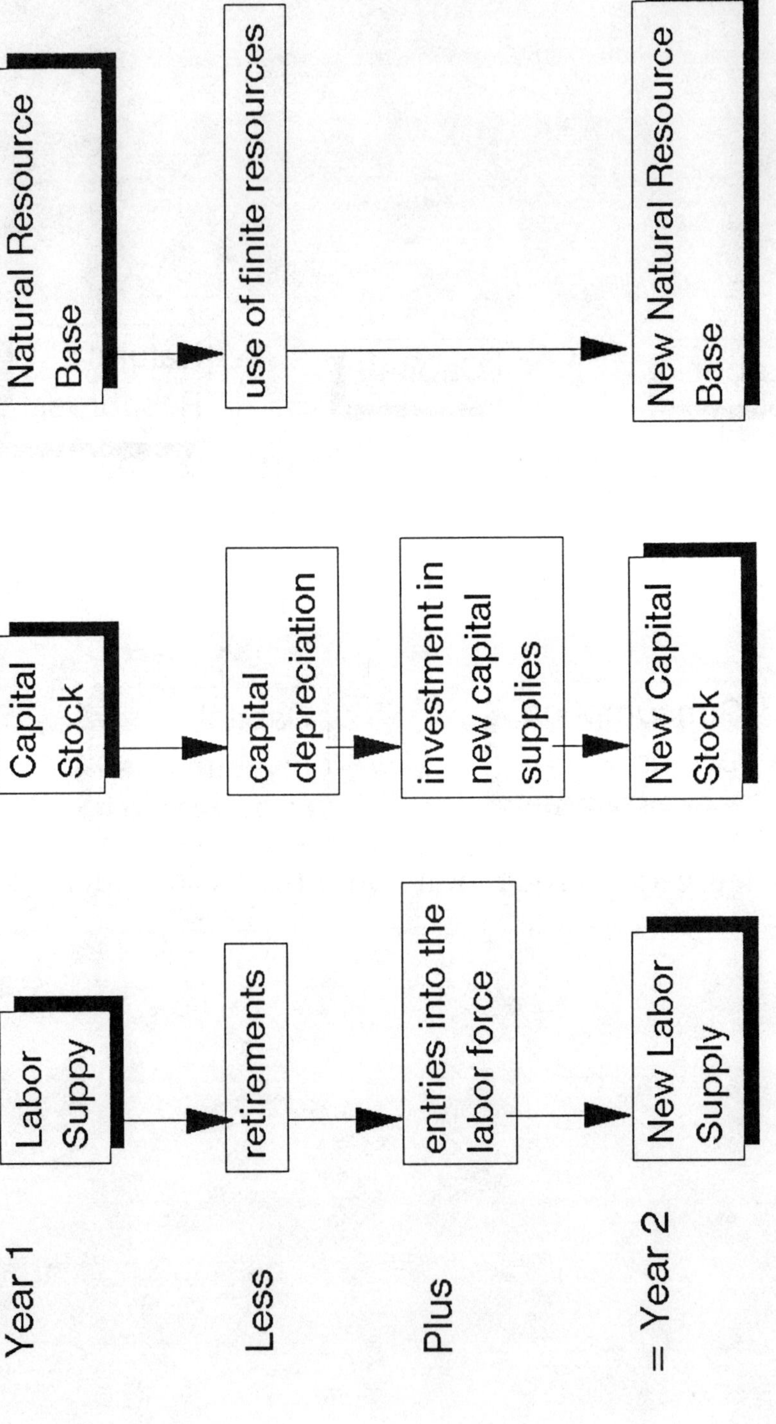

Figure 2: Labor, capital, and natural resource flows under the basic model

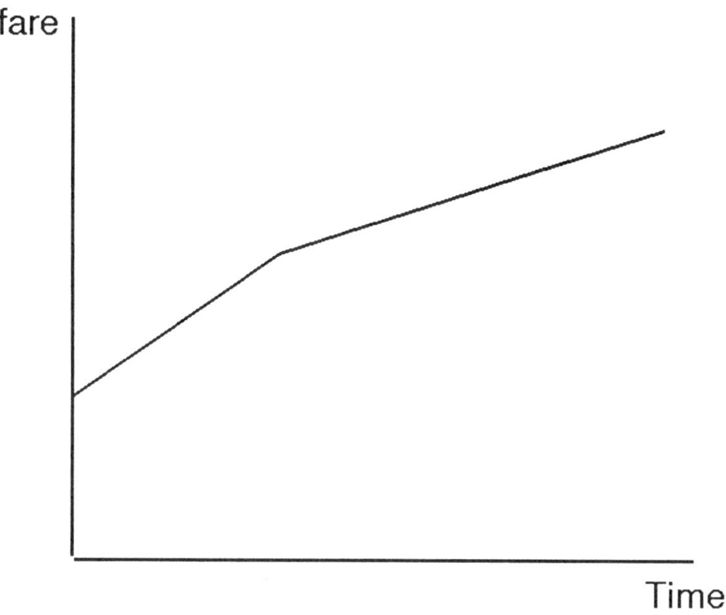

Figure 3: Welfare growth path

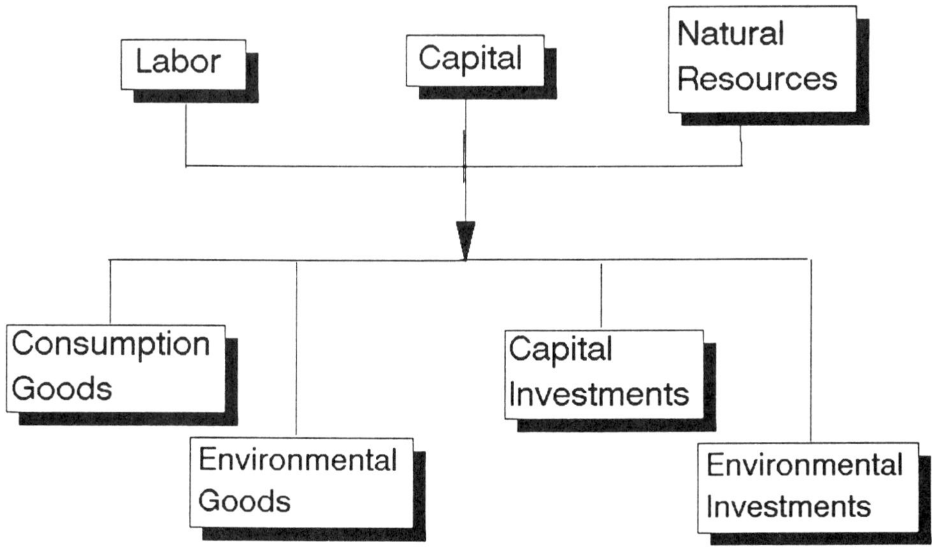

Figure 4: Production under the extended model

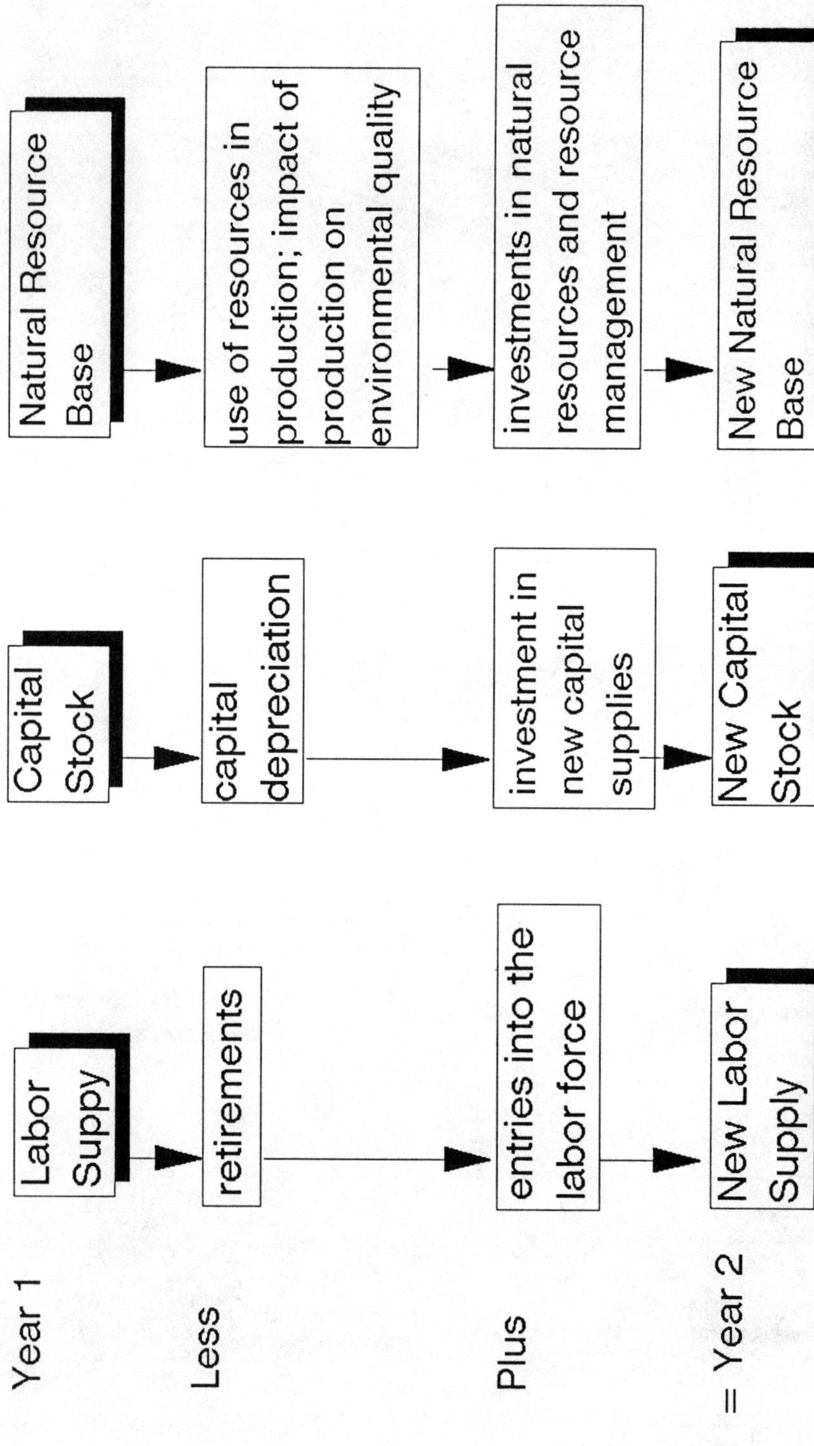

Figure 5: Labor, capital, and natural resource flows under the extended model

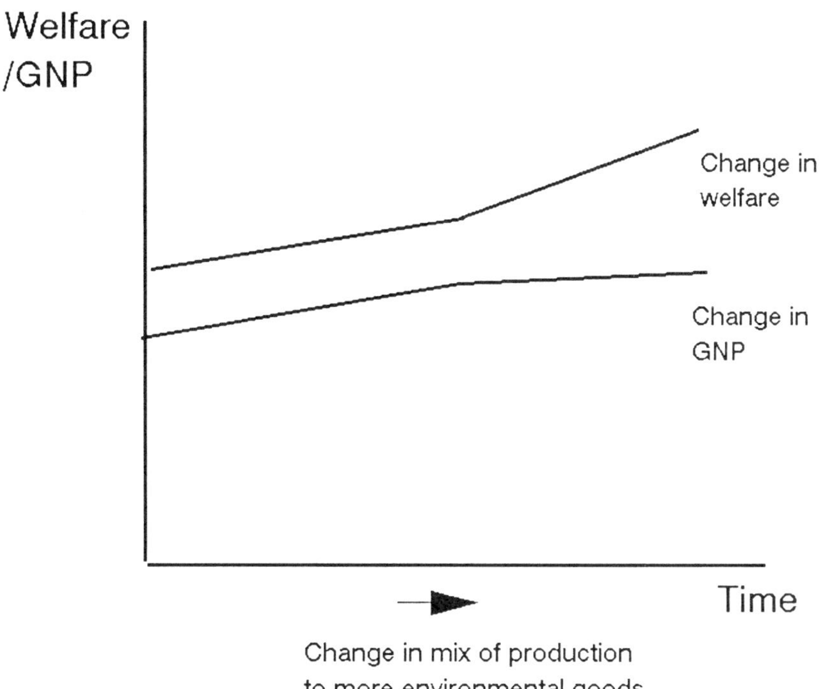

Figure 6: Welfare & GNP effects of environmental goods

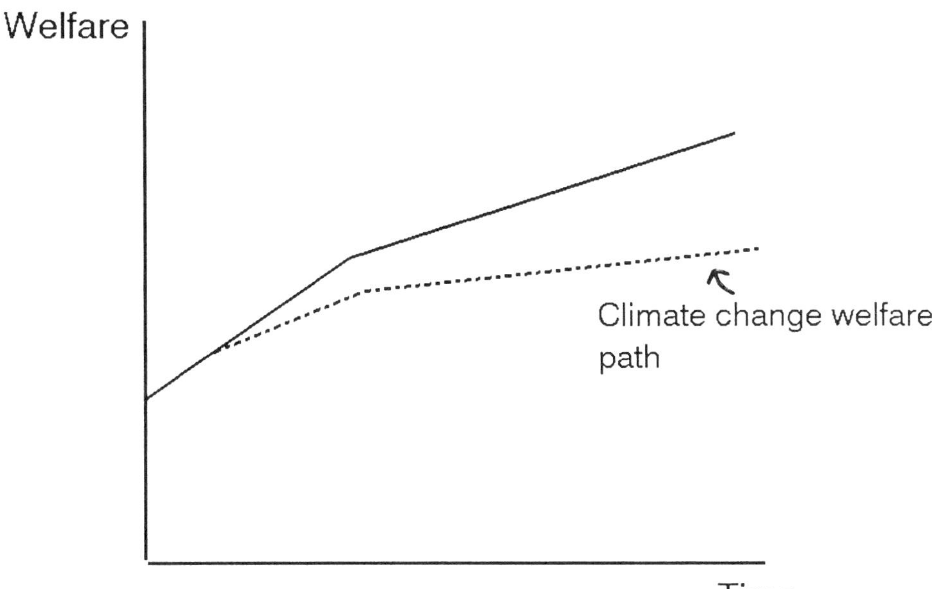

Figure 7: Change in welfare when climate change is taken into account

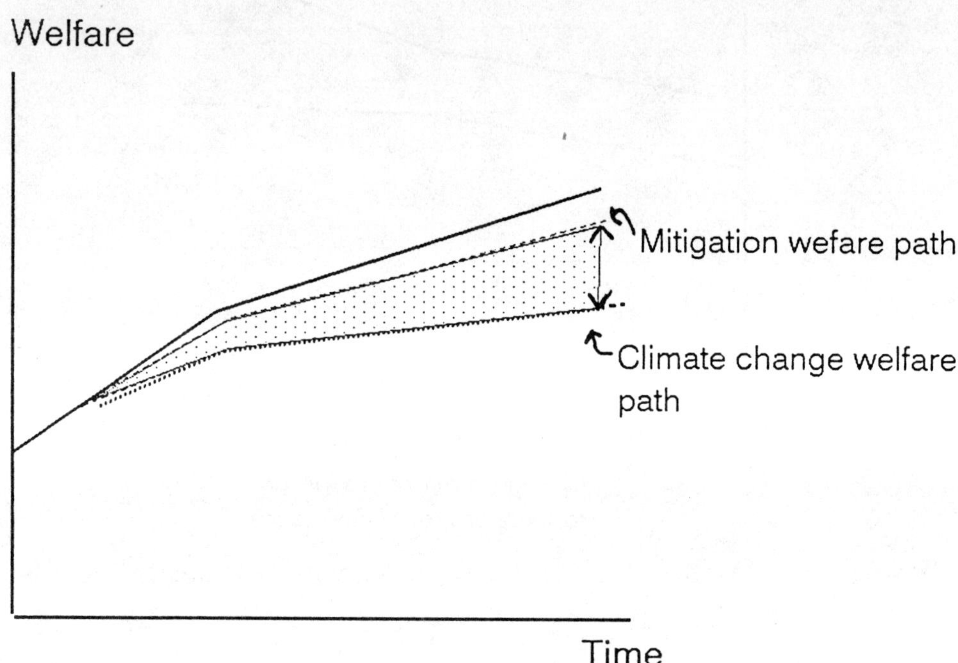

Figure 8: Gross benefits of climate change mitigation

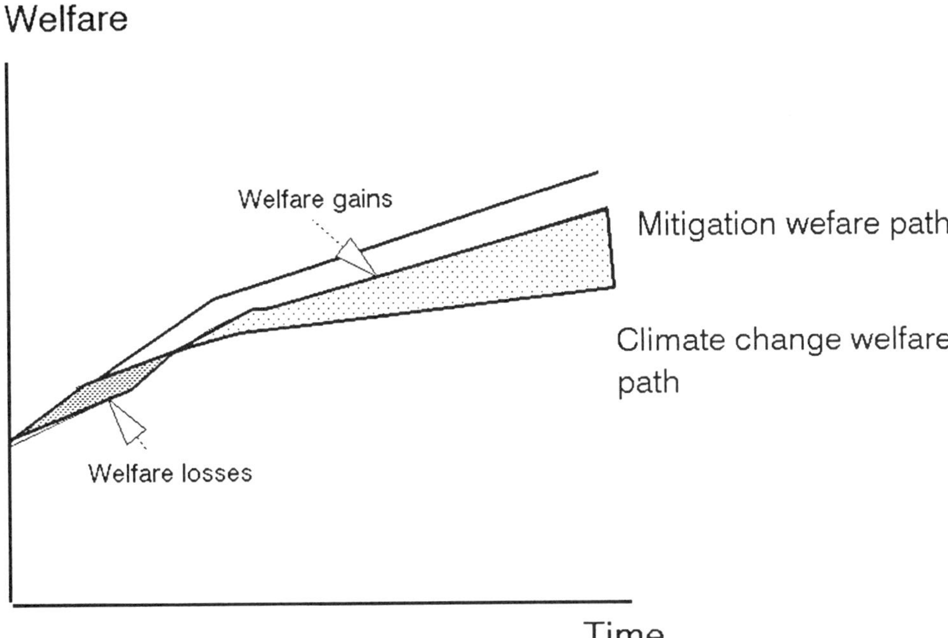

Figure 9: Climate change mitigation benefits when policy costs are taken into account

Notes

1. See Hazilla and Kopp (1990) for an example of a model which does have an endogenous labor market used to assess environmental policy options.

2. This aspect of natural resources is often referred to as "standing value." Standing value actually includes much more, such as protection against future contingencies (one of the most often stated benefits of genetic diversity) and utility derived from knowing that a set of resources will be available to one's children or the next generation.

3. There are, of course, many nonenvironmental external goods, as well as external "bads" of all sorts, which a full model would incorporate. Since we are interested primarily in the welfare effects of climate change and climate change policies, I abstract to a two consumer-good world. I also simply define environmental bads as the loss of an environmental good, rather than the more traditional way of looking at environmental externalities.

4. Consumers actually receive at least as much value from their purchases as the market price would indicate; any additional value is referenced to as "consumer surplus."

5. The total amount of the cost as measured by this integral depends upon the time frame selected.

6. Of course, if the time frame of analysis is sufficiently short to exclude periods of unstable climate, the two welfare paths will be identical. A cost-benefit analysis using such a short time frame, however, cannot hope to inform debate over appropriate policies.

7. It has been suggested that, because of the recent southward migration in the U.S., people might prefer warmer climates. But, of course, to the extent that people have settled in the climates they most prefer, it should be noted that all climates in the U.S. are expected to get warmer, so that no one would end up in their preferred climate without additional population migration!

8. Regulatory incentives have actually been changed in several states to avoid this problem. Most states, however, do not ensure that the least costly means of producing energy services is also the most profitable.

9. See, for example, ICF (1990).

10. Because the U.S. imports nearly half of its oil, reductions in domestic oil consumption can reduce vulnerability to oil supply disruptions and thus increase flexibility in foreign policy and security.

References

Hazilla, Michael and Raymond J. Kopp, "The Social Cost of Environmental Quality Regulations: A General Equilibrium Analysis," *Journal of Political Economy*, Aug. 1990.

ICF Incorporated, *Preliminary Technology Cost Estimates of Measures Available to Reduce Greenhouse Gas Emissions by 2010*, submitted to US EPA, August 1990.

BENEFITS OF MITIGATION AND ADAPTATION

T. C. Schelling

Department of Economics
Room 3105, Tydings Hall
University of Maryland
College Park, MD 20742

I begin with some remarks about adaptation. First, we shall have to adapt to climate change. Mitigate, of course, but adapt certainly. The most ambitious proposals to constrain carbon emissions would do no more than hold constant the annual emissions from developed countries while emissions from China, the Soviet Union, and other developing countries would continue to grow. Thus the concentration of carbon dioxide in the atmosphere is bound to increase, probably at an increasing rate.

Some of the adaptation will be by national governments, but most will be by ordinary people and businesses, or by local and regional governments. There is continuous adaptation even when climate is not changing: we change the technology and the efficacy with which we heat ourselves and cool ourselves and protect ourselves from storms and cope with droughts and floods and dispose of snow. The pace of change will probably be such that people will find themselves adapting to *climate* rather than to *changing climate*.

And if the experience of the past seventy-five years is any guide, changes in the ways that people live and work and clothe themselves and transport themselves, what they eat and what they do for recreation, and changes in health and longevity will be sufficiently great in both developed and developing countries to cloak and perhaps to overwhelm changes in climate. As an experiment, superimposing the kind of climate change forecast for the coming seventy-five years on the seventy-five years just passed would probably not make climate change the most dramatic change in people's lives, even in people's environments.

The benefits of mitigation are consequently conjectural. Being conjectural does not make them inconsequential, only difficult to foresee. The impact of climate change will be on societies vastly changed by the middle of the next century. In the developed countries people and their activities are much less susceptible to weather and climate than they were seventy years ago; in the developing world, people and their economic activities may be much less susceptible to weather and climate after another seventy-five years of development.

Copyright 1991 by Elsevier Science Publishing Company, Inc.
Global Climate Change: The Economic Costs of Mitigation and Adaptation
James C. White, Editor

In assessing the benefits of mitigation, the easiest place to begin is with the impact of climate change on economic activity in the developed countries. With the exception of agriculture, forestry and fisheries, virtually no market-oriented activity in the United States, western Europe, or Japan would be affected. Agriculture itself is a very small and continuously declining part of economic activity, and for half a century agricultural policies in the developed world have been coping with surpluses and keeping acreage out of production. Whatever harm the climate change may do in the developed parts of the world, damage to material well-being is unlikely to be significant. And as per capita income in the developed world will surely double or more than double over the next seventy-five years, any subtraction will be merely a minor reduction from a continual increase.

Developing countries are potentially much more vulnerable. Compared with the two or three percent of GNP devoted to agriculture in the United States, the developing world's dependence on agriculture ranges between one-quarter and three-quarters. While it is not certain that climate change would be adverse even on balance around the world, it certainly could be.

For the developing world the issue of mitigation requires a choice. If the developing world were to invest either resources of its own or resources made available by the wealthy developed world in minimizing the adverse impact of climate change on the welfare of their populations seventy-five years hence, they would have to consider whether more will be accomplished by investing in carbon abatement and slower climate change, or by investing in their own economic development to reduce their vulnerability to climate several generations hence.

I consider it likely that developing countries would elect immediate and continual economic improvement with which to face climate change rather than the long delayed benefits of climate mitigation. Furthermore, much of the developing world, including eastern Europe as well as China, India and Nigeria, is already suffering the poisoning and degradation of its immediate environment with severe impacts on health and child development; moderating climate change may receive lower priority.

More difficult to assess than the impact of climate change on economic activity and material welfare is the likely impact on the natural environment. Again, economic growth and development over the past seventy-five years has had such an impact on the natural environment in most parts of the world that it is hard to assess the relative impacts of climate change and all the other changes that will be occurring the next half century or more. But the possibilities of some discontinuous changes, some sudden surprises, cannot be ruled out, and prudence where climate is concerned is surely a virtue.

My assessment may appear insufficiently alarmist. But that assessment should be compared with the assessments of what it might cost to work out a substantial mitigation of climate change through reductions in carbon emissions. The economic costs of very substantial reduction in carbon emissions can easily sound forbidding when expressed in hundreds of billions of dollars annually worldwide, or tens of trillions of dollars in cumulative future costs discounted to the present. A percent of GNP appears an immense annual investment for the United States in an era when reducing the federal deficit by that amount proved almost politically unmanageable within the past year.

The problem is more political than economic. If the GNP of the United States were to be permanently reduced by 2% below what it would otherwise be, we could plot the curve of per capita GNP from now through the middle of the next century, and while per capita GNP was more than doubling we would find that the 2% reduction due to carbon constraints would be not much more than the thickness of the curve drawn with a number two pencil. To state it differently: if per capita income were to increase by only 1% per year through the middle of the next century, the impact of investing 2% per year of GNP in carbon abatement would mean that the per capita income we would otherwise achieve by the year 2050 would be delayed to the year 2052. I find that comforting.

METHODS OF ANALYSIS

William R. Cline

Institute for International Economics
11 Dupont Circle, Suite 620
Washington, D.C. 20036

My approach is one of social cost-benefit analysis, which in my opinion will be where the finance ministers will eventually have to sign off on greenhouse policy. I would like to submit that the stakes do seem to be very large, perhaps larger on the benefit side than so far might have been suggested.

I would like to suggest a distinction between greenhouse issues and normal environmental issues. Global warming is global rather than local as are many environmental issues, and it is irreversible rather than reversible. Once the carbon is in the air, it basically stays for 300 or 400 years. It is useful to begin from this point.

The first element to establish is the baseline for the build-up of atmospheric carbon dioxide and other trace gases, which is driven by baseline economic assumptions.

We have the base-case emissions (Figure 1) from three of the leading models, Manne and Richels, Nordhaus, and Edmonds-Reilly. They go through the year 2100, and basically go from 5 gigatons a year from fossil fuels to 15 or 20 gigatons (or billion tons) of carbon per year. These are driven by population growth. They are driven by labor productivity, and total production levels with their required energy inputs. This in turn depends on the energy efficiency of production. Those elements all together give the global energy use, and that in turn gives the global emissions.

I'm going to come back to these very long-term projections, but the Intergovernmental Panel on Climate Change, or IPCC, has concluded that we will have a doubling of carbon dioxide equivalent, including other trace gases, by the year 2025. That is fairly soon; nonetheless, that is where the debate has been concentrated.

What are the benefits of action on global warming? As we have heard, they amount to the avoidance of damage. It's what is called a Hobson's choice. It's a question of how much you lose one way or the other.

Let us first consider agriculture (Figure 2). We have reasons to expect lower agricultural output because of decreased soil moisture in continental interiors and because of increased heat stress. On the other hand, there could be a positive contribution to

Copyright 1991 by Elsevier Science Publishing Company, Inc.
Global Climate Change: The Economic Costs of Mitigation and Adaptation
James C. White, Editor

agriculture from carbon dioxide fertilization of plants, which does cause yields to increase.

The Environmental Protection Agency made some earlier estimates that the agricultural effects would be a plus or minus $12 billion a year at 1981 prices, for a doubling of carbon dioxide and a temperature increase globally of three degrees Centigrade. There have been much higher estimates recently by Professor Parry of the United Kingdom who suggests that the United States alone could lose up to $30 billion a year in agricultural output. So it is not necessarily true that there would be a wash on agriculture even for a doubling of carbon dioxide, let alone for higher levels.

The second element of the damage avoided is increased electricity demand for air conditioning. As an offset to that element, non-electric space heating would be reduced. In recent EPA estimates, that electricity demand figure is very large. These are first-cut estimates and I would flag that they are high.

A third element is coastal protection from sea level rise. The IPCC predicts that there will be something like a 60 centimeter sea level rise from a doubling of CO_2. That means the need to defend coastal cities with sea walls and the loss of some agricultural land.

There are other complements of cost or of damage which so far have not really been incorporated in the benefits side – species loss, forest loss, damage from storms. There are estimates that a doubling of CO_2 would cause a 50 percent increase in hurricane damage.

Human disamenity? People may be prepared to pay something for avoiding warmer summer days, especially in warmer areas, above and beyond what they have to pay for increased air conditioning.

One can see that in Figure 2 Nordhaus' bottom line is a damage of only $7 billion in 1981 prices; and he says that the damage avoided, if we could avoid a doubling of CO_2, might be somewhere on the order of one percentage point of GNP.

I like to stress the fact that this is not really a 40-year decision; this is something like a 200- or 300-year decision in my opinion because this is an irreversible process and history is not going to stop in the year 2025. What do we think about the baseline for emissions over much longer horizons? These models, when you project the growth rates of their final periods, give you emissions rates that are in the range of 50 to 70 billion tons of carbon annually by the year 2250, even with very moderate assumptions about population growth after the first half of the next century.

These emissions are compatible with carbon resources. One professor has suggested that, if you looked at a horizon like this, coal would become so scarce that people would be wearing it for earrings, but the numbers do not work out that way. Potential coal resources, when you multiply them by the carbon content, are perfectly adequate to permit this kind of increase in the emissions and still have relatively economical access to the coal resources.

These emissions, of course, translate into concentrations of carbon dioxide that, instead of going from 280 to 560 parts per million, go to the range of 2000 parts per million by the year 2200 to 2250. And that's just carbon.

The IPCC says that, through the year 2100, the radiative forcing from all trace gases including carbon is 1.5 times the radiative forcing from carbon alone. The total radiative forcing of trace gases tends to go up linearly even though the radiative forcing from carbon dioxide alone is logarithmic, in other words, less than linear, because of the saturation of certain bands of the spectrum. The bottom line is that you get a straight-line increase in radiative forcing and, therefore, a straight-line increase in the global warming (Figure 4).

The way I calculate it – and I base my parameters on the IPCC report, although they do not carry out their projections this far – by the year 2250, we are talking about a commitment to mean global warming of 10 degrees Centigrade. That's a far cry from the debate which is currently focused on a 2.5-degree Centigrade warming.

What about the benefits of avoiding global warming if these are the stakes? We would expect the benefits to be nonlinear. Consider, for example, sea level rise. The IPCC estimates of sea level rise assume that the Antarctic is a sink for water rather than a source. The temperature in the Antarctic is lower than the minus 12-degrees Centigrade threshold at which warming causes net melting rather than net buildup because, at those temperatures, the effect of greater snow and more moist air causes glacier increase rather than glacier decrease.

Suppose that we get into the phase of 10-degree warming instead of 2.5-degree warming. It is highly likely that the Antarctic would become a major source rather than a sink of water and that the sea level rise, instead of being 50 centimeters or 60 centimeters, will be closer to two meters or three meters. That's an obviously nonlinear relationship on the benefits of avoiding sea level rise and the attendant costs of sea walls and lost agricultural land.

Consider agriculture. We have positive effects of CO_2 fertilization and we have negative effects of soil moisture and heat stress, but when we get warming on the order of 10 degrees Centigrade we begin to get into a phase where agriculture just shuts down, which occurs somewhere in the range of 40 to 45 degrees Centigrade – somewhat higher for rice and corn, somewhat lower for wheat and barley. When you get an actual shutdown of agriculture in a number of areas, you are really talking about nonlinear costs.

Consider human disamenity. Today, out of 66 major U.S. cities, only two have average July maximum daily temperatures of 100 degrees Fahrenheit or higher. With 10-degrees Centigrade global mean warming and applying the regional coefficients in the three high-resolution models cited in the IPCC, there would be 42 American cities out of the 66 with global mean daily maximum July temperatures in excess of 100 degrees. I suspect that people would be prepared to pay something to avoid that.

The summary point here is that, if the benefits of avoiding global warming for (a) the standard benchmark of a doubling of CO_2 and 2.5 degrees Centigrade of warming are on the order of one percent of GNP; and (b) we expect those benefits to be non-linear, to go up faster than the warming; and (c) that the stakes of warming are closer to 10 degrees Centigrade over a 200-year horizon rather than simply stopping history in the year 2025, then I think the bottom line is that the benefits of avoiding global warming could be probably closer to the range of four, five or six percent of GNP in the very long term. And we must keep that kind of stakes in mind when doing the benefit-cost analysis.

Let's turn to the cost side. We have to look at what are the costs of trying to freeze the current emissions today or bring them back down by 20 percent and freeze them at those levels. In analyzing conceptually what these costs are, essentially we are going to have to use less energy to produce our economic goods, and you get into the law of diminishing returns (Figure 5). We are going to have to combine more labor and more capital with less energy, and that means that we will have lower production. If you look at the righthand panel you will see that there's a bit more labor, a bit more capital, but there's a lot less energy, and that gives you less total production than you had before.

There are means of adjusting to the lack of availability of energy when we have a carbon constraint. One is intra-energy substitution, and at the middle panel you have a shift from carbon energy, coal, oil and gas to a greater percentage contribution from non-carbon sources—nuclear, solar and hydro. You will note, however, that the total energy does go down substantially, obviously depending on technical change.

In the third panel we have product substitution. It's possible to consume fewer of the goods that are carbon-intensive, take fewer jet trips to England, and to consume more of the products which are not carbon-intensive—listen to stereophonic electronics products or have a more elegantly decorated, smaller living quarters instead of spacious living quarters that require more energy.

Most of the studies that we will be hearing about on the costs of limiting carbon are based on the concepts that are represented in this chart. All of them have this production function, law of diminishing returns. All of them have substitution among alternative energy sources. Some of them have substitution among products demand.

There's a third element that some of them also add, and that is productivity growth. And there is the concept of what economists call energy-using technical change. If you have technical change increasing productivity of factors, which is dependent on energy, then you may suppress the rate of productivity growth; you may reduce the rate of productivity growth as you attempt to cut back on energy and make it more expensive.

What are the kinds of results that are coming out of the models? One leading model by Manne and Richels provides a lot of detail on alternative energy sources including new technologies—for example, coal gasification with CO_2 removal by a 100-mile pipeline or nuclear energy with passive safety features. This particular model is not a sectoral model. It has a lot of detail on the energy composition but then it tends to go to an aggregate production function rather than having the 45 different sectors of the economy either in the production functions or in the bottom of the diagram in the product demand.

Manne and Richels conclude that to cut carbon by 20 percent and freeze it costs between 0.8 percent of GNP and 4 percent of GNP into the indefinite future for the United States. The range depends on the autonomous rate of energy efficiency increase, on technological change coming into availability, and on the assumption about what's called the back-stop technology—which is essentially the cost in the long run of noncarbon energy.

They conclude that it would require a carbon tax of $250 per ton of carbon to accomplish this cut-back of 20 percent and freeze, and that's a benchmark that others including Nordhaus have used. Other estimates tend to be somewhat smaller. The Jorgenson, Wilcoxen model—which incidentally does incorporate this productivity suppression so you might think it would be even more costly because productivity growth will not be as high—comes more to the lower end of that range, again, something like one percent of GNP. Part of that is because of very flexible capital substitutability in their model.

But, to some extent, the modelers do not quite yet know what are the differences. They're going to be sitting down in an energy-modeling forum and conducting a "horse race" among the models, and hopefully that process will clarify some of the differences in the estimates. The CBO figure, of something like one to two percent of GNP as the cost to reduce or to freeze carbon emissions for the U.S. economy, does not sound implausible to me.

There are what you could call the process-engineering critics of these kinds of costs, the people who say, "If we simply use state-of-the-art efficiency engineering processes, we could sharply reduce carbon emissions and energy requirements at no cost." The economists' normal response to this argument is, "If that were true, why isn't it already happening through the market process?"

I think there are many explanations economists have. For example, I don't know how many people in this room have substituted their light bulbs with compact fluorescent lights in order to save on energy, even though it might be efficient. The thrust of the engineering process criticism which should be acknowledged is that when we do these range estimates, there is perhaps a reason to consider that the lower ends of these ranges may be a little more germane than we might otherwise think precisely because of some of these opportunities.

Let me emphasize greatly another element in the cost-benefit analysis—the time-discount factor. The costs of adjusting tend to be immediate. The benefits are going to be over the longer range. If you take a five-percent discount rate and look ahead 200 years, one dollar saved 200 years from today at a five percent discount rate is worth one seventeen-thousandth of a dollar today. So you can simply forget it.

Our standard discounting will simply not be appropriate for this issue. What we are going to have to do is to conceptualize, I would submit, the following discounting process. You have two elements in your discount rate. One is the rate of return on investment, and the other is based on the utility of additional income at different income levels. As for the former, we're probably talking about 10 percent annual rates of return, but investment is only 15 percent of the economy. And we have not devised a way to change that much, regardless of all of our policy instruments, so let's take that as given.

For the other 85 percent, we ought to judge the discount rate based on how much richer people will be 200 years from today than they are now. I'm richer than my ancestors were at the time of the Revolutionary War, but I'm not 17,000 times richer. So it would be inappropriate to use a five percent discount rate on that basis, let alone the fact that we are imposing a damage on our descendants rather than making them better off by abstaining.

I would submit that the proper discount rate on the consumption side is something like a half a percent or one percent a year. At a half a percent, my grandson's grandson when he sacrifices a dollar improves my situation by about 30 cents. At a one percent discount rate, it's only about five cents. So those rates are already paying little attention to the distant future.

And if you weigh these two discount rates, the 10 percent for the investment, say, and one percent for consumption, you get a discount rate that's on the order of two percent annually or less. That's an interesting rate because GNP is growing at this order of magnitude. So the scale factor for the benefits, many of which tend to be GNP proportional, can offset the time-discount element in the cost-benefit analysis. This kind of analysis is a rough outline of the kind of thinking we will have to do to make the discounting part of the cost-benefit analysis meaningful.

I would submit that when this is carried out and when the benefits – perhaps on the order of one percent of GNP for a doubling of CO_2 rising to something like five percent of GNP for the much larger warming over the 200-year horizon – are compared with the costs which may be on the order of one to two percent GNP but would probably fall as you have incentive to technical change given by the price system, as you make carbon expensive and you thrust technological change in a carbon-saving direction, then it's quite possible that over a long period the benefit-cost comparison will show that it pays to take aggressive action on the global warming problem.

I will flag two other key issues. One is the international coordination issue. Today, the developing countries only account for about a fifth of carbon emissions. By the year 2075, they are going to account for about 60 percent. We cannot ignore the developing countries when we set up the international regime to deal with the problem. That will mean incentives to bring the developing countries along; that will mean channeling resources to them from the benefits that are thereby obtained.

Another key issue is uncertainty. We hear that uncertainty means we cannot act. It does not really mean that. It probably means that we should wait for five, maybe ten years before we do very massive things, and get as much possible scientific confirmation as we can. But in the end, if there's uncertainty it can also apply to the upside risk: how about these unexpected disasters?

And if policymaking is risk-averse, then it attaches a greater penalty to those large upside risks than it does to the gains of having lower damage and not having to pay quite as much. Uncertainty, then, has to be incorporated into our analysis. I do not think that it will necessarily yield the conclusion of inaction.

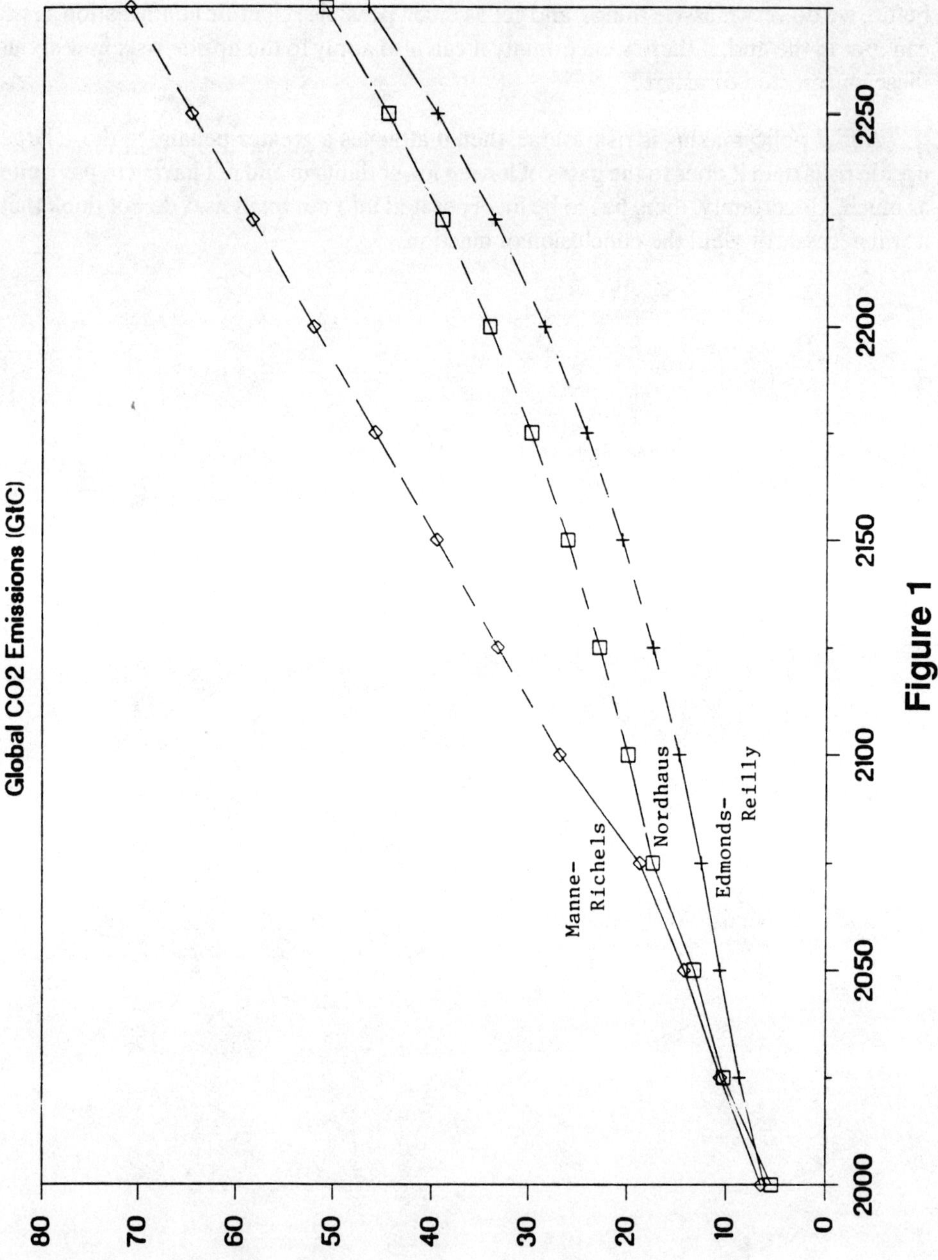
Global CO2 Emissions (GtC)
80
70
60
50
40
30
20
10
0
2000
2050
2100
2150
2200
2250
Manne-
Richels
Nordhaus
Edmonds-
Reilly

Figure 1

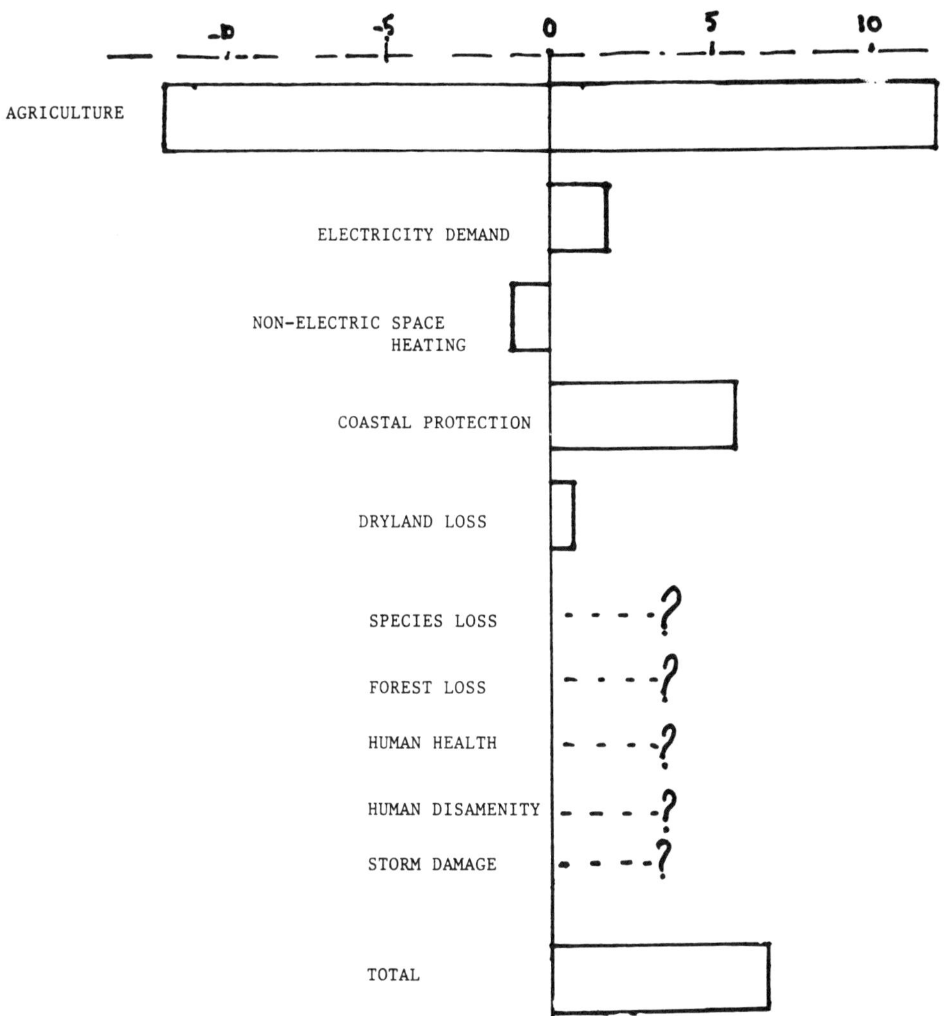

Figure 2

ANNUAL BENEFITS FROM AVOIDANCE OF CO2 DOUBLING (U. S., BILLIONS OF 1981 DOLLARS; NORDHAUS ESTIMATES)

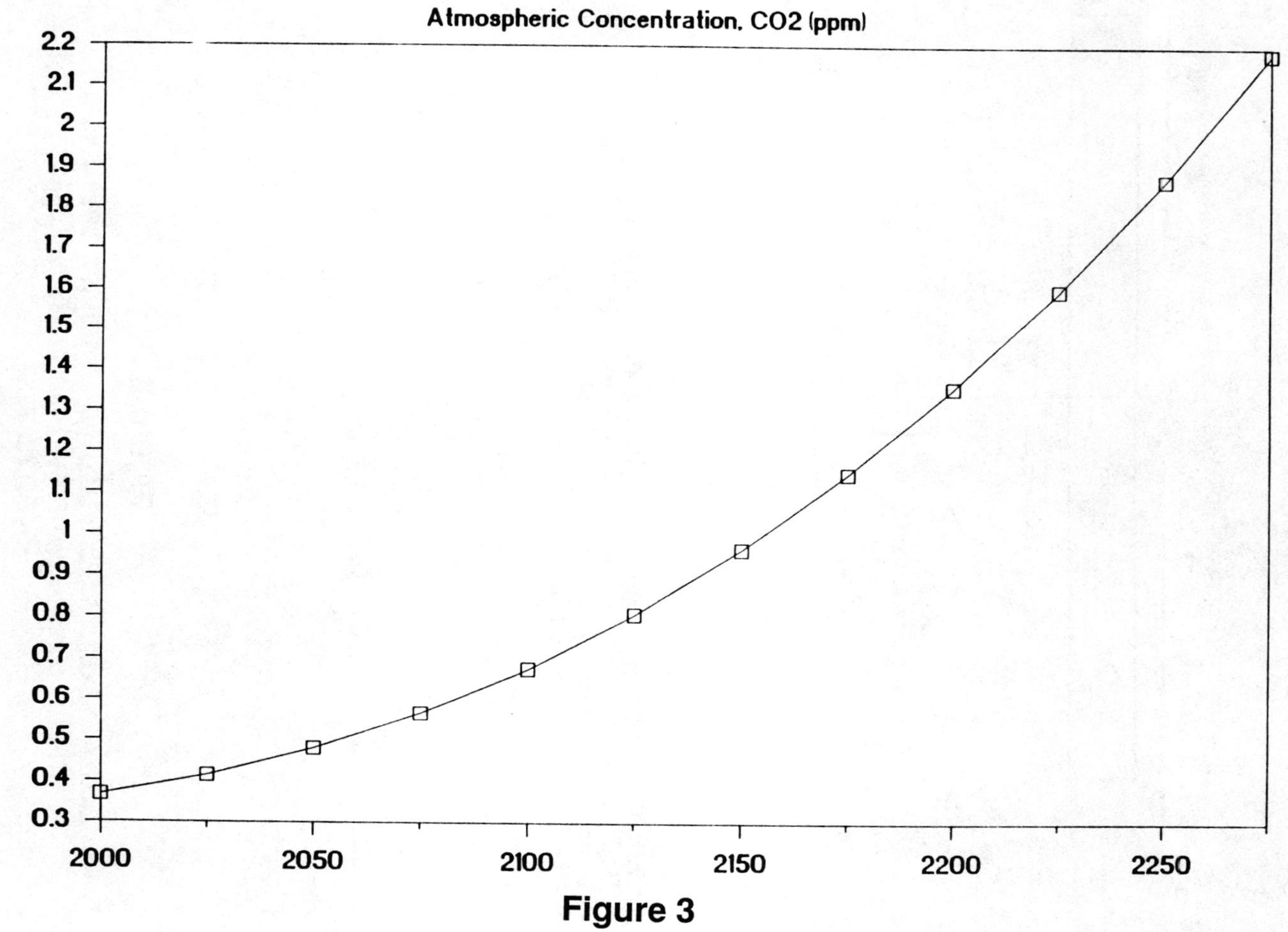
Atmospheric Concentration, CO2 (ppm)
(Thousands)
2.2
2.1
2
1.9
1.8
1.7
1.6
1.5
1.4
1.3
1.2
1.1
1
0.9
0.8
0.7
0.6
0.5
0.4
0.3
2000
2050
2100
2150
2200
2250

Figure 3

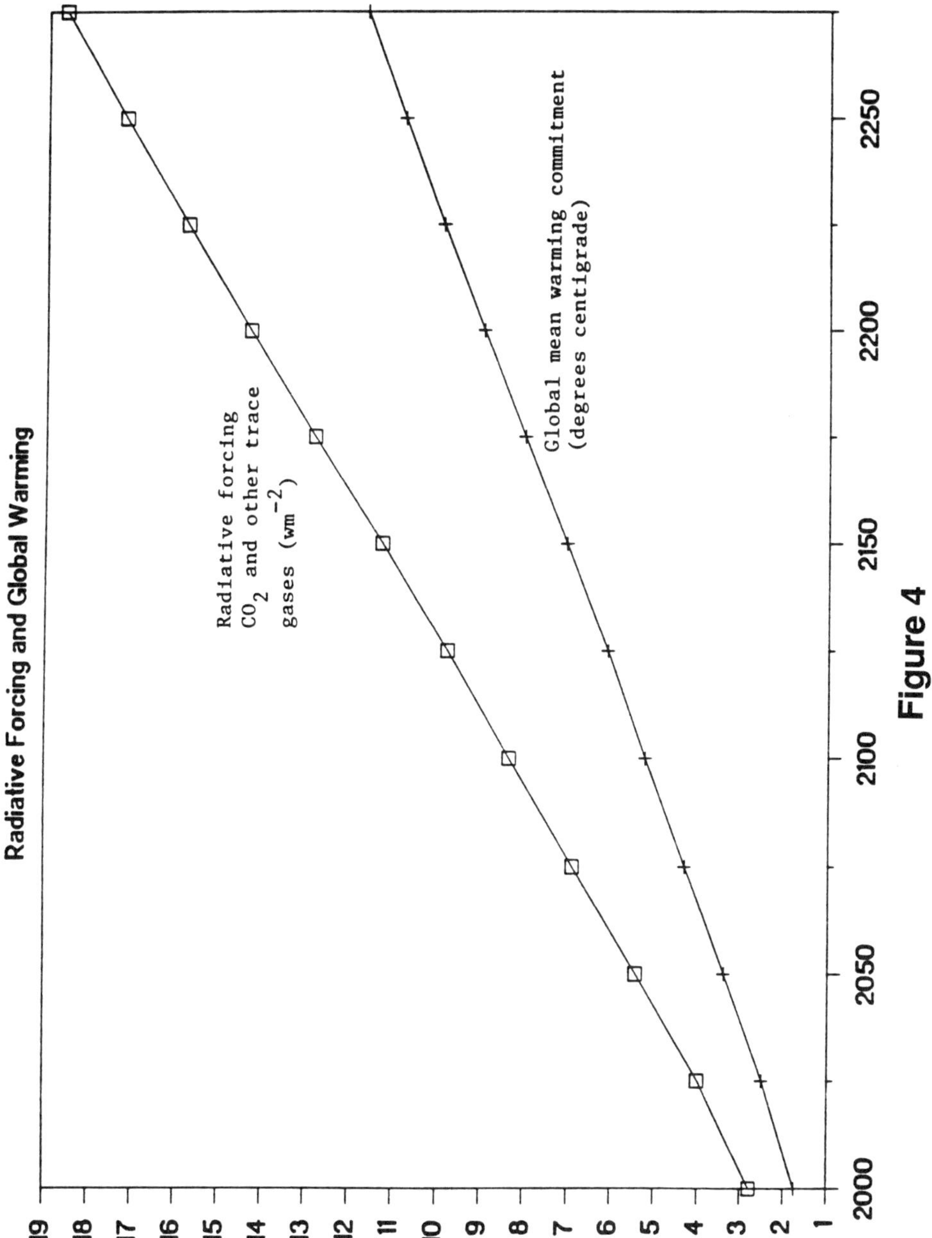
Radiative Forcing and Global Warming
Radiative forcing
CO_2 and other trace
gases (wm^{-2})
Global mean warming commitment
(degrees centigrade)
19
18
17
16
15
14
13
12
11
10
9
8
7
6
5
4
3
2
1
2000
2050
2100
2150
2200
2250

Figure 4

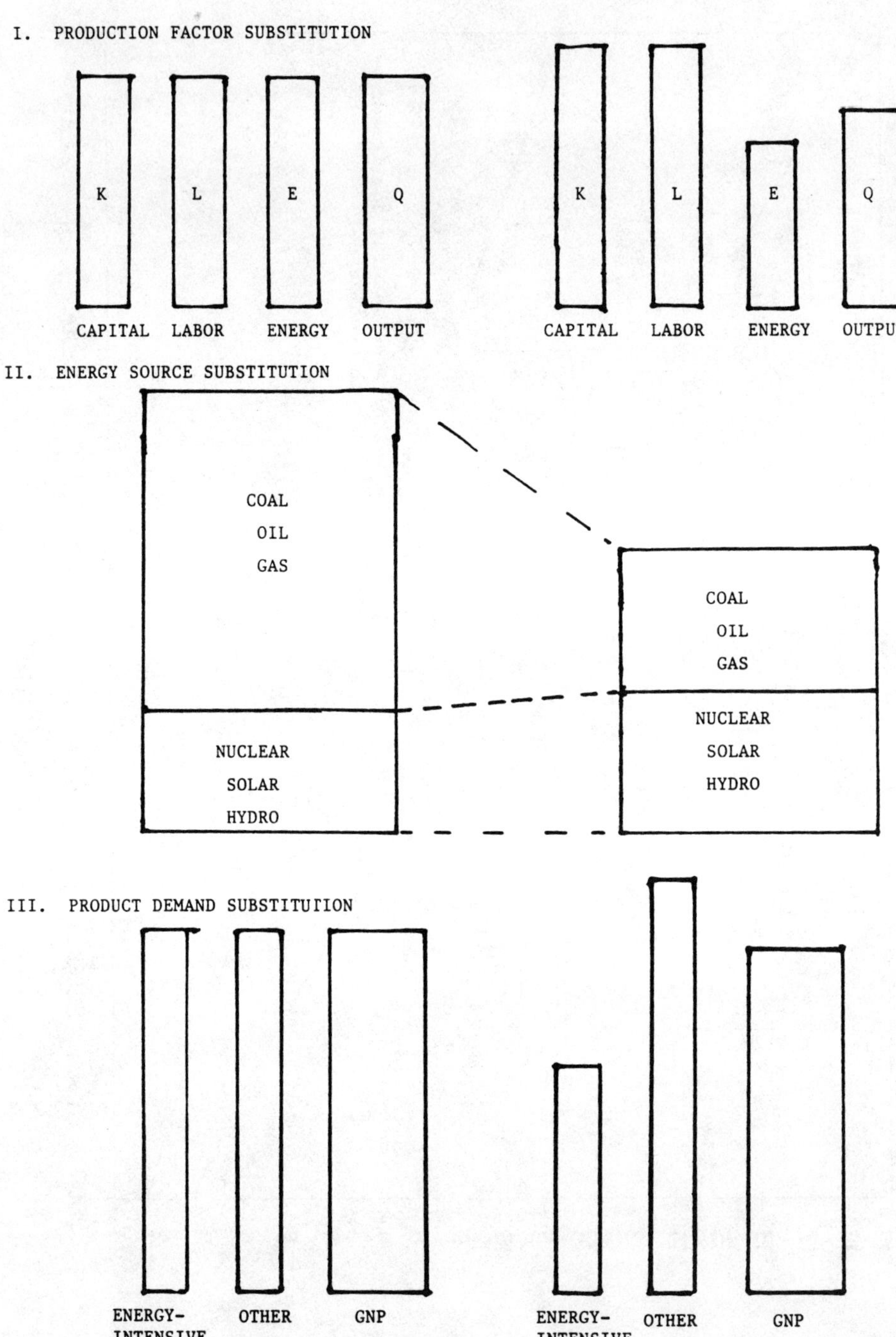

Figure 5
ECONOMIC RESPONSES TO GLOBAL WARMING

CURRENT ANALYSIS OF ECONOMIC COSTS

Lester B. Lave

Graduate School of Industrial Administration
Carnegie-Mellon University
Pittsburg, PA 15213

Uncertainties of Greenhouse Effects

Although the following papers concern economic aspects, you are not going to hear about the dollar benefits and costs of activities to abate emissions of greenhouse gases. Rather, the panel will attempt to clarify a few aspects of the social benefits and costs. No one is able to provide definitive dollar estimates at this time. Among other difficulties, atmospheric scientists cannot tell us what would happen to global mean climate as greenhouse gases accumulate in the atmosphere. There are no confident estimates of the extent of temperature change or whether precipitation will increase or decrease. Estimates of regional climate effects are little better than guesses. Furthermore, in order to calculate the economic implications of global climate change, economists would need to know about regional weather, not just regional climate.

Even if the atmospheric scientists were certain about the extent of changes in the atmosphere, the oceanographers and glaciologists would not know what other kinds of changes to expect in the physical environment or what would be the magnitudes of these changes. But even these uncertainties are dwarfed by uncertainties concerning the changes to expect in local ecologies and in the competition among individual species of plants and animals. Any environmental change that affects one species is likely to affect the whole ecology.

No one knows how humans are going to cope with climate changes and their sequelae. One reaction may be geoengineering, an attempt to reverse the changes in the atmosphere. No one knows within a large range how much petroleum, natural gas, coal, and other fossil fuels exist and what is the cost and environmental effects of recovering them.

Projections about emissions of greenhouse gases are extraordinarily naive, rarely going beyond the assumption that they will increase each year by a given proportional amount. The models assume that population will continue to grow, as will economic activity. But there is little or no technological change in the models or any feedback from climate change to growth rates of population, economic activity, and use of fossil fuels.

Copyright 1991 by Elsevier Science Publishing Company, Inc.
Global Climate Change: The Economic Costs of Mitigation and Adaptation
James C. White, Editor

It is easy to overstate what is known about greenhouse effects. This panel will focus on abatement costs. Although the other authors are likely to agree with my characterizations of how uncertain we are, I want to spell out the quantitative and even qualitative uncertainties.

Is There a Free Lunch?

We are going to hear about enhancing energy efficiency. If all nations could perform the same tasks with 10% or 75% less energy, humans could reduce our use of fossil fuels and put off, or even eliminate, global climate change. Some doubters refer to this prospective energy savings as the "free lunch" hypothesis, asserting that no energy savings is possible without cost. If saving energy could be done with little trouble or cost, why are businesses and consumers so stupid that they are not doing this?

Although I am not a free lunch cynic, I confess to thinking that some of the energy conservation assertions are a bit too simple. For example, look at the ceiling of this meeting room. You see perhaps 1,000 incandescent bulbs. We could light the room with a dozen or so fluorescent bulbs, saving 40,000 watts each hour, not to mention all of the air conditioning required to take away the heat generated by all these lights. Why don't we all march up to the hotel manager's office to announce our profound discovery? If this meeting room is used 12 hours per day, we could save the hotel about 500,000 watts per day in lighting this room – that is worth about $50.00.

Do you think the manager will thank us and immediately hire electricians to remove this beautifully decorative lighting arrangement in favor of a dozen or so fluorescent tubes? I suspect that the manager will tell us that the hotel is aware it could light this room more cheaply, but they are trying to make the room attractive, even beautiful, rather than to minimize electricity consumption. They rent this room for more than $1,000 per day. If prospective tenants find the room unattractive, the hotel management is worried that almost no electricity will be used to light this room.

I have a more personal anecdote to tell. My wife and I have been refurbishing our house. As we were finalizing the plans for reconstruction, I asked the architect to specify energy saving, compact fluorescent bulbs instead of the incandescent bulbs he had scattered so generously in the ceilings, walls, and in free-standing fixtures. I explained that I was concerned about greenhouse effects and wanted to make my contribution to keeping the Earth from warming.

He said, "You won't like it."

I replied, "I'll like it. I haven't complained about the cost of all this. Just do this one thing for me."

He said, "You won't like it. You won't like the color values."

"Color values?" I asked. "We are talking about lighting, not paint."

He said, "Fluorescent bulbs have very different color values from incandescent bulbs."

I said, "I have heard that the fluorescent-bulb people have made great progress in how fast the bulbs generate light when you switch them on, in not flickering, and in warmer color values. Indulge me. Remember, I am the one who writes the checks."

"No," he said, "You won't like it."

There are many reasons why businesses and consumers don't conserve energy. There are subtle dimensions, such as the color values. After losing this fight with the architect, I calculated that we were spending more than $20,000 to make the house look better. Switching to fluorescent lighting might save as much as 20 cents per day. The interest and depreciation on the renovation amount to about $14 per day. It doesn't make sense to have this expensive renovation look slightly less good in order to save on lighting. In a more fundamental sense, we would have saved much more energy by not doing this renovation or by doing a less elaborate renovation.

The point of my example is that lighting uses little electricity and electricity is cheap. If electricity were ten times more expensive, it might have had an appreciable influence on the lighting decision.

Business also has other matters to worry about, like trying to make a product or service that consumers want, trying to provide the requested ancillary services, hiring, training and promoting workers, complying with laws against employment discrimination, providing safe and environmentally sound products and working conditions, etc. For most firms, energy is such a small part of their costs that they give energy conservation little or no attention. It is difficult for energy saving technologies to get tried and have a chance to prove themselves. Without a demonstrated record of reliability and savings, few firms are willing to adopt an energy conservation technology.

The Unintended Effects of Technologies

I want to say a few words about unintended effects of new technologies. The current technologies for producing electricity, powering our cars, and heating our houses are marvelous. They are reliable, relatively inexpensive and have relatively small unintended side effects. In aggregate, the automobiles in Washington emit enough oxides of nitrogen, carbon monoxide, and unburned hydrocarbons to cause ozone problems on some summer days. But the emissions of these gases is a trivially small proportion of the private and social issues associated with automobiles (from their cost and reliability, to their safety and congestion). Who could have know when automobiles were being developed that such great concern would be focused on ozone resulting from these emissions? Should government leaders have stopped the development of the automobile because of concern that emissions might someday produce unhealthy smog levels in the summer?

Throughout the United States, there are thousands, indeed millions, of inventors and entrepreneurs developing new ideas and technologies to solve problems and fulfill consumer demands. The successful technologies are used by millions of Americans and have some unwanted side effects. One will generate air pollution, another water pollution. One will be a safety hazard while another causes environmental problems when it is thrown away. In some cases, this unforeseen side effect will be so important that we will reject the technology, e.g., three-wheel recreational vehicles. More generally, we need to recognize that the largest generator of problems is past solutions.

The technologies being developed and proposed for energy conservation have the same problems. Some will not turn out to be attractive, even though they now look promising, e.g. early heat pumps in the north. Some will turn out to have side effects so bad that we will ban the technology. A very small proportion of the technologies that now look attractive will be adopted and be successful.

Without wanting to be discouraging about new energy conserving technologies or technologies for using renewable energy resources, I want to warn that few of the currently attractive technologies will be commercially successful. Even a technology that works great in one location will turn out to not work well in another location.

I want to encourage the millions of inventors and entrepreneurs; they are what give the U.S. economy its zip. But, in analyzing any single idea, we do not want to be naive about its chances of being successful in accomplishing what we want.

Furthermore, it is likely to take much longer to implement a new technology that the inventor thinks is reasonable. As a wonderful new technology is implemented, the kinks are worked out, it becomes more reliable, and often less expensive. Those who invest in it first get a less satisfactory version than those who hold off to let the product be improved.

My point is that there are a host of reasons for expecting that most of the energy conserving technologies that are now being proposed will turn out to be less attractive once all the bugs have been worked out; even the attractive ones will take a long time to implement.

Flexibility

I began by describing the pervasive uncertainties associated with global climate change. When a situation is so uncertain and there are large costs associated with making the wrong investments, flexibility is extremely valuable. For example, should the United States begin a crash program to build nuclear reactors in order to phase out all coal use and cut down on the use of petroleum and natural gas? Many people would regard such a step as potentially costly. However, if this were the only strategy that would head off disrupting global climate change, it would be accepted.

Before implementing this strategy, we must be aware not only of the uncertainties concerning the magnitude and effects of global climate change, but also the uncertainties concerning the current generation of nuclear reactors, the characteristics of new technologies, and other energy sources that don't generate carbon dioxide. It would be foolish to stake our energy future on these technologies because of the problems with the current nuclear technology. One way of gaining flexibility would be a research program to develop advanced nuclear technologies that don't have the problems of the current technologies. Even if advanced nuclear technologies are never built, this investment is worthwhile since nuclear technologies could give us flexibility in case other technologies don't pan out.

Unfortunately, as suggested by this example, flexibility often is expensive. Resources that are invested in research on advanced nuclear technologies are resources that won't be available for research on fusion or solar power. Another example is that a new electricity generation plant could be built to burn any fuel, from coal to natural gas to biomass. Unfortunately, a boiler designed to handle such a wide range of fuels would be expensive and inefficient. It would be better to build a more specialized boiler, knowing that it might be obsolete before the end of its physical life.

Biases in Well Documented Analyses

You will be pleased with the papers in this session. All are well documented analyses. Unfortunately, there is a bias in well documented analyses. If a researcher is doing a careful job, not making statements that can't be documented, the analysis will be a conservative one that favors the status quo. I don't know which renewable energy technology will be most attractive, but I am willing to bet that there will be several that will be attractive. Since the attractiveness of any one technology cannot be documented, a careful analyst will tend to underestimate the contribution that will be made by renewable energy sources in the future.

More generally, in a well documented analysis, the costs will tend to be overestimated and the benefits will tend to be underestimated. Such an analysis will suggest that taking action isn't very attractive and that less should be done. I am certainly not suggesting that you ought to disregard well documented analyses. Quite the contrary. But you should understand that they have a bias toward not doing anything.

In particular, the best current analyses suggest that a major abatement of carbon dioxide emissions would cost about 2% of the nation's economic output. One reaction, a reaction of the European greens and Europeans more generally is: "Is that all?" I agree, but point out that total expenditures for all environmental programs in the United States is about 2% of GNP. The recent negotiations over reducing the nation's budget deficit were over about 2% of GNP. If devoting these resources to preventing global climate change would save the Earth, no one would demure. But, if the extent of climate change is highly uncertain, and if the effects of warming might be good as well as bad, there are many pressing needs for our nation's resources.

I am concerned about changing the climate of the Earth, but current uncertainties are overwhelming. In my judgment, the United States should be devoting $1-2 billion per year to improving our knowledge of greenhouse effects and how to adjust to a different climate and taking reasonable steps to conserve energy and reduce greenhouse gas emissions. It is premature to devote $100 billion per year to massive abatement of greenhouse gas emissions.

THE COST OF REDUCING GREENHOUSE GAS EMISSIONS IN THE UNITED STATES[1]

Alex Cristofaro

Air and Energy Policy Division
US EPA
401 M Street, SW
Washington, DC 20460

Summary

The costs of greenhouse gas mitigation strategies for the United States are sensitive to the baseline rates of growth for such gases. This paper builds upon the previous work of EPA (e.g., "Policy Implications of a Comprehensive Greenhouse Gas Budget," by Cristofaro and Scheraga) and establishes a projection of future emissions of major greenhouse gases. The focus is on the next twenty years, a period in which the U.S. will reduce emissions of a number of greenhouse gases because of implementation of the Clean Air Act and the CFC Protocol. Future CO_2 emissions are expected to rise, however, and two possible emission scenarios are presented. Possible emission control scenarios are then discussed in the context of a comprehensive greenhouse gas approach, and preliminary cost estimates are given for a limited set of policy options.

Base Case Emissions

To determine the costs of reducing U.S. emissions of greenhouse gases one must first estimate the magnitude of emissions that will likely result if no policy action is undertaken. This represents the "base case" from which emission reductions must be undertaken in order to meet emission targets.

While much attention has focused on carbon dioxide limits in the context of global warming mitigation, it is but one (albeit an important one) of a number of greenhouse gases. Others include methane (CH_4), chlorofluorocarbons (CFCs), nitrous oxide (N_2O) and ozone (O_3). Of additional concern are gases that are not themselves greenhouse gases, but are chemically active in the atmosphere and influence the concentration of greenhouse gases. Examples of these are volatile organic compounds (VOCs), carbon monoxide (CO), and nitrogen oxide (NO_x).

Copyright 1991 by Elsevier Science Publishing Company, Inc.
Global Climate Change: The Economic Costs of Mitigation and Adaptation
James C. White, Editor

The ideal goal would be to construct a base case emission scenario for all gases contributing to global warming and determine the least-cost strategy for keeping warming within acceptable limits. To do this, it is useful to translate gases into a common metric that takes into account the contribution each gas makes to potential global warming. This metric would take into account the relative lifetimes of individual gases, their potency in trapping heat, and the interaction among gases as they chemically react in the atmosphere. There is no universally accepted methodology for developing this kind of global warming index. Nonetheless, the Intergovernmental Panel on Climate Change (IPCC) in its *Science Report* (1990) did present preliminary values for gas indices. These appear in Table I.

In this table, the potential of each gas to warm the earth is indexed relative to CO_2. Thus, over a 100-year time horizon, one ton of methane has the potential to cause 21 times the warming as one ton of CO_2 emitted in the same year.

This information can play an important role in allocating the burden of control measures among gases, as it allows decision makers to maximize the impact of control expenditures in reducing potential warming.

Using the information in Table I, it is possible to determine the total greenhouse gas "budget" of the United States for an historical year. Table 2 depicts U.S. greenhouse gas emissions for 1987, a prototypical year of the latter 1980s.[2] Emissions are expressed in terms of "carbon equivalents" using the 100-year global warming potential (GWP) factors developed by the IPCC. As can be seen from the table these totalled 2,328 million tonnes, with CO_2 accounting for 56%, followed by CFCs (16%), methane (10%), NO_x (9%), nitrous oxide (3%) and others (5%).

These historical emissions can be projected into the future to develop a base case scenario. Such a projection must take into account the effect of economic and population growth, the vintaging of capital equipment, and the government policies in place that are likely to affect future greenhouse gas emissions. Foremost among the latter are the Clean Air Act, the "America the Beautiful" tree planting initiative and, potentially, the National Energy Strategy.

Future Rates of Emissions

Ideally, future emissions would be estimated using a consistent (general equilibrium) framework that incorporates all gases, macroeconomic variables, and existing government policies. Unfortunately, no such modeling structure exists at this time. However, it is possible to piece together a view of likely emission trends from studies underway at EPA and existing studies from a variety of sources. In what follows, a base case emissions scenario is constructed by combining the results of several studies – some of which rely on general equilibrium models, some partial equilibrium analyses. Some employ capital vintaging models, others rely on alternative methodologies. In some cases, assumptions concerning economic growth rates may vary slightly from one study to another. Nonetheless, if one wants to assess the costs of meeting greenhouse gas emission targets, one must first surmise what emissions will be in the absence of such targets given existing government policies. Moreover, if the ultimate goal is to limit possible warming, it seems prudent to focus on all gases that contribute to warming, no matter how analytically messy this may be.

The base case scenario focuses on the next twenty years, a time frame that may be viewed as too short in the context of global warming. If current atmospheric models prove to be correct, it will be necessary to sustain emission limitations over decades to appreciably limit warming. Cost estimates over the next twenty years should be viewed in this context. However, our knowledge of the technologies that will be available to mitigate greenhouse gases twenty years hence, as well as their associated costs, is extremely limited. It therefore seems that an examination of the next twenty years can be instructive despite these limitations.

Carbon Dioxide (CO_2)

Estimates of future CO_2 emissions in the United States are taken from two sources: the official U.S. submission to the Energy and Industry Subgroup (EIS) of the IPCC, and from a recent analysis by Jorgenson and Wilcoxen. The former is based on the output of an energy sector model, Fossil 2, employed by the Department of Energy, the details of which are referenced in the EIS report. It estimates future U.S. CO_2 emissions of approximately 1550 million tpy in the year 2000 and 1850 million tpy in 2010. Jorgenson and Wilcoxen employ an econometric model of the U.S. fitted with data on interindustry transactions extending from 1947 to 1985. The model contains 35 industrial sectors and 672 types of households. Producer and consumer behavior is based upon "rational expectations" (i.e., both are forward looking) and each sector's production cost function incorporates an endogenous determination of the rate of productivity growth. Using this model, Jorgenson and Wilcoxen estimate that, in the absence of changes in government policy, CO_2 emissions will grow from their 1987 level of 1310 million tonnes to 1506 million tonnes in the year 2000 and 1638 million tonnes in 2010.

However, neither projection takes into account several changes in government policy that will affect future carbon emissions. These include the following:

Tree Initiative

The "America the Beautiful" reforestation initiative has set a goal of planting one billion trees per year on 1.5 million acres for ten to twenty years, and of improving forest management practices on an additional 100,000 acres per year. This program is being proposed for reasons quite apart from global warming mitigation, but would have the ancillary benefit of sequestering 9 million tons of carbon in the year 2000 and 50 million tons in 2010 – if fully funded. EPA is currently analyzing the impact of the program as it emerged in the Farm Bill.

It should be noted, however, that while this program will sequester additional carbon, estimates of total net sequestration for the U.S. for the next 20 years do not exist. Such estimates would consider the impact of projected air pollution improvements on forest health, as well as changes in land use. In the absence of such estimates, this paper assumes that the only net change to carbon sequestration will be the "America the Beautiful" program.

DOE Initiatives

While the Department of Energy (DOE) has yet to complete the development of its National Energy Strategy, in 1990, it did undertake a number of initiatives that will affect carbon emissions. These include promulgation of energy efficiency standards for appliances, the development of model building codes to be adopted by states, a program to further promote state least-cost utility planning efforts, and an initiative to expand hydroelectric power.

DOE has issued estimates of the energy savings associated with these initiatives for the year 2000 and these have been converted into carbon dioxide emission reductions. To calculate the impact for the year 2010, EPA has assumed that many of the programs will be as effective in the years 2000-2010 as DOE has calculated for the years 1990-2000. For example, due to stock turnover and retrofits, HUD adoption of DOE building standards will result in a given energy savings in 2000. Continued stock turnover and retrofits will result in a nearly equal energy savings in the next ten years, through 2010. When more detailed analysis was available, as with appliance standards, those energy savings estimates were used. A few initiatives, such as expanded hydro power, were assumed to remain flat beyond 2000, i.e., to exhibit no further growth. The energy savings estimates from DOE programs yields carbon emission reductions totalling 36 million tonnes in 2000 and 65 million tonnes in 2010.

The Clean Air Act

The recently enacted Clean Air Act (CAA) Amendments will affect carbon emissions in a number of ways. First, acid rain controls will raise the cost of electricity – especially coal-based electricity – causing some amount of fuel switching (away from coal or oil to natural gas) and some amount of conservation. Using ICF's Coal and Electric Utilities Model and assuming a demand elasticity of -1.0 (while taking into account the energy penalty associated with sulfur dioxide scrubbing technologies) results in carbon emission reductions of 16 million tonnes in 2000 and 19 million tonnes in 2010.

The Act will affect carbon emissions in other ways the magnitude of which has not been estimated by the Agency. Tailpipe standards will raise the costs of new cars and slow the rate at which newer, more fuel-efficient cars replace older, less fuel-efficient cars. Standards regulating the quality of gasoline and other transportation fuels will raise fuel prices, causing reductions in total vehicle miles travelled and increasing the demand for fuel-efficient vehicles (offsetting the effect of tighter tailpipe standards). Many control technologies for stationary sources rely on the incineration of pollutant emissions and would therefore result in increases in carbon emissions. Finally, the Act, which entails annual control costs of $25 billion, will affect total national income growth and investment, which will in turn affect carbon dioxide emissions.

Total Future CO_2 Emissions

The total effect of new government initiatives are summarized in Table 3. The question is should these be subtracted from the carbon totals generated by the DOE and Jorgenson/Wilcoxen models to yield revised base case projections?

There seems to be a strong case for adjusting the estimates of the models. The 1947-1985 period over which the Jorgenson and Wilcoxen model was estimated, while containing a number of important government interventions concerning energy and environmental regulation, does not reflect the future roles of such interventions. Environmental regulation will be playing a larger role in the economy, with environmental expenditures estimated to grow from their 1972 level of 0.9% of GNP to 2.7% by the year 2000. It would seem reasonable to take these trends into account.

That said, it is another matter as to how best to do so. The government estimates of the impacts of these programs are generated using partial equilibrium analyses, meaning that important behavioral feedbacks may be omitted. In addition, there is a potential for double counting. For example, the endogenous technological change component of the Jorgenson and Wilcoxen model may have already factored in the high efficiency refrigerators mandated by DOE's rulemaking. The conservation induced by the Acid Rain program may be the same conservation DOE is targeting with its Least Cost Electricity Planning Program. Given these problems, the approach taken in this paper is to use the EIS (Energy and Industry Subgroup of the IPCC) submission (less policy initiatives) as the upper value of base case carbon emissions and the Jorgenson/Wilcoxen estimates (less policy initiatives) as the lower value. This results in future carbon emissions of 1401 to 1503 million tonnes in the year 2000 and 1498 to 1627 in 2010. These compare to 1989 emissions of approximately 1322 million tonnes.

CFCs

Future emissions of CFCs were estimated using an EPA capital stock vintage model while taking into account recent revisions to the Clean Air Act and the Montreal Protocol – both of which call for a phase-out of CFC production by the year 2000. This model has been used by EPA in regulatory proceedings to calculate the impact of proposed rulemakings. The model takes into account the lag between the production of CFCs and their release into the atmosphere. For example, it may be several years between the time CFCs are produced for use in refrigeration and the time they are released into the atmosphere. In addition to accounting for the projected increased demand for compounds with CFC attributes, the model also takes into account the phase-in of CFC substitutes as CFCs are phased out. Many substitutes are themselves greenhouse gases, although of less potency than CFCs. Estimates from this model indicate that CFCs and their substitutes will account for 256 million tonnes of "carbon equivalents" in the year

2000 and 188 million tonnes in the year 2010. This compares with annual emissions of approximately 367 million tonnes in the 1985-90 time period.

Methane

Future U.S. emission rates of methane were taken from the U.S. submission to EIS with one exception: landfill emission estimates were updated to reflect likely future solid waste disposal rates and the imposition of environmental controls to capture landfill off-gases. The latter are being established under the new source performance standards provisions of the Clean Air Act. The goal of the regulation is to capture conventional pollutants (such as volatile organic compounds). However, an ancillary benefit will be the control of methane. Because of this regulation, total U.S. emissions of methane are expected to decline from current levels of 235 million tonnes of carbon equivalents in 1987 to 208 million tonnes in 2000 and 212 in 2010.

Other Pollutants

Emission trends of nitrous oxide (N_2O) have not been the subject of scrutiny and the simplifying assumption has been made that they will remain unchanged in the future. Since they account for only 3% of U.S. greenhouse gas emissions, this is unlikely to affect overall results.

Future emissions of VOCs, NO_x, and CO were estimated using capital stock vintage models developed by EPA and employed during Clean Air Act deliberations. The analysis takes into account the effect of the new CAA on emissions of these gases. The models employed include the ERCAM-VOC model developed by E.H. Pechan and Associates, EPA's Mobile 4 model, as well as ICF's Coal and Electric Utility model. In this analysis, the control measures and timetables included in the Senate version of the Act were analyzed. To the extent that the final version of the Act differs from the Senate version, emission reduction estimates associated with these gases will diverge from those presented here. In addition, this analysis generally assumes timely and full implementation of the Act on the part of EPA and the states and full compliance with theoretically achievable emission rates on the part of sources. Total projected future emissions of these pollutants (expressed in terms of carbon equivalents) are depicted in Tables 4 and 5.

Sulfur Dioxide and Sulfates

An additional pollutant worthy of discussion in a comprehensive analysis of greenhouse gases is sulfur dioxide (SO_2). SO_2 is emitted from a number of industrial sources (about two thirds of current U.S. emissions are from powerplants) and reacts with other gases to form sulfate aerosols in the atmosphere. Sulfate particles, in turn, serve as cloud condensation nuclei. That is, they form the nuclei around which water vapor condenses to form clouds. Clouds, in turn, are important determinants of the earth's albedo – i.e., they are important in determining how much of the sun's energy reaches the earth and how much is reflected back into space. The chain is thus as follows: sulfur dioxide emissions form sulfates; sulfates help form clouds; and clouds (depending on the altitude) may reflect solar energy back into space, keeping the earth cooler than would otherwise be the case. The IPCC identified this chain as a potentially important mechanism in global warming, although it provided no quantitative estimates relating SO_2 to warming. Given the new Clean Air Act's requirement to reduce SO_2 by 50%, closer examination of this possible phenomenon may be warranted in the future as it may have implications for future rates of warming.

Total Future Emissions of Greenhouse Gases

Tables 4 and 5 depict total U.S. greenhouse gas emissions given current government policies for the years 1987, 2000, and 2010. These tables differ only in their projection of CO_2 – with Table 4 relying on the adjusted EIS submission and Table 5 employing the adjusted Jorgenson and Wilcoxen estimates. As is evident in the figure, subject to the caveats previously discussed, the U.S. contribution to global warming will be approximately the same in the year 2000 as it was in 1987. The year 2010 shows a slight increase (4%) over these levels in Table 4 and virtually no change in Table 5.

This base case scenario indicates that an increase in carbon emissions (14-24%) will be offset by decreases in other gases regulated by the Clean Air Act. Base case scenarios such as this are sensitive to assumptions. Many of these have been discussed earlier. For example, these results could change if the implementation of environmental or energy policies does not materialize at assumed rates or if the economy grows at a rate faster than expected. Generally, in the analysis of Clean Air Act issues, the Agency has assumed that historical trends in such factors as vehicle miles travelled will continue in the future and that economic growth will be in the range of 2.0-2.5% per year. The output of Jorgenson and Wilcoxen is somewhat different in this respect. In their model, economic growth is an output rather than input of the model, and is determined by econometrically derived coefficients of consumer and producer behavior. Their model projects a 2% growth in GNP over the 1990-2000 period, and a slight decrease thereafter.

The Costs of Greenhouse Gas Targets

Given this base case emissions scenario, the question becomes what can be learned about the costs of meeting various greenhouse gas targets?

First, if the goal is to cap the overall level of weighted greenhouse gases during this time period, it may be that only a limited amount of additional government intervention is needed. This is because current government energy and environmental policies are sufficient to keep total greenhouse gas emissions (expressed as carbon equivalents) below 1987 levels in the year 2000. In 2010 total emissions are projected to be only 4% above current levels in one scenario and essentially the same as 1990 levels in another.

There are differences in the trends of individual gases, however. CO_2 is expected to grow in both scenarios. Thus, if the goal is stabilization of the absolute level of CO_2, additional government control programs would be necessary. The rate of increase associated with CO_2 levels in the low scenario, however, is less than the rate of projected U.S. population growth (currently 0.7% per year). Thus, existing government programs may be sufficient to limit per capita CO_2 emissions to today's levels in the years 2000 and 2010.

Opportunities for Additional Reductions

There are a number of opportunities for achieving additional reductions in the U.S. contribution of greenhouse gases. In evaluating these opportunities, it is important to keep in mind that (1) different gases have different greenhouse potentials and (2) government actions to further reduce greenhouse gases typically will entail other environmental benefits that should be factored into any judgment about whether such actions are cost-beneficial.

A simple example can illustrate the importance of these considerations. According to Table 1, NO_x emissions are 40 times more potent at trapping heat than CO_2 emissions (using a 100-year time horizon). This means that NO_x control costs of $400 per ton would be as cost effective in reducing warming as CO_2 controls costing $10 per ton. In addition, NO_x controls provide other important environmental benefits. NO_x deposition has been identified as an important contributor to nitrogen loadings in water bodies and NO_x plays a role in the formation of tropospheric ozone, visibility impairment, and acid deposition. These benefits should be taken into account in analyzing the economic efficiency of reducing greenhouse gases.

A number of control opportunities exist that have not been systematically analyzed in the context elucidated above. The discussion below focuses on options of particular relevance to the United States for further study. The list is not intended to be inclusive and is for illustrative purposes only. Benefits of control are not calculated in this paper and should be considered in making any control decision.

Methane

According to Table 1, methane is 21 times more potent than CO_2 in its contribution to the greenhouse effect (over a 100-year period). Thus, methane control should not be overlooked in a greenhouse gas policy. Control opportunities include the following:

1. Animal Waste: Methane from animal waste is a major source of greenhouse gases. In addition, animal wastes have been identified as a major source of nonpoint

water pollution and groundwater contamination. A promising technology for managing such waste at sites with large animal herds such as feedlots and dairy farms is decomposition in anaerobic digesters. Full methane recovery at these sites would reduce emissions by approximately 50%, and can be used to generate electricity at a cost of five to seven cents per kwh.

2. Coal mining: Methane released during coal mining can be recovered, processed, and injected into natural gas pipelines.

3. Landfills: EPA hopes to soon issue regulations controlling landfill emissions. In addition to methane, these emissions consist of volatile organic compounds and air toxins. The issue of the size of landfills to be subject to regulation is likely to be the subject of public comment. Extending the size cutoff to smaller sources results in additional methane control and other air quality benefits. Such benefits, however, must be weighed against increased control costs.

4. Sewage Sludge: Anaerobic decomposition of sewage sludge is also a source of methane in the United States. However, the feasibility and cost of controlling this source has not been the subject of detailed study.

It should be noted that the lack of data on methane control could be rectified in the near future. The Clean Air Act Amendments of 1990 require EPA to issue a report to Congress on methane emission inventories and control costs. Obviously, further examination of the net benefits of controls is needed before their use should be mandated.

Nitrous Oxide (NO_x)

Recall that NO_x has a greenhouse potency factor 40 times that of CO_2. Thus, NO_x control has the potential to make a significant contribution to a mitigation strategy. The 1990 Clean Air Act will result in over 2 million tons of NO_x reductions. However, several NO_x control opportunities remain. For example, low NO_x burner controls could be extended to existing industrial boilers and small electric utility boilers. Stationary internal combustion engines and turbines could be controlled, and selective catalytic reduction controls could be required of major stationary sources.

Carbon Monoxide (CO)

CO has a warming potential 8 times that of CO_2. The current Clean Air Act requires CO control through the use of tighter tailpipe standards (i.e., "cold start CO") and oxygenated fuel requirements. The latter are required only during winter months except for severe ozone nonattainment areas. The coverage of such controls could be broadened both geographically and temporally. In addition, raising the oxygenate percentage requirements would further reduce CO. Again, additional analysis of total program cost and benefits should be examined before such controls are undertaken.

Carbon Dioxide

A number of studies have been undertaken on the costs of controlling CO_2 and the estimates of such control vary widely. A number of these are reported in a DOE White Paper entitled *The Economics of Long Term Global Climate Change: A Preliminary Assessment*. Estimates of such costs in the 1990-2020 time frame range from a *net savings* of 75% over the energy system costs that would otherwise have been incurred in the absence of CO_2 emission targets to 5% of GNP for a 0-20% reduction.

Recently, EPA retained Jorgenson and Wilcoxen to analyze CO_2 control costs. Using the Dynamic General Equilibrium Model (DGEM) briefly described earlier, they analyzed alternative CO_2 emission targets.

These targets were achieved using three different policy instruments: taxes on the carbon content of primary fuels, the energy (BTU) content of such fuels, and an ad valorem tax. Some preliminary results from the carbon tax analysis are presented in Table 6. They indicate that emissions can be maintained at 1990 levels through the imposition of a carbon tax of approximately \$17 per ton. This amounts to a \$11.12 tax per ton of coal, a \$2.34 tax per barrel of oil; and a \$0.28 tax per thousand cubic foot of natural gas. It should be noted, however, that the Jorgenson and Wilcoxen analysis did not take into account the impact of the Clean Air Act, DOE initiatives, or tree planting programs mentioned earlier. These programs will reduce the amount of tax necessary to maintain CO_2 emissions at 1990 levels. It should be noted that the results of the model are preliminary and may change as refinements are made.

Accommodative Monetary and Fiscal Policies

The Jorgenson and Wilcoxen analysis examined revenue neutral energy tax programs. Revenues generated through energy taxes were offset by proportional reductions in income and business taxes. The results of their analysis indicate the possible medium-term (i.e., 15-20 years) impacts of carbon emission targets. However, such targets would entail serious economic impacts in the immediate to short term if not accompanied by accommodative monetary and fiscal policies.

Energy taxes would send a surge of inflation through the economy as consumers pay higher prices for energy. This inflationary effect could be reduced if income taxes were simultaneously reduced. Reductions in income taxes alone, however, would not be sufficient to totally offset this inflation if government revenue (or the deficit) is to be held constant. Monetary policy could be tightened to control this inflationary surge, but this would cause an adverse impact on GNP growth. What is therefore needed is a government policy that can both offset the inflationary impacts of energy taxes while maintaining total national income. This policy must simultaneously reduce business costs by the same magnitudes that they are increased by energy taxes. This can be accomplished either by subsidizing business activities or by reducing other business taxes.

To illustrate the importance of this phenomenon, EPA retained Data Resources, Inc. and had them analyze a gasoline tax sufficient to maintain light duty vehicle carbon emissions at 1989 levels through 2010. Such a tax would amount to $0.59 per gallon (in 1990 dollars) in the year 2010. Gasoline tax collections would average $61 billion annually from 1991 to 2010. However, the effect of such a tax on GNP growth depends largely on what other government policies accompany the tax. Table 7 illustrates two alternative cases.

In case I, the gasoline tax revenues are accompanied with reduced income taxes. Although this policy is both revenue and deficit neutral, total national income (as measured by GNP) is reduced by 0.4% in the year 2000. In Case II, the gasoline tax revenues are accompanied with a reduction in payroll taxes (paid by businesses). In the scenario, GNP remains virtually unchanged from baseline levels. The lessons from these scenarios are twofold:

1. Mitigation strategies that rely on taxes by definition involve changes in fiscal policies. In reviewing the results of studies of such strategies, it is useful to distinguish between impacts that are related solely to the fact that fiscal policy has changed from those that relate to the particular focus on greenhouse gases. Accommodative policies can be designed to alleviate the former.

2. In this context, it is useful to conceive of mitigation strategies that shift the tax base towards greenhouse gases and away from other business taxes. Such strategies are likely to minimize the impact on total national income although tax burdens on different sectors of the economy will change.

Utility Sector CO_2 Mitigation Strategies

A variety of studies in recent years have highlighted the large potential for electric and natural gas end-use efficiency gains in the residential, commercial, and industrial sectors. Estimates of technical potential for savings in the electric utility sector range from 24% to 75% of today's consumption level; estimates for natural gas consumption fall within the same range. Differences in estimates of technical potential are largely due to different definitions. Typically, technical potential is defined as the "overnight implementation" of all currently commercially available and technically applicable technologies which provide similar or increased levels of energy service. Cost-effectiveness is not a limiting criterion. Some estimates use a broader definition that includes measures still on the drawing board or in early stages of commercial development.

Notwithstanding the differences, all the estimates show a significant potential for gains in end-use efficiency. A goal of policy makers should be to design strategies that allow the capture of as large a share of the cost-effective technical potential for energy savings as possible.

EPA has recently funded an effort with RCG/Hagler, Bailly, Inc. to develop a modeling system capable of assessing the impact of various utility sector policy options. The modeling system – known as the Electric and Gas Utilities Modeling System

(EGUMS) – is specifically designed to focus on near-term demand-side alternatives that can be used to reduce the consumption of energy and lessen its concomitant adverse environmental impacts, particularly CO_2 emissions. Three policy options have been analyzed to date: (1) an incremental tightening of the federal household appliance standards, (2) an incremental tightening and augmentation of the federal commercial lighting efficiency standards, and (3) a federal commercial building energy code.

Policy option (1) assumed that federal appliance standards were updated to include all cost-effective technologies based on a 7% real societal discount rate. Average national utility prices and appliance utilization rates were used to calculate cost effectiveness. The tightened appliance standards were assumed to take effect beginning in 1995. EGUMS projects that the tightened standards could save 2.3% of total base case electricity consumption by 2000, and 6.3% by 2010. Natural gas consumption savings amount to 1.4% in 2000, and 4.1% in 2010. Carbon emissions savings for both electric and natural gas measures totaled approximately 21.5 million tonnes in 2000, and 83.0 million tonnes in 2010.

Policy option (2) assumed an electronic ballast and fluorescent lamp standard. Energy savings were also projected for reduced utilization of building HVAC systems. EGUMS projects electricity consumption savings of 1.2% in 2000, and 1.6% in 2010. Carbon emissions savings are approximately 9.0 million tonnes in 2000, and 17.2 million tonnes in 2010.

Policy option (3) combines the lighting standards of option (2) with the performance-based ASHRAE Standard 90.1 which applies only to new commercial buildings. EGUMS projects electricity consumption savings of 1.4% in 2000, and 2.5% in 2010. Natural gas consumption savings total 0.6% in 2000, and 1.6% in 2010. Carbon emissions savings are approximately 13.1 million tonnes in 2000, and 29.2 million tonnes in 2010.

Other policy options and market scenarios to be modeled with EGUMS include: (1) tax restructuring/shifting initiatives, including carbon taxes; (2) growth of utility-sponsored demand-side management programs; (3) growth of unregulated energy service companies; and (4) alternative rates of market penetration and product promotion of energy efficient end-use appliances.

Efforts to Encourage More Use of Biomass

There is heightened interest in using biomass as a source of energy supplies in place of fossil fuels. From the standpoint of global warming, the advantage of biomass is that it is a "sink" for CO_2 in comparison to fossil fuels. Thus, CO_2 emissions resulting from the combustion of the biomass originate in the atmosphere and are merely regenerated when the fuel is burned.

On the surface, using biomass would appear to cause no net environmental increase in CO_2. The problem with this reasoning is that fossil fuels are used to produce biomass. In order to compute the net CO_2 benefits of using biomass, one must account for the "imbedded (fossil fuel) energy" that is used in the production of the biomass.

One logical place to encourage greater use of biomass is by blending ethanol into gasoline. We have attempted to calculate the net reductions in CO_2 from substituting ethanol for gasoline, accounting for the ramifications of greater use of ethanol upon corn production upon existing farm support programs. CO_2 reductions range from roughly 25-50% on an energy equivalent basis. These are larger than some other studies have found. We calculate that this substitution will occur at a cost effectiveness per ton of carbon reduced of greater than $1000/ton. Compared to other options to reduce CO_2, this is not cost effective – although the effect of ethanol on other emissions (e.g., CO) has not yet been taken into account. Current research effort is being devoted to looking at technologies that can reduce the costs of producing ethanol and in the long run to using alternative feedstocks (e.g., short rotation woody crops).

Table 1

Trace Gas	Estimated Lifetime, years	Global Warming Potential Integration Time Horizon, Years		
		20	100	500
Carbon Dioxide	*	1	1	1
Methane - inc indirect	10	63	21	9
Nitrous Oxide	150	270	290	190
CFC-11	60	4500	3500	1500
CFC-12	130	7100	7300	4500
HCFC-22	15	4100	1500	510
CFC-113	90	4500	4200	2100
CFC-114	200	6000	6900	5500
CFC-115	400	5500	6900	7400
HCFC-123	1.6	310	85	29
HCFC-124	6.6	1500	430	150
HFC-125	28	4700	2500	860
HFC-134a	16	3200	1200	420
HCFC-141b	8	1500	440	150
HCFC-142b	19	3700	1600	540
HFC-143a	41	4500	2900	1000
HFC-152a	1.7	510	140	47
CCl_4	50	1900	1300	460
$CH_3 CCl_3$	6	350	100	34
CF_3Br	110	5800	5800	3200
INDIRECT EFFECTS				
Source Gas	Greenhouse Gas Impacted			
CH_4	Tropospheric O_3	24	8	3
CH_4	CO_2	3	3	3
CH_4	Stratospheric H_2O	10	4	1
CO	Tropospheric O_3	5	1	0
CO	CO_2	2	2	2
NO_x	Tropospheric O_3	150	40	14
NMHC	Tropospheric O_3	28	8	3
NMHC	CO_2	3	3	3

Table 2

Current Emissions
1987
(100 Year GWP Factors)

Carbon Equivalents – Millions of Tonnes

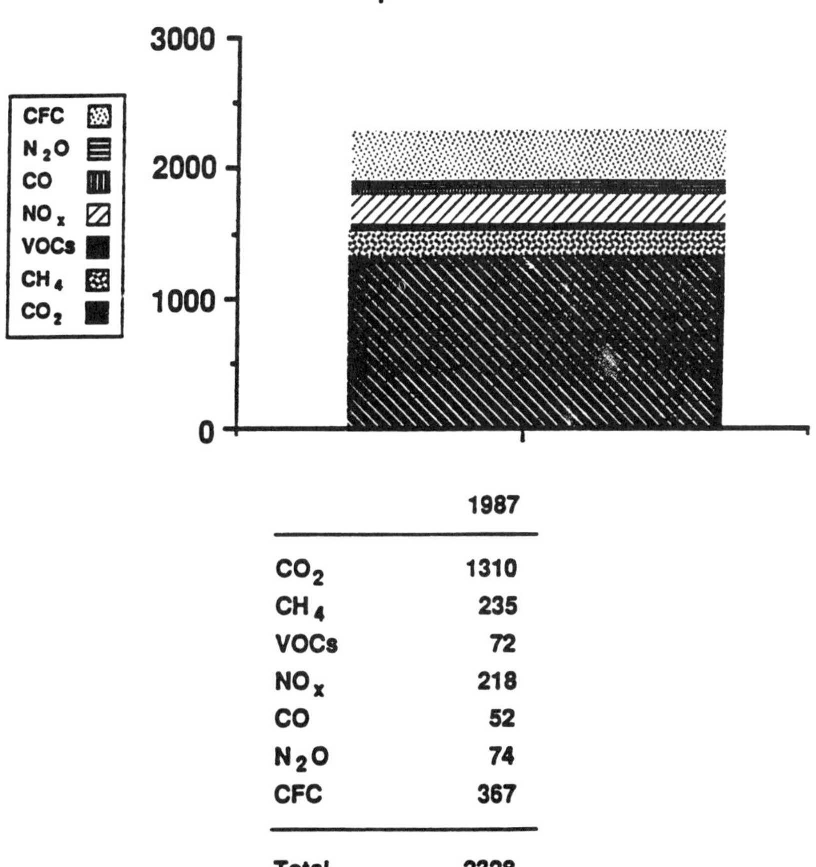

	1987
CO_2	1310
CH_4	235
VOCs	72
NO_x	218
CO	52
N_2O	74
CFC	367
Total	2328

Table 3
Effect of Recently Announced Government Program on Carbon Emissions (Reductions In Tonnes of Carbon)

ACTION/PROGRAMS	2000	2010
DOE Appliance Standards (Refrigerators, Washers, Dryers, Dishwashers)	4.4	4.9
DOE Efficiency Initiatives	27.8	55.7
Federal Building Lighting	1.4	2.8
Commercial Buildings Lighting	2.5	5.0
Promote State LCUP	9.0	18.0
State Adoption of Interim Building Standards	8.2	16.4
Energy Analysis and Diagnotic Centers	6.0	12.0
HUD Adoption of DOE Building Standards	0.8	1.6
DOE Renewable Initiatives	4.1	4.1
Expanded Hydro Power	3.5	3.5
Photovoltaic Technology	0.5	0.5
Clean Air Act		
Acid Rain	16.4	19.1
Transportation Biofuels	0.8	1.6
GRAND TOTAL	53.5	82.6

Table 4

Emissions
(With Current Commitments)
(100 Year GWP Factors)

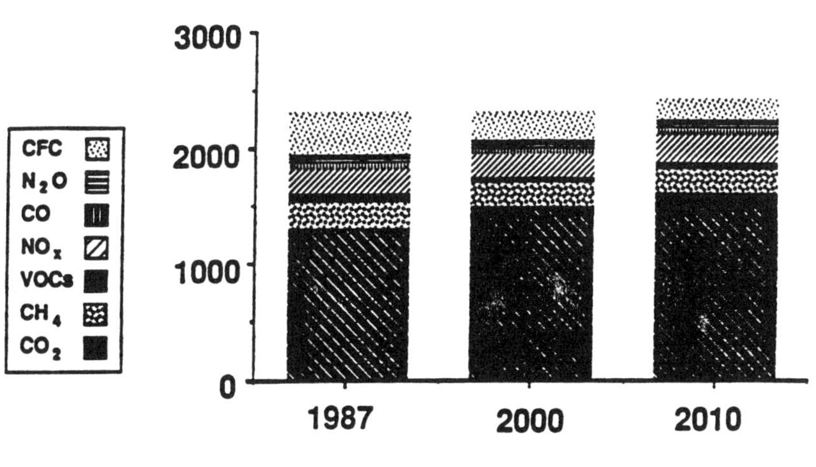

	1987	2000	2010
CO_2	1310	1503	1627
CH_4	235	208	212
VOCs	72	48	50
NO_x	218	199	235
CO	52	45	46
N_2O	74	74	74
CFC	367	256	188
Total	2328	2332	2430

Table 5

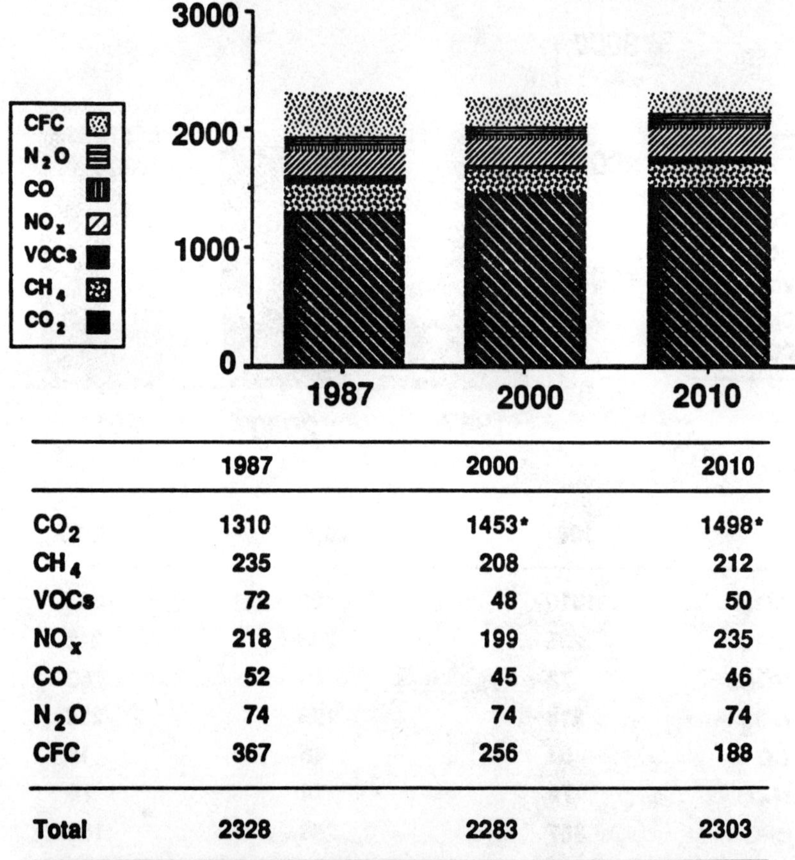

	1987	2000	2010
CO_2	1310	1453*	1498*
CH_4	235	208	212
VOCs	72	48	50
NO_x	218	199	235
CO	52	45	46
N_2O	74	74	74
CFC	367	256	188
Total	2328	2283	2303

* CO_2 estimates derived from baseline emissions projections in Jorgenson and Wilcoxen (1990)

Table 6

LONG-RUN EFFECTS OF A CARBON TAX

(JORGENSON-WILCOXEN)

SCENARIO: HOLD CO2 EMISSIONS IN ALL FUTURE YEARS AT 1990 LEVELS

VARIABLE	*LEVELS IN 2020*
CARBON TAX	*$17.13*
TAX ON COAL	*$11.12*
TAX ON OIL	*2.34*
TAX ON GAS	*0.28*
CARBON EMISSIONS	*-19.6%*
REAL GNP	*-0.5%*
TAX REVENUE	*$20.7 billion*
EFFECTS ON COAL INDUSTRY:	
* *PRICE OF COAL*	*+37.8%*
* *QUANTITY OF COAL*	*-34.5%*

KEY RESULTS

- *A CARBON TAX IS THE LEAST EXPENSIVE INSTRUMENT FOR A GIVEN TARGET*
- *CONSIDERABLE REVENUE RAISED BY ANY TAX*
- *MEDIUM RUN GNP GROWTH EFFECTS ARE SMALL BUT ADD UP*
- *COAL MINING WILL BE HARD HIT UNDER ANY SCENARIO*

Table 7

EFFECTS OF ALTERNATIVE TAX
SCEMES ON GNP
(DRI)

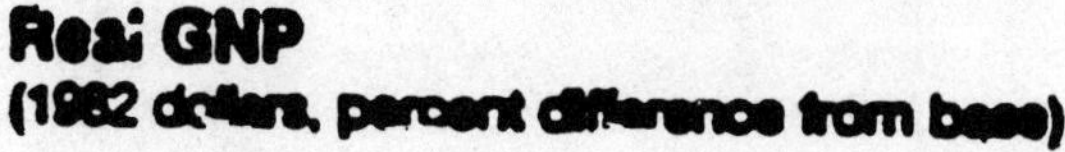

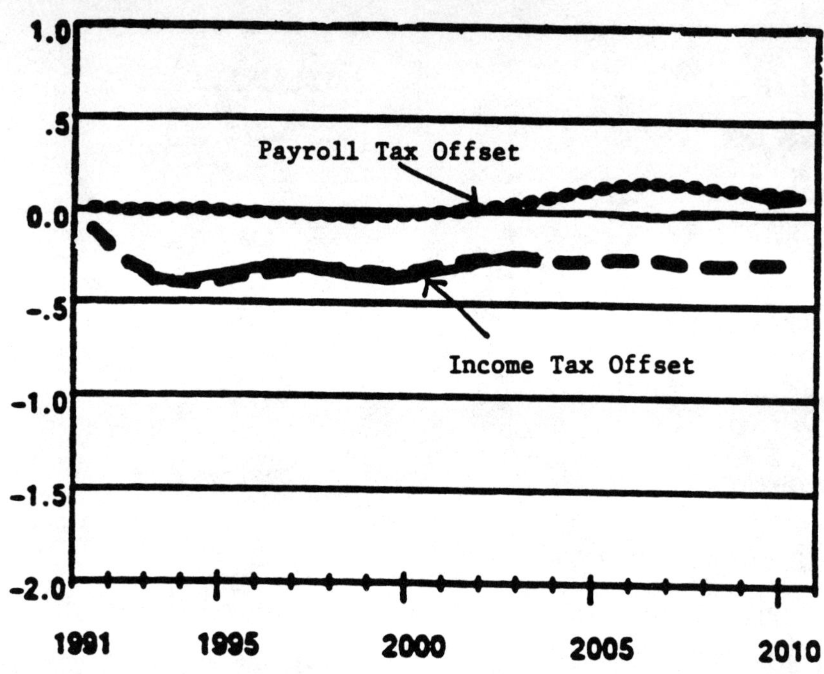

Notes

1. Data and analyses in this paper are the work of the Energy Branch of EPA's Office of Policy Analysis, its contractors and other EPA Offices. Much of this information represents work in progress and does not represent an official statement on the part of EPA.

2. Due to limitations in data, emissions for some gases are taken from other years in the late 1980s.

SAVING MONEY AND REDUCING THE RISK OF CLIMATE CHANGE THROUGH GREATER ENERGY EFFICIENCY

Howard S. Geller

American Council for an Energy-Efficient Economy
1001 Connecticut Ave., N.W.
Washington, DC 20036

Abstract

Improving the efficiency of specific energy-using devices such as automobiles or appliances can save consumers and the country money while reducing emissions of carbon dioxide and other pollutants. In the case of automobiles, the efficiency improvements that occurred during 1975-89 saved consumers over $20 billion and cut U.S. energy consumption by nearly 5 quads (6%) in 1989. The net cost of avoided carbon emissions was approximately negative $200 per metric ton. Improvements in the efficiency of refrigerators during the past 15 years were even more cost-effective, with a social cost-benefit ratio of about 8 and net cost of avoided carbon emissions of approximately negative $300 per metric ton. For both end uses, a further doubling in efficiency is both technically feasible and cost-effective.

Policies and programs aimed at increasing energy efficiency also can save money and reduce carbon dioxide emissions. Appliance efficiency standards and utility conservation programs are two examples of what can be done to capture the benefits of greater energy efficiency. Conservation and load management programs operated by the New England Electric System are expected to save consumers $88 million per year and lower carbon emissions over 5% by 1999. A broad set of new energy efficiency initiatives could further reduce energy service costs and carbon emissions in the U.S. The initiatives include new financial incentives, minimum efficiency standards, R&D ventures, and information programs. Adopting the initiatives could cut U.S. energy demand in 2000 by 15-20% from levels otherwise projected, provide consumers with a net savings of $75 billion per year, and reduce carbon emissions in 2000 to 90% of the level in 1988. The policies are needed to overcome barriers inhibiting widespread efficiency improvements.

Copyright 1991 by Elsevier Science Publishing Company, Inc.
Global Climate Change: The Economic Costs of Mitigation and Adaptation
James C. White, Editor

Introduction

Some recent studies profess to show that there are economic penalties associated with achieving a substantial reduction in carbon dioxide emissions in the United States or in the world (1, 2). These studies utilize macroeconomic models and apply carbon or energy taxes as the means for reducing carbon dioxide emissions. Higher energy taxes and prices, without compensating reductions in other taxes, lead to loss of income, less economic activity, and a lower GNP.

The macroeconomic studies are flawed on a number of counts. As pointed out by Williams (3), they fail to adequately account for ongoing reductions in the energy intensity of industrialized countries. Falling energy intensities, which have been occurring for decades, are related to structural changes (e.g., less dependence on energy-intensive materials) and technological innovations. Both of these factors are likely to continue to have a large impact on energy use in the United States and elsewhere (4). Technological innovations in another area, renewable energy technologies, could greatly reduce dependence on carbon-based fossil fuels. This is important since the economic studies extend 20-110 years into the future, a time period when a variety of renewable energy technologies are expected to become cost-competitive and widely implemented (5).

Another serious flaw in the economic models is their simplistic treatment of economic costs and benefits. The models ignore the complex environmental impacts associated with climate change, for example. Global warming is expected to intensify hazards such as tropical pests, hurricanes, and urban air pollution (6). Regional variations in impacts also could lead to major economic and human dislocations. Ignoring these difficult to quantify but important effects limits the value of the economic projections.

Of greatest relevence to this chapter, the economic models fail to account for the full impact of end-use energy efficiency measures such as more efficient automobiles, buildings, appliances, or industrial processes. End-use efficiency improvements usually provide energy savings of greater present value than the initial cost for the conservation measures; i.e., saving energy usually costs less than supplying energy. Consequently, there is a net economic savings for consumers and the economy while energy use and corresponding pollutant emissions decline. When energy is obtained from fossil fuels, there is a reduction in carbon dioxide emissions at a negative economic cost to society. End-use efficiency improvements can be stimulated through market mechanisms such as revenue-neutral fees and rebates, efficiency regulations, or other means.

Technologies for Improving Energy Efficiency

Two examples of the technological possibilities for cost-effective end-use efficiency improvements are presented in this section. Both past accomplishments and near-term opportunities for improving the efficiency of automobiles and refrigerators are addressed. In addition, a few comprehensive studies of energy conservation potential are summarized.

Automobiles

The average fuel economy of new cars sold in the U.S. increased from 15.8 miles per gallon (MPG) in 1975 to 28.6 MPG in 1988 based on the Environmental Protection Agency's (EPA) combined city-highway test procedure (Figure 1). Efficiency gains resulted from a wide range of technical modifications including engine improvements, the shift to front-wheel drive, better aerodynamics, and use of lighter materials (7). Starting in 1978, efficiency gains were achieved without significant reductions in average car size (as measured by interior volume) or performance (as measured by acceleration).

Fuel economy rose during the early 1980s in spite of a steady decline in gasoline price. Corrected for inflation, the average price of gasoline in 1988 was at the lowest level in 40 years (8). The Corporate Average Fuel Economy (CAFE) standards established by the U.S. Congress and Ford Administration in 1975 were the main impetus for fuel economy improvements during 1975-88 (9). The fuel economy of cars produced by U.S. manufacturers stayed especially close to the CAFE standards during the 1980s.

It is possible to calculate the energy, economic, and carbon savings associated with previous automobile efficiency improvements (Table 1). Automobile efficiency experts estimate a cost in the range of $200-400 per car at the retail level for the cumulative fuel economy improvements made between 1978 and 1985 when domestic cars increased from 18.7 to 25.8 MPG on average (10). Also, U.S. auto manufacturers have estimated the average cost for fuel economy improvements during this period by size class. Averaging across companies and size classes, the manufacturers estimate an average cost premium of $397 (10). Based on a car being driven 10,400 miles per year on average (11) and a 15% penalty between the EPA composite rating and on-road efficiency (12), fuel economy improvements between 1978 and 1985 on average saved 180 gallons of gasoline per year.

With estimates of the extra first cost and annual energy savings for the efficiency improvements during 1978-85, it is possible to calculate a "cost of saved energy" (CSE) using the following formula:

$$CSE = (FC \times CRF + AOC)/AES$$

where FC is the extra first cost for the conservation measures, CRF is the capital recovery factor, AOC is any additional non-energy operating cost due to the conservation measures on a yearly basis, and AES is the annual energy savings. The capital recovery factor is given by:

$$CRF = ((1 + D)^L \times D)/((1 + D)^L - 1)$$

where D is the real discount rate and L is the lifetime of the conservation measures. In effect, CSE is the average cost for saving a unit of energy over the lifetime of the efficiency measures.

In this example, assuming a cost premium of $400, no additional non-energy operating cost because of the conservation measures, a real discount rate of 6% (approximately the interest rate or opportunity cost for consumers), and a ten-year lifetime, the cost of saved energy equals $0.30/gal. For comparison, gasoline typically costs consumers in the U.S. $0.86/gal as of 1989 excluding taxes (11). Using this gasoline price, the efficiency improvements show a benefit-cost ratio of 2.9.

Regarding aggregate energy savings, the average on-road fuel economy for the entire passenger car fleet in 1989 was approximately 20.5 MPG (11). For comparison, the fleet-average fuel economy in 1975 was 13.5 MPG. If the entire passenger car fleet in 1989 (143 million vehicles) had only achieved 13.5 MPG, total gasoline consumption would have increased by 37.6 gallons assuming no change in miles driven. This savings is equivalent to 2.45 million barrels of oil per day (MBD), more oil than the U.S. imported from Arab OPEC nations in 1989. The aggregate energy savings, 4.7 quads, also equalled 21% of transportation sector energy use and 6% of total U.S. energy use in 1989. Moreover, the energy savings will grow as older, less efficient cars are removed from the vehicle fleet.

If all gasoline savings in 1989 resulted from efficiency improvements with a cost of conserved energy of $0.30/gal, then the total annualized cost for efficiency improvements in the 1989 car fleet was $11.3 billion. Since the value of the gasoline savings was $32.3 billion (excluding taxes), consumers realized a net economic savings of $21 billion in 1989. This estimate is conservative in that it assumes no change in energy price as a function of gasoline use. In fact if U.S. gasoline use was 2.45 MBD higher than the actual value in 1989, there would have been upward pressure on world oil markets and higher oil and gasoline prices.

Regarding carbon dioxide emissions, the direct reduction in emissions associated with reducing gasoline use by 2.45 MBD in 1989 was about 102 million metric tons (expressed in terms of carbon). This is equivalent to about 7% of actual carbon emissions from fossil fuel combustion in the U.S. that year. Based on a net economic savings of $21 billion, the net "cost" for avoiding this amount of carbon emissions was about negative $206 per metric ton.

Manufacturers have by no means exhausted all cost-effective options for improving automobile efficiency. There are a number of proven technologies such as front-wheel drive, use of four valves per engine cylinder, and better aerodynamics which have been used in some but certainly not all vehicle models.

There are other efficiency measures such as new intake valve control systems and continuously variable transmissions which are just starting to be implemented. Table 2 lists a host of technologies that are available for increasing the efficiency of new cars during the 1990s. All of these measures increase efficiency without reducing vehicle performance or comfort.

Ledbetter and Ross analyzed the savings potential and cost effectiveness of these efficiency measures (13). The results are presented in terms of a "conservation supply curve", where individual measures are displayed according to their cost of saved energy and potential efficiency improvement (Figure 2). Each horizontal line in Figure 2 is a separate conservation measure. The efficiency improvements account for the fraction of new vehicles still eligible for each measure. Widespread adoption of cost-effective measures, i.e., measures with a cost of saved energy that is less than the projected gasoline price, could increase the average rated fuel economy of the new car fleet in 2000 to about 44 MPG. Compared to a scenario with no efficiency improvements, total energy use would drop 2.1 quads (1.1 MBD) by 2000 and 4.9 quads (2.6 MBD) by 2010. The average cost of saved energy for all the cost-effective efficiency measures is only about $0.55 per gallon. Since the average cost of saved energy is still much less than the projected cost of fuel, carbon dioxide emissions continue to be reduced at a negative net cost.

Even greater efficiency improvements are possible with more advanced technologies such as direct-injection diesel engines, two-stroke engines, engine shut-off during deceleration, and aluminum engines or bodies. Prototype vehicles with these features have EPA combined fuel economy ratings of 65-100 MPG (14). Although these prototype vehicles are not yet suitable for mass production, they demonstrate that much higher efficiencies can be achieved. As Ross has stated (15), "The cost and performance that cars like these would have if designed for the market and mass produced remain to be determined. There is, however, every expectation that cars with very high fuel economy and good space and performance characteristics can be built, perhaps without a substantial cost penalty beyond the manufacturer's initial tooling investment."

Refrigerators

Total energy use per household fell approximately 24% during 1973-87 (16), indicating significant conservation of fuels and electricity in the residential sector. Refrigerators, which account for about 15% of electricity use in housing, provide a good example of the technological changes and energy efficiency improvements in this sector. The typical new refrigerator sold in 1989 consumed about 930 kWh/year according to the official test procedure, 46% less than that for the typical new refrigerator sold in 1972 (Figure 3). In addition, today's mix of new refrigerators are larger and have more features such as greater use of automatic defrost. In terms of "energy factor" (a measure of efficiency based on adjusted volume per unit of electricity use), refrigerators sold in 1989 were over twice as efficient as those sold in 1972 (17).

In order to improve energy performance, refrigerator manufacturers switched from fiberglass to polyurethane foam insulation, increased the efficiency of motors and compressors, used larger heat exchangers, reduced internal heat sources, and made other changes (18). The doubling in efficiency occurred without major innovations or radical product redesign. Efficiency gains in refrigerators and other appliances were stimulated by market forces as well as minimum efficiency standards. Standards were first adopted in California in the mid-1970s. After California strengthened its standards and other states passed similar legislation, consensus national appliance efficiency standards were adopted in 1987 (19). The national standards for refrigerators went into effect in 1990.

As was the case for automobiles, the standards caused efficiencies to rise in spite of real energy prices falling in recent years.

Improving the efficiency of refrigerators is very cost-effective (Table 3). Assessments by contractors for the Department of Energy (DOE) as well as independent analysts indicate that a one-third reduction in electricity use from commonplace efficiency measures costs about $40 at the retail level (20, 21). Using a 6% real discount rate, the cost of saved energy is only $0.009/kWh. Based on the 1989 average residential electricity price of $0.076/kWh (11), efficiency improvements adopted in recent years show a societal benefit-cost ratio of 8.4. With these values, consumers realized a net economic savings of about $3.2 billion in 1989.

Around 48 billion kWhs were saved in 1989 due to efficiency improvements in new refrigerators during 1972-89. This is equivalent to nearly 6% of total residential electricity use or the power supplied by about eleven very large (1000 MW) power plants. Splitting up avoided power generation in proportion to the total utility fuel and generation mix, refrigerator efficiency improvements led to approximately 10 million metric tons of avoided carbon emissions in 1989. Given the extreme cost-effectiveness of these improvements, carbon emissions were reduced at a net cost of negative $308 per ton.

The electricity consumption of new refrigerators will continue to decline through greater implementation of available technologies, development of new technologies, and tightening of efficiency standards. Regarding new standards, DOE issued "second-tier" efficiency requirements for new refrigerators and freezers in 1989 (22). The new standards, which take effect in 1993, will ensure a further reduction in average electricity use of around 25% (see Figure 3). In addition, DOE is required to evaluate the technical and economic feasibility of tougher standards in the future.

The 1993 standards do not represent the upper limit on refrigerator efficiency. New insulating materials are under development that, unlike polyurethane foam, do not contain chlorofluorocarbons (CFCs). In particular, various vacuum insulation concepts provide greater resistance to heat flow without use of CFCs (23). Researchers at the Natural Resources Defense Council have estimated the energy savings and cost effectiveness of one type of vacuum insulation material along with other advanced efficiency measures (Table 4). Their analysis starts at the level of efficiency and electricity use mandated by the 1993 standards. Use of the aerogel vacuum insulation material and other measures could lower total electricity use for a typical two-door refrigerator- freezer to around 200 kWh/yr. It is estimated that electricity use of around 260 kWh/yr could be achieved up to a marginal cost of saved energy of $0.075/kWh (24).

Other Technologies

Analysis of a wide range of energy-efficiency measures yields the same results as those portrayed above – saving energy costs less than supplying energy and carbon dioxide emissions are reduced at a negative net cost. A study of the electricity conservation potential in New York State found that it is technically and economically feasible to reduce total electricity demand by 35% (25). The conservation measures that can yield the largest savings include heating, ventilation and air conditioning system retrofits in the commercial sector, more efficient refrigerators and freezers, reflectors for fluorescent light fixtures, compact and other high-efficiency fluorescent lamps, and variable speed motor drives (Table 5). The average cost of saved energy for all conservation measures considered in this study is $0.025/kWh. For comparison, the statewide average electricity price is about $0.090/kWh.

Researchers from Lawrence Berkeley Laboratory have shown that efficiency improvements in all major end uses in buildings can cut carbon emissions at a negative net cost (26). The electricity savings potential and conservation measure costs in this analysis were derived primarily from a study sponsored by the Electric Power Research Institute (27). The analysis indicates that electricity use in U.S. buildings and corresponding carbon dioxide emissions can be cut by 45% (Table 6). All twelve major conservation measures are cost-effective and reduce carbon emissions at a negative net cost based on a cost-effectiveness threshold of $0.075/kWh, the average cost of electricity in buildings.

Energy efficiency improvements in the industrial sector tend to be very economical as well. One study of conservation opportunities in 15 major manufacturing companies found that the cost of saved energy was typically in the range of $1-2/MBtu (28). The conservation measures included boiler improvements, heat recovery equipment, control systems, cogeneration projects, recovery of waste gases, and process modifications. For comparison, industrial consumers typically pay about $3/MBtu for natural gas and $4.70/MBtu for petroleum products (11).

Barriers Inhibiting Implementation

If energy-efficiency measures are so cost-effective, it is reasonable to ask why they are not more widely implemented and why policies and programs are needed to encourage greater adoption. The answer to these questions is that a variety of structural and behavioral factors inhibit full exploitation of cost-effective energy-efficiency measures (29). As evidence of these barriers, consumers in all sectors implicitly require paybacks of three years or less when making tradeoffs between initial costs and reduced operating costs. The resulting problem, often referred to as the "payback gap", is that there is a large difference in the investment criteria used for energy efficiency and energy supply investments. For the latter, paybacks of ten years or more are accepted. Several factors contribute to the payback gap:

- Consumers often lack credible information on the performance of energy-efficient technologies. Consumers usually do not know how much energy is used and how much it costs to operate different pieces of equipment. Although the MPG rating of automobiles is well understood, there is inadequate information about the energy-use characteristics of other products. Even if the energy efficiency ratings of new products are readily available, most consumers do not know how to evaluate the return associated with purchasing an energy-efficient, higher first-cost product.

- Many energy-efficient technologies were developed and commercialized within the past decade. Consumers may view them as risky investments with uncertain savings. There may be questions about non-energy-related performance. Also, there may be a lack of vendors and limited availability of relatively new energy-efficient technologies.

- Certain types of consumers such as low-income households and cash-constrained businesses often lack money for investing in energy conservation measures. Businesses tend to devote investment capital to improving their products or expanding their market share, not to reducing operating costs. One study found that even energy-intensive industries typically require a payback of two years or less for cost-saving investments (28).

- In some cases, those who select and purchase energy-consuming equipment do not pay the operating cost. This is often the case for buildings under construction and for rental properties. In large companies, those in charge of specifying and purchasing equipment are often disconnected from those paying the operating costs. Where split incentives exist, equipment purchasers tend to minimize first cost, which leaves the user with inefficient buildings, appliances, etc.

- Outside of some energy-intensive industries, energy represents a small fraction of the total cost of owning and operating a household, business, or vehicle. In office buildings, for example, the annual cost to an employer for a typical worker is approximately $150 per square foot, while energy costs are typically $1 to $2 per square foot (30). Low relative costs along with competing demands on decision makers' time limit consideration of energy efficiency opportunities.

- Government spending, tax, and regulatory policies are strongly tilted in favor of energy production as opposed to greater efficiency. In the mid-80s, energy supply industries received tax breaks and other subsidies totalling around $40 billion per year (31). And over 90% of the federal government's energy R&D budget is devoted to supply options. At the state level, regulation of electric utilities usually makes it more profitable for utilities to promote electricity consumption rather than to conserve in the short run.

- In addition to the payback gap problem, energy prices do not fully reflect the environmental, security-related, and social costs associated with energy production and use. For example, the costs of acid rain, urban smog, and protecting oil imports are not reflected in the prices of fossil fuels or electricity. One study estimates that the environmental costs related to electricity production from fossil fuels are in the range of $0.01-0.07/kWh, depending on fuel source and power plant type (32). Also, electricity and fuel prices are normally based on average costs, not marginal costs. Average-cost pricing and the failure to incorporate "externalities" into energy prices contribute to underinvestment in energy efficiency from the perspective of minimizing the total cost society pays for energy services.

Policy Instruments

A wide range of policies and programs can be adopted to overcome the barriers described above (33). This section examines two such efforts, appliance efficiency standards and the electricity conservation programs conducted by one particular utility. In addition, a broad set of policies for reducing energy use, economic costs, and carbon dioxide emissions in the United States are presented.

Appliance Efficiency Standards

As mentioned above, the National Appliance Energy Conservation Act of 1987 (NAECA) set minimum efficiency requirements for major residential products and required DOE to consider tightening the standards in the future (19). The legislation was developed jointly by conservation advocates and appliance manufacturers with support from utilities, consumer advocates, and environmental groups. The standards were virtually unopposed because consumers save money, manufacturers sell better-quality products and face uniform regulations, utilities need to construct fewer power plants, and less fuel is burned by eliminating the production, purchase, and use of inefficient appliances. It is truly a "win-win" proposition.

The appliance standards cover a wide range of products including furnaces, air conditioners, water heaters, refrigerators, freezers, and ranges. It is estimated that the initial appliance standards along with "second tier" standards on refrigerators and freezers will reduce electricity use by 54 TWh and total primary energy use by nearly 1.0 quad in 2000 (34). These estimates are based on comparing a standards scenario to a market forces scenario. In the latter, efficiencies improve more gradually. Saving 1 quad in 2000 is equivalent to about 5% of projected primary energy use by the residential sector that year (35). Assuming electricity savings are in the same proportion as total utility fuel shares, direct and indirect fuel savings from the appliance standards lower carbon emissions in 2000 by about 16 million metric tons.

The appliance standards are expected to increase the initial cost of affected products slightly, but on average the operating cost savings will be three times as great as the increase in first cost (34). Consumers should save about $2.7 billion per year on a net basis by 2000 and realize a total savings of $28 billion (nearly $300 per household) over the lifetime of products sold through 2000. Thus the appliance standards should reduce carbon emissions at a net cost of around negative $170 per metric ton.

Additional energy and carbon savings would result by strengthening the efficiency requirements on products already covered by NAECA as well as by adopting efficiency requirements on other products. In 1988, national efficiency standards on fluorescent lighting ballasts were added. The initial requirements, which took effect in 1990, are expected to reduce electricity use by about 28 TWh/yr and save businesses nearly $11 billion by 2000 (36). The benefit-cost ratio for these standards is around five. Other products for which minimum efficiency requirements are technically and economically feasible include lamps, light fixtures, and motors. Total electricity savings of about 72

TWh/yr could occur ten years after the adoption of efficiency standards on these products (37).

Utility Conservation Programs

A growing number of utilities in the United States are investing in end-use efficiency, thereby cutting load growth and reducing the need to build new power plants. The most aggressive utilities are spending around 2-4% of their gross revenues on conservation and load management (C&LM) programs (38). Utilities that invested large sums in C&LM during the 1980s experienced load growth that was about 1%/yr below that of other utilities (39).

In the late-1980s, the New England Electric System (NEES) developed one of the most comprehensive and well-funded electricity conservation programs in the nation. NEES spent about $65 million on C&LM programs in 1990 (about 4% of its revenues) and plans to increase its budget to $85 million in 1991 (40). NEES offers rebates and other financial incentives to stimulate the adoption of conservation measures by all types of consumers. NEES's 1990 program was expected to reduce peak demand by 105 MW and power production costs by $150 million.

Regulatory commissions in states where NEES operates (MA, RI, and NH) have provided the utility with incentives for achieving the maximum electricity savings as cost-effectively as possible. If NEES meets its 1990 savings and cost targets, its shareholders will get to keep about 10% of the net benefits to society from its C&LM programs after recovery of program costs (40). Over a twenty-year period starting in 1990, NEES expects to reduce its peak demand by about 1250 MW and obtain 32% of its "new resources" from C&LM (Figure 4) (41).

Fossil fuels provide about two-thirds of the power generated by NEES. It is possible to estimate the avoided carbon emissions and the cost effectiveness of carbon avoidance due to NEES's C&LM programs (Table 7). This analysis is based on the total cost for the conservation measures, i.e., the cost to the utility as well as the consumers. The average cost of saved energy for the C&LM programs was $0.037/kWh in recent years and is projected to reach about $0.052/kWh in the future. In both cases, the cost of saved energy is less than half the avoided cost of power (41). A net economic savings of about $850 million is projected as a result of all C&LM efforts through 1999.

C&LM programs underway at NEES are expected to cut NEES's carbon emissions in 1999 by about 250,000 metric tons, equivalent to over 5% of projected carbon emissions that year. The net cost of avoided carbon emissions, negative $140-230 per ton, is similar to that for other types of conservation measures and programs. In 1990, NEES was in the process of increasing its long-run electricity savings targets (42). Boosting energy savings efforts should lead to even greater carbon avoidance.

National Energy Efficiency Platform

The 74.3 quads of energy consumed in the United States in 1986 was almost identical to the amount of energy use in 1973. GNP rose 36% during this period, so national energy intensity declined 2.3%/yr on average. About three-quarters of the reduction in energy intensity was due to efficiency improvements, the remainder was due to structural change and interfuel substitution (16). Since 1986, however, energy use and carbon emissions have increased nearly as fast as GNP. Both U.S. energy use and carbon emissions in 1989 were 9% higher than in 1986 (11). Considering automobiles, for example, the average efficiency of new cars began to decline in 1989 (Figure 1).

In order to get the U.S. back on the "energy efficiency track", a national energy efficiency platform was developed in 1989 by four energy conservation research and advocacy organizations (43). The platform combines economic incentives, efficiency standards, R&D initiatives, and information programs (Table 8). Of course, the list is not exhaustive. Other policies such as a carbon-based fuels tax or additional efficiency standards could be adopted in order to achieve even further savings. The energy efficiency initiatives are briefly described and analyzed as follows:

- The original fuel economy standards are now ineffective and outdated. The standards could be gradually increased to 45 MPG for cars and 35 MPG for light trucks over the next ten years. New standards could either require each manufacturer to achieve a specified overall average efficiency (i.e., an extension of the current approach), require equal percentage efficiency improvements from all manufacturers, or require specified average efficiency levels for each size class.

- Only very low fuel-economy luxury cars are now subject to the federal gas guzzler tax. The tax could be extended to other inefficient cars (in proportion to MPG rating) and the new tax revenue could be used to provide rebates to purchasers of highly efficient cars. Rebates could be offered for the best cars in each size class.

- The market price for gasoline does not reflect its real cost to the nation, i.e., considering externalities such as environmental impacts and national security costs associated with oil imports. Furthermore, the gasoline tax in the U.S. is far below that in most other industrialized nations. Substantially raising the gasoline tax would rekindle consumer interest in fuel economy, complement new fuel economy standards, and help to limit growth in vehicle usage.

- Most utilities are penalized when they operate successful energy efficiency programs due to the loss of sales revenue in the short run. To remedy this problem, energy efficiency measures could be allowed to compete fairly with energy supply options under the Public Utility Regulatory Policies Act (PURPA). Also, state regulatory authorities could provide utilities with financial incentives for pursuing energy efficiency and least-cost energy services.

- New power production and distribution technologies such as combined-cycle power plants, advanced steam-injected gas turbines, and amorphous-alloy distribution transformers are more efficient than conventional technologies. Utilities could be given financial and regulatory incentives such as accelerated depreciation in order to stimulate the adoption of these new technologies. Also, R&D and demonstration programs related to electricity supply could emphasize options that increase overall efficiency.

- Acid rain legislation adopted in 1990 caps total emissions of sulfur dioxide by providing each affected utility with an annual emissions allowance. An individual utility can sell some of its allowances should its actual emissions fall below its emissions cap. Also, utilities can receive extra allowances for early emissions reductions due to investments in energy efficiency and renewable energy technologies. This legislation, if rigorously implemented, will give utilities in the Midwest and elsewhere an incentive to increase the efficiency of electricity supply and end use.

- The Department of Energy is required to review the appliance efficiency standards on a regular basis and promulgate more stringent standards if deemed technically and economically feasible. DOE could tighten the minimum efficiency requirements on air conditioners, water heaters, and other products covered in the original NAECA law. Also, new efficiency standards could be adopted for incandescent lamps, fluorescent lamps, fluorescent light fixtures, heating and air conditioning equipment used in commercial buildings, showerheads, and motors.

- Home energy rating and labeling programs have been successfully implemented in some parts of the country (44). Ratings and labels could be required on all new homes. For existing homes, their use could be promoted at the time of sale. This will help consumers to identify an efficient home, encourage builders to exceed minimum building code requirements, and make it easier for lending agencies to offer larger mortgages for buyers of very efficient homes.

- The federal mortgage lending agencies (FHA, Fannie Mae, etc.) can offer larger mortgages to buyers of homes that meet strict efficiency guidelines because residents of such homes are able to afford a higher monthly loan payment. A program along these lines was started, but it has had little impact so far. The program needs to be streamlined, better promoted, and expanded.

- The energy efficiency requirements in most state building codes are outdated and below cost-effective levels. DOE could encourage and assist states that are interested in upgrading their codes, using new standards that are mandatory for federally owned buildings but voluntary for the private sector as a model (45). Also, new homes financed by the federal government (e.g., public housing and homes receiving FHA loans) could be required to meet more stringent efficiency standards such as the Model Energy Code issued by the Council of American Building Officials in 1989 (46).

- Without exception, the federal government should purchase energy-efficient lighting products, motors, heating and cooling equipment, etc. when the operating savings exceed the extra first cost. A large revolving fund could be established to finance such investments.

- DOE's conservation R&D program was cut by two-thirds during 1980-89 in spite of DOE advancing the development and commercialization of numerous energy-efficient technologies during the 1970s (47). Funding was modestly increased for 1990-91, but many worthwhile projects are still unfunded or underfunded. The program could be doubled within a few years through redirecting money from other parts of the agency. Also, DOE could reinstitute demonstration of new energy-conserving technologies and expand efforts to transfer new technologies to the private sector.

- The federal government has stopped collecting data on annual energy use from large industries. This limits awareness of energy use and energy efficiency trends. The Energy Information Administration could reinstitute an annual survey of energy use by major manufacturers. Also, data on the extent of implementation of efficiency measures could be collected.

- In order to encourage innovation in the industrial sector, joint government-industry research centers could be established for energy-intensive industrial processes. The centers could conduct basic and applied research, striving for advances that provide energy savings along with other benefits.

National Energy Efficiency Platform – Potential Impacts

Adopting the energy efficiency platform could dramatically lower national energy use, expenditures on energy services, and carbon emissions by 2000 (Table 9). The savings are calculated relative to the energy use forecast issued by DOE in 1989 (48). To avoid double counting, savings already assumed in the DOE forecast are excluded. To facilitate the analysis, some of the policy measures are grouped together. No energy savings are directly attributed to the R&D proposals.

Adopting the entire platform could cut projected energy use in the year 2000 by nearly 16 quads (18%). Consumers could realize a net savings of about $75 billion per year taking into account the initial cost of efficiency measures. Carbon emissions in 2000 could fall by over 350 million metric tons, with a net cost of carbon avoidance of negative $210 per metric ton on average. Additional energy, economic, and carbon savings will occur after the turn of the century as the full impact of the policies is felt.

Table 10 compares energy use, energy services cost, and carbon emissions in 1988 with the respective values in 2000 from a frozen efficiency scenario, the 1989 DOE forecast, and a high efficiency scenario represented by implementing the platform. In the high efficiency scenario, there would be a modest drop in absolute energy use between 1988 and 2000. Assuming GNP increases 2.5%/yr, national energy intensity (E/GNP) would fall 3.0%/yr on average during 1988-2000 in the high efficiency scenario. This rate of energy intensity reduction is moderately greater than the average rate that prevailed during 1973-86. With the energy efficiency initiatives, carbon emissions fall by 21% compared to the DOE forecast and 11% compared to actual emissions in 1988.

The potential reduction in carbon emissions from the proposed set of energy efficiency initiatives is consistent with the goal of achieving a 20% reduction in CO_2 emissions from 1988 levels by 2000. Efficiency improvements provide most but not all of the CO_2 reductions necessary to meet this goal. A modest increase in renewable energy sources, shifting from more carbon-intensive fuels to natural gas, or further conservation will be needed in order to reduce CO_2 emissions 20% by the turn of the century.

This analysis is consistent with other studies of the potential to cut national CO_2 emissions through increasing energy efficiency. In Sweden, it is estimated that aggressive efficiency improvements can lower projected carbon emissions in 2010 by one-third or more while reducing the overall cost of energy services (49). In Canada, it is estimated that drastically reducing carbon emissions in 2005 from projected levels through efficiency improvements could provide a net economic benefit of about $60-110 billion (50). Both of these studies indicate that strong policy initiatives will be needed in order to realize such large savings.

Conclusion

Examination of specific energy end uses as well as conservation policies and programs indicates that it is possible to greatly reduce carbon dioxide emissions while reducing the cost of energy services and increasing economic growth. In fact, consumers and our nation can save money while taking actions to reduce the threat of global warming. The key to obtaining these benefits is to focus on increasing the efficiency of energy end use. Increasing the efficiency of energy supply can yield similar results (51). In other words, both environmental and economic benefits result from a more rational balance between investments in energy efficiency and energy supply.

Given the existence of cost-effective efficiency opportunities, any thorough analysis of the potential cost of carbon dioxide emissions avoidance should contain a negative cost portion (Figure 5). The challenge for analysts is to identify the maximum amount of energy and carbon that can be saved at a negative cost. Likewise, policymakers and those implementing energy policies should strive to maximize the amount of carbon avoidance achieved with a net economic benefit.

A variety of structural and behavioral barriers inhibit the adoption of cost-effective energy efficiency measures. In order to overcome these barriers, a mix of policies including economic incentives, efficiency regulations, R&D initiatives, and educational programs can be adopted. Past experience with automobile and appliance efficiency standards, for example, demonstrates that some conservation policies have been highly successful. But policies must be carefully designed and implemented in order to have the desired impacts (33). Given the potentially catastrophic impacts of global warming, it is all the more important that the United States and other nations adopt comprehensive strategies for accelerating the implementation of cost-effective energy efficiency measures.

Economic models examining the cost of reducing CO_2 emissions have failed to include a negative cost portion (1, 2, 52). Those engaged in modeling national and international energy systems should incorporate end-use efficiency improvements directly into their models. Ignoring the reductions in energy use, energy service costs, and pollutant emissions that result from end-use efficiency improvements misses a large target of opportunity. Also, ignoring end-use efficiency improperly characterizes the total systems used for obtaining energy services. Models that fail to account for end-use efficiency and potential efficiency improvements should not be used to estimate the net economic impact of reducing greenhouse gas emissions.

Table 1

ENERGY, ECONOMIC, AND CARBON SAVINGS ASSOCIATED WITH PREVIOUS AUTO EFFICIENCY IMPROVEMENTS

INITIAL COST FOR INCREASING FUEL ECONOMY FROM 18.7 TO 25.8 MPG:

Duleep - $200-400
U.S. auto companies - $397 on average

AVERAGE GASOLINE SAVINGS FROM INCREASING FUEL ECONOMY FROM 18.7 TO 25.8 MPG:

180 gallons per year

OTHER ASSUMPTIONS:

Discount Rate - 6% real
Lifetime - 10 years

COST OF SAVED ENERGY (based on $400 cost premium):

$0.30/gal = $2.40/MBtu

1989 AVERAGE GASOLINE PRICE (excluding taxes):

$0.86/gal

SOCIETAL BENEFIT-COST RATIO:

$0.86/$0.30 = 2.9

TOTAL GASOLINE SAVINGS IN 1989 FROM AUTO EFFICIENCY IMPROVEMENTS DURING 1975-89:

37.6 billion gallons = 2.45 MBD = 4.7 Quads

NET ECONOMIC SAVINGS IN 1989 DUE TO EFFICIENCY IMPROVEMENTS DURING 1975-89:

$32.3 billion - $11.3 billion = $21.0 billion

AVOIDED CARBON EMISSIONS IN 1989 FROM AUTO EFFICIENCY IMPROVEMENTS DURING 1975-89:

102 million metric tons

NET COST OF AVOIDED CARBON EMISSIONS:

- $206 per ton

Table 2

SELECTED AUTOMOBILE FUEL ECONOMY TECHNOLOGIES

Technology	Potential Fuel Economy Improvement (%)
Roller cam followers	1.5
Overhead cam engines	6.0
Intake valve control	6.0
Front wheel drive	10.0
Four valves per cylinder	6.8
Improved aerodynamics	4.6
Improved engine accessories	1.7
Torque-converter lock-up	3.0
Four-speed auto. transmission	4.5
Five-speed auto. transmission	2.5
Electronic transmission controls	1.5
Multi-point fuel injection	3.0
Advanced friction reduction	2.0
Continuously variable transmission	2.5
Improved lubricants and tires	1.0
Aggressive transmission management	8.0
Engine idle off	9.0

Source: Ledbetter and Ross, 1990 (Ref. 13).

Table 3

ENERGY, ECONOMIC, AND CARBON SAVINGS ASSOCIATED WITH REFRIGERATOR EFFICIENCY IMPROVEMENTS

INITIAL COST FOR REDUCING ELECTRICITY USE (1989 $):

DOE - $31 for 431 kWh/yr reduction
ACEEE - $54 for 414 kWh/yr reduction

AVERAGE ELECTRICITY SAVINGS BETWEEN NEW MODELS PRODUCED IN 1972 AND 1989:

Typical 1972 model - 1726 kWh/yr; 3.84 Energy factor
Typical 1989 model - 934 kWh/yr; 7.78 Energy factor
Savings - 792 kWh/yr

OTHER ASSUMPTIONS:

Discount Rate - 6% real
Lifetime - 19 years

COST OF SAVED ENERGY (based on $0.101/kWh/yr cost premium):

$0.009/kWh = $0.78/MBtu

1989 AVERAGE RESIDENTIAL ELECTRICITY PRICE:

$0.076/kWh

SOCIETAL BENEFIT-COST RATIO:

$0.076/$0.009 = 8.4

TOTAL ELECTRICITY SAVINGS IN 1989 FROM REFRIGERATOR EFFICIENCY IMPROVEMENTS DURING 1972-89:

48.3 billion kWhs = 0.56 Quads

NET ECONOMIC SAVINGS IN 1989 DUE TO EFFICIENCY IMPROVEMENTS DURING 1972-89:

$3.67 billion - $0.44 billion = $3.23 billion

AVOIDED CARBON EMISSIONS IN 1989 FROM REFRIGERATOR EFFICIENCY IMPROVEMENTS DURING 1972-89:

10.4 million metric tons

NET COST OF AVOIDED CARBON EMISSIONS:

- $308 per ton

Table 4

REFRIGERATOR CONSERVATION SUPPLY CURVE (1)

Efficiency Measure	Elect. Use (kWh/yr)	Elect. Savings (kWh/yr)	Cost of Saved Energy Marginal (c/kWh)	Cost of Saved Energy Average (c/kWh)
Baseline	677	--	--	--
Condensor anti-sweat heater	572	195	1.3	1.3
Adaptive defrost	487	85	3.0	2.0
EER=5.3 compressor	466	21	6.4	2.5
0.75" aerogel insulation	374	92	6.8	3.8
1.00" aerogel insulation	320	54	3.5	3.8
1.25" aerogel insulation	285	35	5.4	3.9
1.50" aerogel insulation	260	25	7.5	4.1
1.75" aerogel insulation	242	18	10.5	4.4
2.00" aerogel insulation	228	14	13.4	4.7
Two-compressor system	188	40	20.0	5.9

(1) 18 cubic foot refrigerator/freezer with automatic defrost.

Source: Goldstein, et al., 1990 (Ref. 24).

Table 5

COST-EFFECTIVE ELECTRICITY CONSERVATION
POTENTIAL IN NEW YORK STATE

Conservation Measure	Savings Potential (GWh/yr)	Cost of Saved Energy ($/KWh)
Reflectors for fluorescent fixtures	4140	0.010
High eff. refrigerators and freezers	5280	0.011
Residential infiltration reduction	590	0.017
HVAC retrofits in commercial buildings	6850	0.020
Commercial bldg. variable speed drives	3473	0.024
Energy saving incandescent lamps	880	0.028
High eff. industrial lighting	470	0.028
Occupancy sensors in comm. buildings	500	0.033
High eff. commercial fluor. lighting	2190	0.036
Industrial variable speed drives	2550	0.040
Compact fluorescent lamps	2020	0.040
Infrared reflecting lamps	810	0.044
Daylighting in commercial buildings	1660	0.047
Heat pump clothes dryer	860	0.065
Other	2070	--
All measures	34,340 (1)	0.025

(1) The savings potential refers to 1986, when total electricity consumption in the region under study equalled about 99,000 GWh.

Source: Miller, Eto, and Geller, 1989 (Ref. 25).

Table 6

COST OF SAVED ELECTRICITY AND CARBON DIOXIDE THROUGH GREATER ENERGY EFFICIENCY IN BUILDINGS

Conservation Measure (1)	Cost of Saved Energy (c/kWh)	Potential Elect. Savings (TWh/yr)	Potential Carbon Savings (Mton/yr)	Net Cost of Saved Carbon (2) ($/ton)
Reducing urban heat islands	0.5	45	8.7	-362
Residential lighting	0.9	56	10.6	-349
Residential water heating	1.3	38	7.1	-332
Commercial water heating	1.4	10	1.9	-321
Commercial lighting	1.5	166	31.9	-312
Commercial cooking	1.5	7	1.3	-310
Commercial cooling	1.9	115	22.1	-291
Commercial refrigeration	2.2	22	4.1	-284
Residential appliances	3.3	103	19.6	-221
Residential space heating	3.7	105	20.2	-198
Commercial space heating	4.0	22	4.1	-188
Commercial ventilation	6.8	45	8.7	- 36
All measures (3)	2.4	734	140.3	-265

(1) Specific conservation measures are aggregated for each major end use.

(2) The net cost of saved carbon is calculated using the 1989 average electricity price for the buildings, $0.075/kWh.

(3) For reference, the buildings sector consumed 1627 TWh of electricity in 1989 and generating this power resulted in about 311 million metric tons of carbon emissions.

Source: Rosenfeld, et al., 1990 (Ref. 26).

Table 7

ELECTRICITY SAVINGS, COST EFFECTIVENESS, AND AVOIDED CARBON EMISSIONS FROM CONSERVATION PROGRAMS SPONSORED BY THE NEW ENGLAND ELECTRIC SYSTEM (NEES)

Parameter	1987–89 Program	Projected 1987–99 Program
Electricity savings in last year (GWh/yr)	182.8	1,292.5
Electricity savings in last year as a fraction of electricity sales (%)	0.8	4.5
Conservation program cost (Million $)		
Utility	56.2	560.0
Society (i.e., utility plus consumers)	65.2	647.9
Average cost of saved energy for society ($/kWh)	0.037	0.052
Average avoided cost in last year (1989 $/kWh)	0.079	0.120
Societal benefit–cost ratio	2.15	2.32
Net economic savings in last year (Million $)	7.7	87.9
Net economic savings over the lifetime of the conservation measures (Million $)	75.0	855.2
Avoided carbon emissions in last year (Ktons)	34.9	246.9
Avoided carbon emissions in last year as a fraction of total carbon emissions (%)	0.9	5.3
Net cost of avoided carbon emissions ($/ton)	- 143	- 231

Source: New England Electric System, 1990 (Refs. 41 and 42).

Table 8

ENERGY EFFICIENCY PLATFORM FOR THE UNITED STATES

1. Raise car and light truck fuel economy standards, expand the gas guzzler tax, and establish gas sipper rebates so that new cars average 45 mpg and new light trucks average 35 mpg by 2000.

2. Raise the federal gasoline tax by 50 cents per gallon within five years and spend part of the revenue on mass transit and energy efficiency programs.

3. Reform federal utility regulation to foster investment in end-use energy efficiency and cogeneration systems.

4. Increase the efficiency of electricity supply through development, demonstration, and promotion of advanced generating technologies.

5. Implement new acid rain legislation that encourages more efficient production and use of electricity.

6. Strengthen federal appliance efficiency standards and adopt new efficiency standards on lamps, light fixtures, showerheads, and motors.

7. Reduce energy use in residential and commercial buildings through home energy ratings, mortgage-based incentive programs, and new building standards.

8. Reduce federal energy use through life-cycle cost-based purchasing.

9. Reduce industrial energy use through research, demonstration, and reporting programs.

10. Increase government-sponsored energy efficiency R&D and reinstitute demonstration programs.

Source: Geller, 1989 (Ref. 43).

Table 9

POTENTIAL BENEFITS FROM THE ENERGY EFFICIENCY PLATFORM

Policy Proposal	----------SAVINGS IN 2000--------- Energy (Quads)	Money (Billion $)	Carbon Emissions (1) (Megatons)
Raise vehicle efficiency standards and gas guzzler tax/sipper rebates	2.0	12	41
Increase the gasoline tax	0.9	10	19
Reform utility regulation	3.8	14	94
Increase the efficiency of electricity supply	0.9	--	20
Implement new acid rain legislation	1.9	9	48
Equipment efficiency standards			
existing standards	1.2	6	25
new standards	1.1	5	25
Reduce energy use in buildings through home energy ratings, mortgage programs, and standards	1.2	5	22
Reduce federal energy use through life-cycle cost-based purchasing	0.2	1	4
Reduce industrial energy use through R&D, demonstration, and reporting	2.7	13	58
Increase conservation R&D	--	--	--
TOTAL	15.9	75	356

(1) Units are million metric tons of carbon.

Source: Geller, 1989 (Ref. 43).

Table 10

OVERALL ENERGY USE, COST, AND CARBON EMISSIONS
FOR DIFFERENT EFFICIENCY SCENARIOS

Scenario	Energy Use (Quads)	Energy Services Cost (1) (Billion $)	Carbon Emissions (Megatons)
Actual 1988	80.2	416	1503
Frozen efficiency 2000	107.2	735	2010
EIA Base Case 2000	90.6	621	1699
High Efficiency 2000 (2)	74.7	546 (3)	1343

(1) Annual energy services cost expressed in 1988 dollars.

(2) Based on savings estimates from the policy proposals shown in Table 8.

(3) Includes the levelized cost of additional conservation measures relative to the EIA base case, but excludes any tax impacts from the gasoline tax or economic impacts from the initiative to increase the efficiency of electricity supply.

Sources: EIA, 1989 (Ref. 48) and Geller, 1989 (Ref. 43).

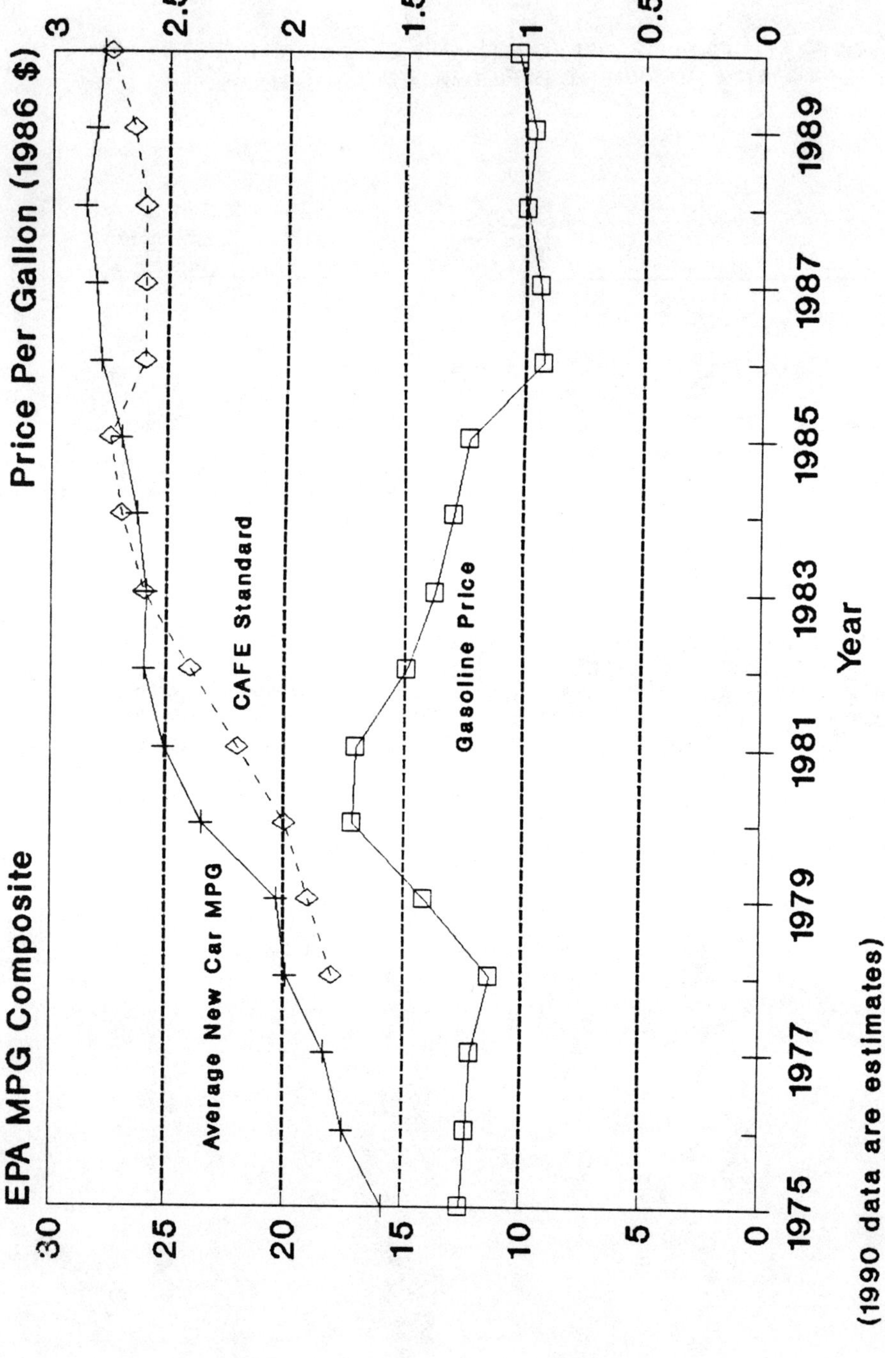

Figure 1. Trends in Automobile Fuel Efficiency and Gasoline Price

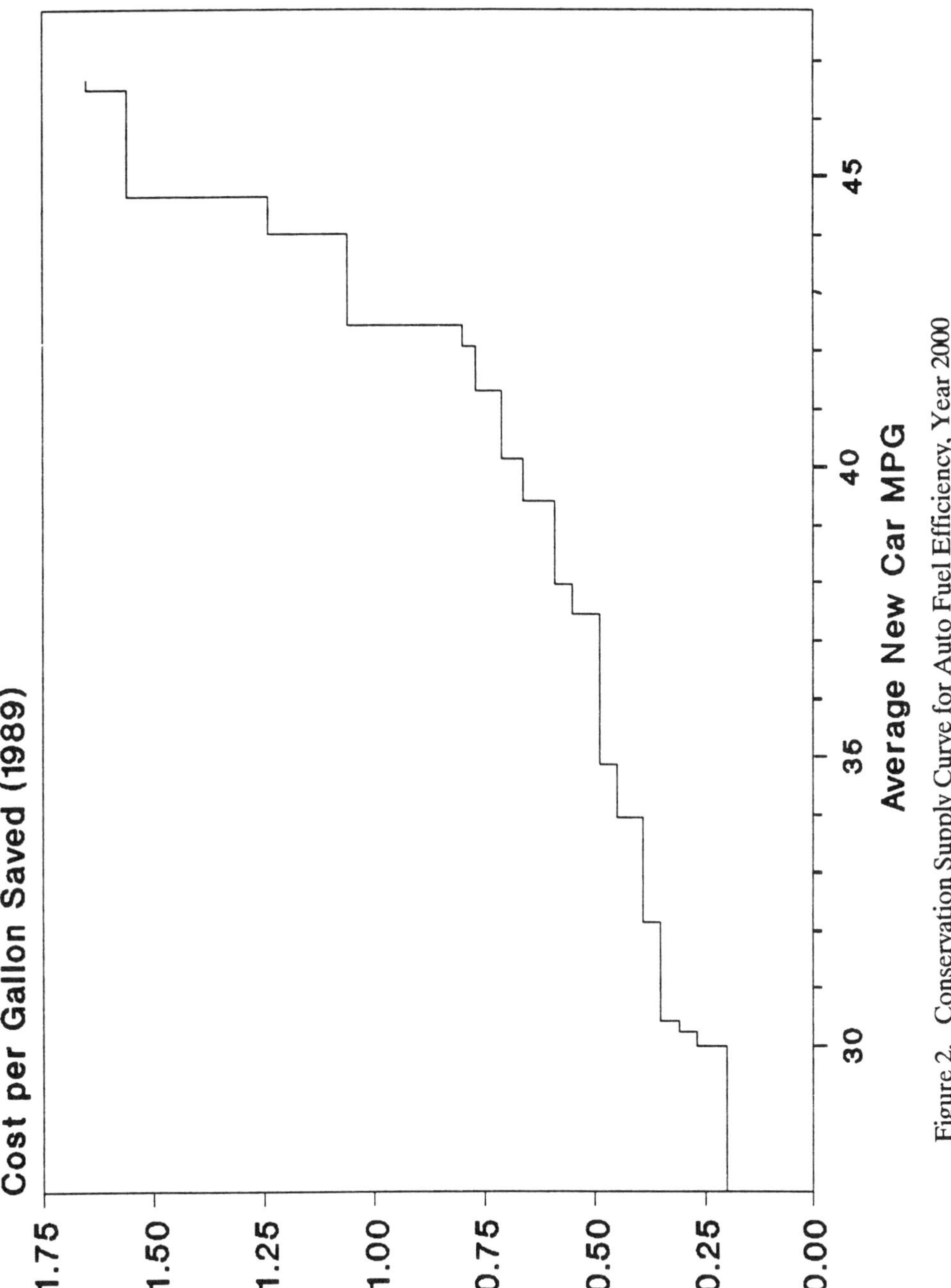

Figure 2. Conservation Supply Curve for Auto Fuel Efficiency, Year 2000

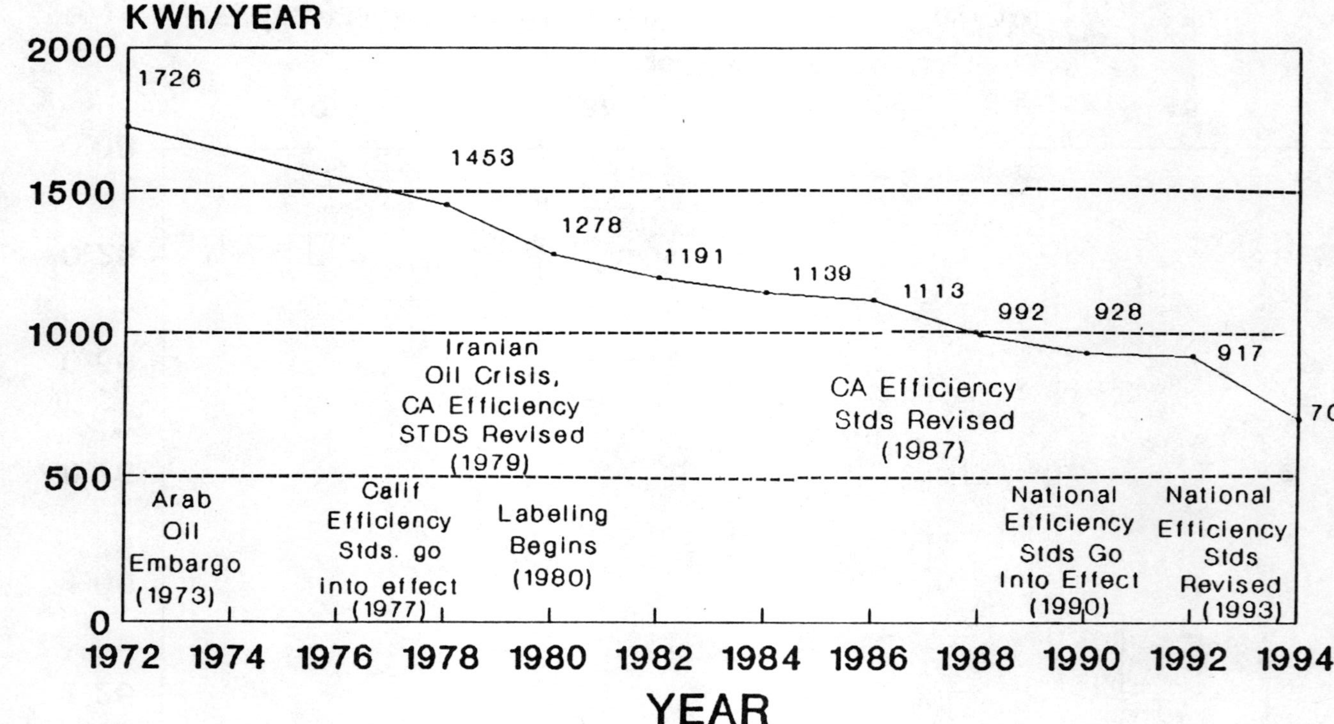

Source: Association of Home Appliance Manufacturer Historic Data, DOE Projections

Figure 3. Average Electricity Use of New Refrigerators Sold in the United States

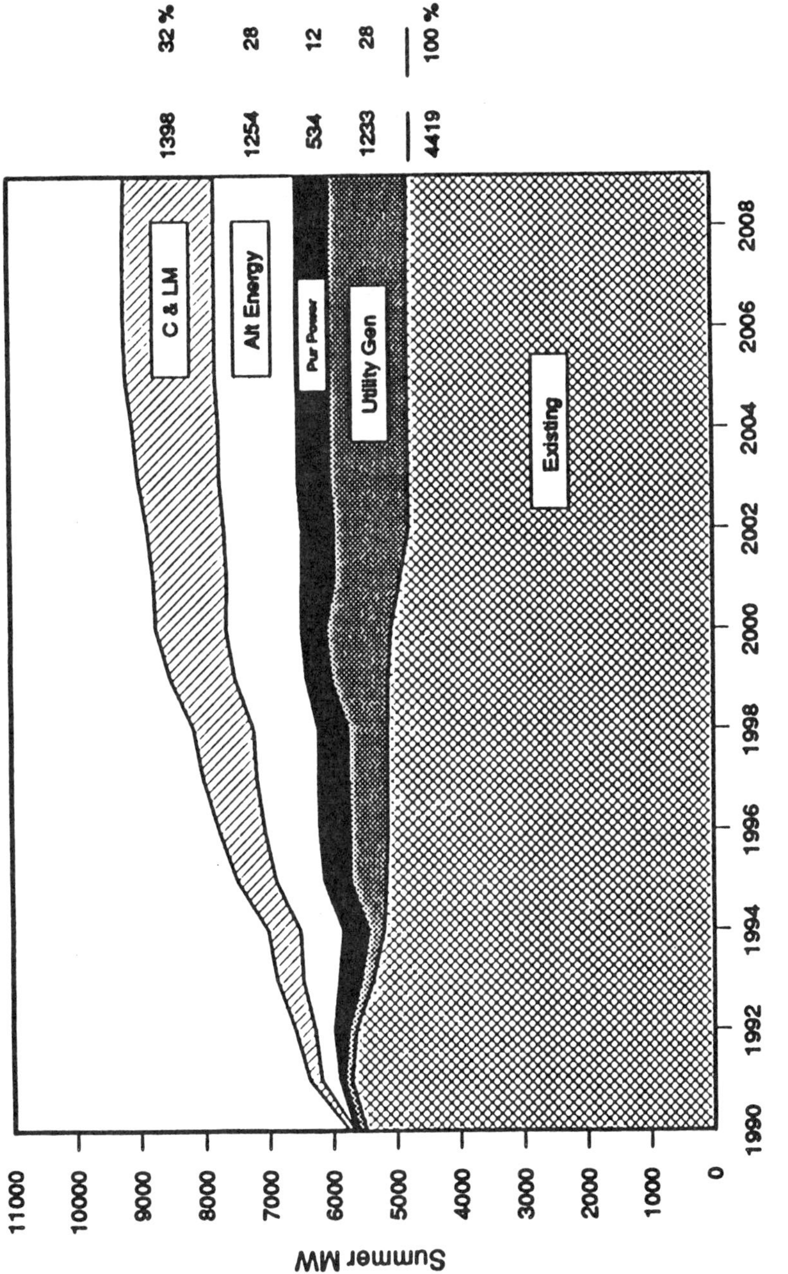

Figure 4. New England Electric System 1990 Resource Plan

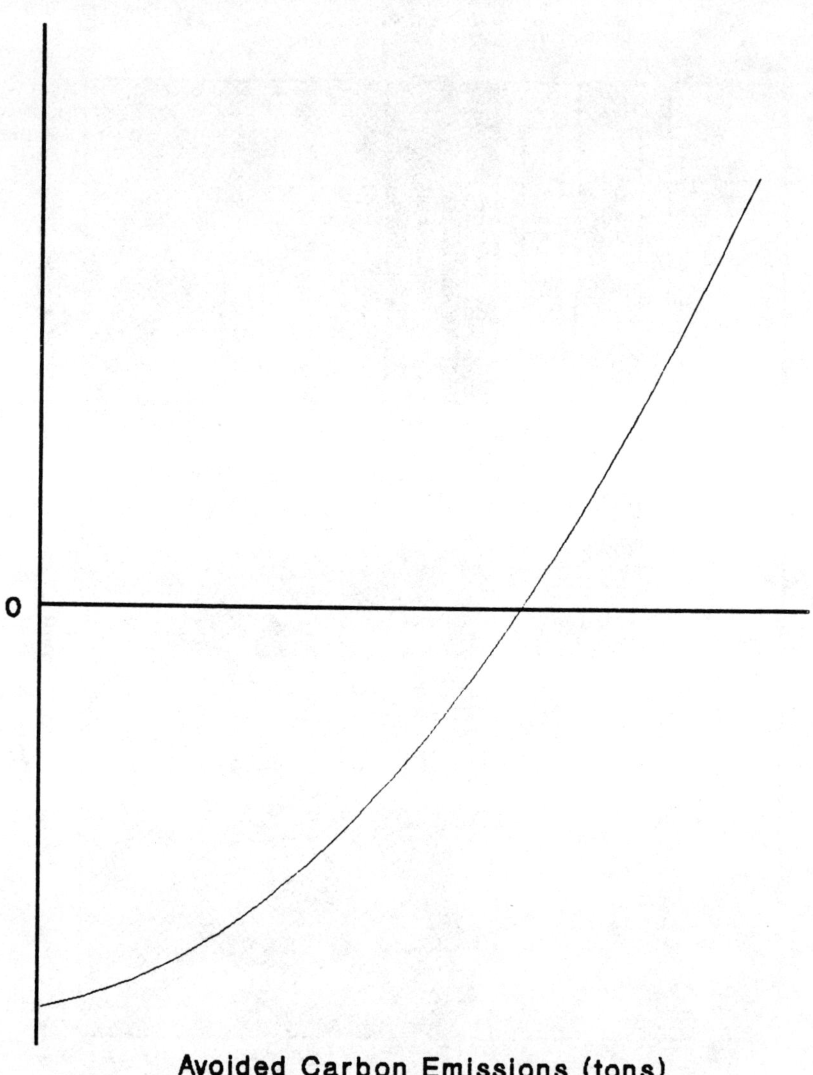

Figure 5. Cost Curve for Carbon Emissions Avoidance

Acknowledgment

Preparation of this chapter was made possible by a grant from the John D. and Catherine T. MacArthur Foundation.

References

1. A. Cristofaro, *The Cost of Reducing Greenhouse Gas Emissions in the United States*, Air and Energy Policy Division, U.S. EPA, Dec. 1990. Also, see this volume.

2. A.S. Manne and R.G. Richels, "Global CO_2 Emissions Reductions – The Impacts of Rising Energy Costs," *The Energy Journal* 11(3), 1990. Also, see this volume.

3. R.H. Williams, "Low-Cost Strategies for Coping with CO_2 Emission Limits," *The Energy Journal* 11(3), pp. 35-59, 1990.

4. M.H. Ross and D. Steinmeyer, "Energy for Industry," *Scientific American* 263(3), pp. 88-101, Sept. 1990.

5. C.J. Weinberg and R.H. Williams, "Energy from the Sun," *Scientific American* 263(3), pp. 146-155, Sept. 1990.

6. A. Miller, I. Mintzer, and P.G. Brown, "Rethinking the Economics of Global Warming," *Issues in Science and Technology*, pp. 70-73, Fall 1990.

7. D.B. Bleviss and P. Walzer, "Energy for Motor Vehicles," *Scientific American* 263(3), pp. 103-109, Sept. 1990.

8. J.J. MacKenzie and M.P. Walsh, *Driving Forces: Motor Vehicle Trends and their Implications for Global Warming, Energy Strategies, and Transportation Planning*, World Resources Institute, Washington, D.C., Dec. 1990.

9. D.L. Greene, "CAFE OR PRICE?: An Analysis of the Effects of Federal Fuel Economy Regulations and Gasoline Price on New Car MPG, 1978-89," *The Energy Journal* 11(3), 1990.

10. D.L. Greene and J.T. Liu, "Automotive Fuel Economy Improvements and Consumers' Surplus," *Transportation Research – A*, 22A(3), pp. 203-218, 1988.

11. U.S. Department of Energy, *Monthly Energy Review August 1990*, DOE/EIA-0035(90/08), Energy Information Administration, Washington, D.C., Nov. 1990.

12. F. Westbrook and P. Patterson, "Changing Driving Patterns and their Effect on Fuel Economy," paper presented at the SAE Government/Industry Meeting, Washington, D.C., May 2, 1989.

13. M. Ledbetter and M. Ross, *Supply Curves of Conserved Energy for Automobiles*, American Council for an Energy-Efficient Economy, Washington, D.C., March 1990.

14. D.L. Bleviss, *The New Oil Crisis and Fuel Economy Technologies: Preparing the Light Transportation Industry for the 1990s*, Quorum Books, New York, 1988.

15. M. Ross, "Energy and Transportation in the United States," *Annual Review of Energy* 14, pp. 131-171, 1989.

16. L. Schipper, R.B. Howarth, and H. Geller, "United States Energy Use from 1973 to 1987: The Impacts of Improved Efficiency," *Annual Review of Energy* 15, pp. 455-504, 1990.

17. Data on refrigerator energy efficiency and consumption trends, Association of Home Appliance Manufacturers, Chicago, IL, 1990.

18. H.S. Geller, *Residential Equipment Efficiency: A State-of-the-Art Review*, American Council for an Energy-Efficient Economy, Washington, D.C., May 1988.

19. National Appliance Energy Conservation Act, Public Law 100-12, March 17, 1987.

20. U.S. Department of Energy, *Consumer Products Efficiency Standards Engineering Analysis Document*, DOE/CE-0300, Washington, D.C., March 1982.

21. D.B. Goldstein, P.M. Miller, and R.K. Watson, *Developing Cost Curves for Conserved Energy in New Refrigerators and Freezers*, American Council for an Energy-Efficient Economy and Natural Resources Defense Council, Washington, D.C., Jan. 1988.

22. U.S. Department of Energy, *Energy Conservation Program for Consumer Products: Energy Conservation Standards for Two Types of Consumer Products*, Final Rule, Docket No. CE-RM-87-102, Office of Conservation and Renewable Energy, U.S. Department of Energy, Washington, D.C., Nov. 13, 1989.

23. I. Turiel and M.D. Levine, "Energy-Efficient Refrigeration and the Reduction of Chlorofluorocarbon Use," *Annual Review of Energy* 14, pp. 173-204, 1989.

24. D. Goldstein, R. Mowris, B. Davis, and K. Dolan, "Initiating Least-Cost Energy Planning in California: Preliminary Methodology and Analysis," testimony presented by the Natural Resources Defense Council and The Sierra Club to the California Energy Commission, Docket No. 88-ER-8, San Francisco, CA, May 10, 1990.

25. P.M. Miller, J.E. Eto, and H.S. Geller, *The Potential for Electricity Conservation in New York State*, Report 89-12, New York State Energy Research and Development Authority, Albany, NY, Sept. 1989.

26. A.H. Rosenfeld, et al., *A Compilation of Supply Curves of Conserved Energy for U.S. Buildings*, Center for Building Science, Lawrence Berkeley Laboratory, Berkeley, CA, Sept. 1990.

27. A. Faruqui, *Efficient Electricity Use: Estimates of Maximum Energy Savings*, CU-6746, Electric Power Research Institute, Palo Alto, CA, March 1990.

28. Alliance to Save Energy, *Industrial Investment in Energy Efficiency: Opportunities, Management Practices, and Tax Incentives*, Washington, D.C., July 1983.

29. M.A. Brown and E. Hirst, "Closing the Efficiency Gap: Barriers to Improving Energy Efficiency," in *Energy Efficiency: How Far Can We Go*, ORNL/TM-11441, Oak Ridge National Laboratory, Oak Ridge, TN, Jan. 1990.

30. E. Hirst, J. Clinton, H. Geller, and W. Kroner, *Energy Efficiency in Buildings: Progress and Promise*, American Council for an Energy-Efficient Economy, Washington, D.C., 1986.

31. H.R. Heede, R.E. Morgan, and S. Ridley, *The Hidden Costs of Energy*, Center for Renewable Resources, Washington, D.C., 1985.

32. R.L. Ottinger, et al., *Environmental Costs of Electricity*, Oceana Publications, Inc., New York, 1990 (available from ACEEE).

33. W.U. Chandler, H.S. Geller, and M.R. Ledbetter, *Energy Efficiency: A New Agenda*, American Council for an Energy-Efficient Economy, Washington, D.C., 1988.

34. H.S. Geller, *Energy and Economic Savings from National Appliance Efficiency Standards*, American Council for an Energy-Efficient Economy, Washington, D.C., March 1987.

35. U.S. Department of Energy, *Annual Energy Outlook 1990*, DOE/EIA-0383(90), Energy Information Administration, Washington, D.C., Jan. 1990.

36. H.S. Geller and P.M. Miller, *1988 Lighting Ballast Efficiency Standards: Analysis of Electricity and Economic Savings*, American Council for an Energy-Efficient Economy, Washington, D.C., Aug. 1988.

37. S. Nadel, *Efficiency Standards for Lamps, Motors and Lighting Fixtures*, American Council for an Energy-Efficient Economy, Washington, D.C., Aug. 1990.

38. "New England Electric, Led by Incentives, Tops Conservation Spending," *The Electricity Journal* 3(10), p. 2 Dec. 1990.

39. H.S. Geller and S.M. Nadel, "Electricity Conservation: Potential vs. Achievement," Paper prepared for the NARUC Least-Cost Utility Planning Conference, American Council for an Energy-Efficient Economy, Washington, D.C., Oct. 1989.

40. J.W. Rowe, "Making Conservation Pay: The NEES Experience," *The Electricity Journal* 3(10), pp. 18-25, Dec. 1990.

41. New England Electric System, *New England Electric Conservation and Load Management Annual Report*, Westborough, MA, May 1990.

42. Personal communication with Mark Hutchinson, New England Electric System, Westborough, MA, Dec. 1990.

43. H.S. Geller, *National Energy Efficiency Platform: Description and Potential Impacts*, American Council for an Energy-Efficient Economy, Washington, D.C., Aug. 1989.

44. E. Vine, B.K. Barnes, R. Ritshcard, *Implementation of Home Energy Rating Systems*, LBL-22872, Lawrence Berkeley Laboratory, Berkeley, CA, Feb. 1987.

45. "Energy Conservation Voluntary Performance Standards for Commercial and Multi-Family High Rise Residential Buildings; Mandatory for New Federal Buildings; Interim Rule," *Federal Register* 54, pp. 4538-4720, Jan. 30, 1989.

46. Council of American Building Officials (CABO), *Model Energy Code, 1989*, Falls Church, VA, 1989.

47. H. Geller, J.P. Harris, M.D. Levine, and A.H. Rosenfeld, "The Role of Research and Development in Advancing Energy Efficiency: A $50 Billion Contribution to the U.S. Economy," *Annual Review of Energy* 12, pp. 357-395, 1987.

48. U.S. Department of Energy, *Energy Information Administration Annual Energy Outlook 1989*, DOE/EIA-0383(89), Washington, D.C., Jan. 1989.

49. B. Bodlund, E. Mills, T. Karlsson, and T.B. Johansson, "The Challenge of Choices: Technology Options for the Swedish Electricity Sector," in T.B. Johannson, B. Bodlund, and R.H. Williams, eds., *Electricity: Efficient End-Use and New Generation Technologies, and Their Planning Implications*, Lund University Press, Lund, Sweden, 1989 (available from ACEEE).

50. Erik Haites, "Canada," in W.U. Chandler, ed., *Carbon Emissions Control Strategies*, World Wildlife Fund and the Conservation Foundation, Washington, D.C., 1990.

51. R.H. Williams and E.D. Larson, "Expanding Roles for Gas Turbines in Power Generation," in T.B. Johannson, B. Bodlund, and R.H. Williams, eds., *Electricity: Efficient End-Use and New Generation Technologies, and Their Planning Implications*, Lund University Press, Lund, Sweden, 1989 (available from ACEEE). Also, see R.H. Williams, this volume.

52. W.D. Nordhaus, "Greenhouse Economics: Count Before You Leap," *The Economist*, pp. 21-24, July 7, 1990.

Global CO_2 Emission Reductions: The Impacts of Rising Energy Costs

Alan S. Manne

Stanford University
Palo Alto, CA 94305

Richard G. Richels

(Presenter)
Environmental Risk Analysis
Electric Power Research Institute
Box 10412
Palo Alto, CA 94303

(Published in The Energy Journal, *Vol. 12, No. 1, 1991)*

Introduction

In recent years there has been growing concern that the increasing accumulation of greenhouse gases in the earth's atmosphere will lead to undesirable changes in global climate. This concern has resulted in a number of proposals, both in the U.S. and internationally, to set physical targets for reducing greenhouse gas emissions.

With CO_2 believed to be responsible for approximately half the problem, the energy sector plays an important role in strategies to delay climate change. Two bills introduced in the U.S. Congress during 1989 (the National Energy Policy Act of 1989 and the Global Warming Prevention Act) called for a national energy policy to reduce global warming. Each bill would establish national goals of reducing carbon dioxide emissions by 20% by the year 2000.

Calls for reducing CO_2 emissions have been a common theme at international conferences on global warming. For example, the final statement from the June 1988 Toronto conference on "The Changing Atmosphere" called for 20% worldwide reductions of CO_2 emissions by the year 2005. The Hamburg Conference (November 1988) called for a 30% reduction by the year 2000. In each instance the major share of the reduction would be borne by the industrialized countries.

Published 1991 by Elsevier Science Publishing Company, Inc.
Global Climate Change: The Economic Costs of Mitigation and Adaptation
James C. White, Editor

In the diplomatic arena, perhaps the most notable group is the Intergovernmental Panel on Climate Change (IPCC). This is being sponsored by the World Meteorological Organization and the United Nations Environment Programme. In addition to developing a comprehensive initial assessment of the scientific evidence and impacts of climate change, the IPCC will be exploring strategies for policy responses. The analysis is scheduled for completion in August 1990.

As the global climate debate moves toward the consideration of specific legislative initiatives and policy options, international negotiations are likely to take on increasing importance. The greenhouse effect is inherently a global problem. Most countries are taking the position that, if significant measures are required to reduce CO_2 emissions, they should only be taken in the context of an international agreement.

The negotiation of such an agreement would be extraordinarily complex. Major reductions in carbon emissions will be expensive. The difficulty in achieving a given target is likely to vary greatly among nations. Some nations might incur enormous costs in order to achieve relatively modest reductions, while the converse might hold for others. See Grubb (1989). For a set of limits to be broadly accepted, they must be perceived as equitable, enforceable and based on gradual rather than abrupt changes in the status quo. Economic efficiency is likely to be a secondary criterion, but could be achieved through international markets in carbon emission rights. For projecting the evolution of such markets, it is important to understand how the costs of emissions abatement vary among regions. This requires an explicit analysis of how the abatement costs might be distributed among the potential signatories to any treaty.

In the following analysis, we seek to improve our understanding of how the costs of a CO_2 emissions limit will vary depending on the stage and pattern of economic development, the fuel mix, and the initial endowment of hydrocarbon resources. The analysis is based on Global 2100, an analytical framework for estimating the economy-wide impacts of rising energy costs. We will explore how emissions are likely to evolve in the absence of a carbon limit, and how the regional pattern is likely to shift during the next century. We will then examine alternative strategies to limit global emissions, calculate the impact of higher energy costs upon conventionally measured GDP and indicate the size of the carbon tax that would be required to induce individual consumers to reduce their dependence on carbon-intensive fuels. Finally, differences in the time path of carbon taxes among regions will be analyzed to identify potential opportunities for trade in emission rights. We have not attempted to estimate the *benefits* of slowing down the rate of climate change through a reduction in worldwide CO_2 emissions. Our analysis is confined to the impacts of carbon emission limits upon the *cost* of energy – and the resulting effects upon the economy as a whole. This measurement of costs is only a part of the story. It is a far more formidable task to estimate the benefits from reduced emissions, and this is well beyond the scope of the present analysis. Clearly, policy makers will need to balance *both* the benefits and the costs in order to arrive at an overall judgment.

The Global 2100 Model

The name Global 2100 has been adopted in order to emphasize both the global nature of the carbon emissions problem and also the need for a long-term perspective. There are long time lags inherent in the buildup of CO_2 and in the transition away from carbon-based fuels. Our model is benchmarked against 1990 base-year statistics, and the projections cover 11 ten-year time intervals extending from 2000 through 2100. This is an intertemporal rather than a recursive model. It is assumed that producers and consumers will be sufficiently farsighted to anticipate the scarcities of energy and the environmental restrictions that are likely to develop during the coming decades. In its present form, Global 2100 is based on parallel computations for five major geopolitical groupings: (1) the USA, (2) other OECD nations (Western Europe, Canada, Japan, Australia and New Zealand), (3) the USSR and Eastern Europe, (4) China, and (5) ROW (rest of world). In defining these regions, we have attempted to employ the minimal level of disaggregation necessary to provide meaningful insights into how the costs of a carbon constraint may vary among nations.

Our regional categories have been based upon the following two considerations: (1) any solution to the climate problem is likely to require differentiated responses by industrialized and developing countries; and (2) over the long term, the CO_2 problem is primarily a coal problem, and 97% of the world's coal resources are contained in the OECD, USSR, Eastern Europe and China. See Clark (1982).

A distinction should be drawn between the first four regions and the ROW. The latter represents a catch-all category containing all those countries not included in any other region. It is required to maintain a consistent global balance of energy and carbon flows. However, the countries in ROW are expected to pursue their own individual interests rather than the welfare of the group as a whole. It would be misleading to treat ROW as a homogeneous entity. Had we been specifically concerned with oil and gas energy issues, we would have separated the ROW between OPEC and non-OPEC nations.

Within each region, the analysis is based on ETA-MACRO, a model of two-way linkage between the energy sector and the balance of the economy. This is a merger between ETA (a process model for energy technology assessment) together with a macroeconomic growth model providing for substitution between capital, labor and energy inputs. For a technical description of this model, see Manne and Richels (1990b).

ETA-MACRO is a tool for integrating long-term supply and demand projections. It is designed to compare the options that are realistically available to each region as the world moves away from its heavy present dependence upon oil and gas resources toward a more diversified energy economy.

At some point in the future, we hope to adopt a computable general equilibrium (CGE) framework. As an initial step in this direction, we make a series of assumptions on the future path of international crude oil prices, and place bounds on the willingness of each region to import or export oil. A CGE framework would allow us to deal explicitly with issues such as trade in carbon quota rights, trade in carbon-intensive commodities and the impact of carbon quotas upon the international division of labor.

Key Demand Parameters

Three of the demand parameters are: (1) potential gross domestic product (GDP) growth, (2) elasticity of price-induced substitution (ESUB) and (3) autonomous energy efficient improvements (AEEI). They are crucial to the debate over energy and environmental futures. There is no easy way to estimate these coefficients econometrically. The values adopted here have been determined so that our model will track closely with the conventional wisdom expressed, for example, by the median poll responses of the International Energy Workshop. See Manne and Schrattenholzer (1990).

One key estimate is the rate of GDP growth. This rate depends both upon population and per capita productivity trends. In parallel with the slowdown of population growth during the 21st century, there will be a diminishing rate of growth of GDP, and hence a slowdown in the demand for energy. Figure 1 shows our assumptions about the rate of potential GDP growth within each region. Typically, these represent extrapolations of performance during the past two decades. Such extrapolations are subject to enormous uncertainty. If, for example, China moves back toward more centralized control of its economic system, it is unlikely to enjoy the high growth rates that are suggested by Figure 1.

Because of energy-economy interactions, the *potential* GDP growth rates do not uniquely determine the realized rates. Energy costs represent one of the claims upon an economy's output. Tighter environmental standards and/or an increase in energy costs will reduce the net amount of output available for meeting current consumption and investment demands. The potential will then exceed the *realized* GDP.

Energy consumption need not grow at the same rate as the GDP. Over the long run, they may be decoupled. In Global 2100, these possibilities are summarized through two parameters. One is termed ESUB (the elasticity of price-induced substitution), and the other is AEEI (autonomous energy efficiency improvements).

There is a good deal of possible substitutability between the inputs of capital, labor and energy. The degree of substitutability will affect the economic losses from energy scarcities and price increases. One example of such a trade-off would be insulation to replace heating fuels in homes and other structures. A second example would be the increased use of heat exchangers and of cogeneration within industry. In the aggregate, the ease or difficulty of these trade-offs is summarized by ESUB. The higher the value of ESUB, the less expensive it is to decouple energy consumption from GDP growth during a period of rising energy prices. When energy costs are a small fraction of total output, ESUB is approximately equal to the absolute value of the price elasticity of demand. In ETA-MACRO, this parameter is measured at the point of secondary energy production – electricity at the busbar, crude oil and synthetic fuels at the refinery gate.

Here the numerical value of ESUB has been taken to be .40 for the USA and other OECD nations (OOECD). These countries have already demonstrated their ability to use the price mechanism as an aid in decoupling energy from GDP growth. Elsewhere, price-induced substitution is more problematical, and it depends upon structural economic changes. We have therefore set ESUB at .30 for these other regions.

AEEI (costless energy efficiency improvements) are even more difficult to estimate than those that are induced by price increases. For a sensitivity analysis of the importance of this parameter, see Manne and Richels (1990a). Econometric investigations of the U.S. post-1947 historical record show no evidence for autonomous time trends of this type. See Brown and Phillips (1989), Hogan (1988) and Jorgenson and Wilcoxen (1989). Indeed, Hogan and Jorgenson (1990) suggest that the AEEI for the U.S. might have been negative during this historical period. Technologically oriented end-use analysts, however, have suggested that non-price efficiency improvements may be induced by deliberate changes in public policy, e.g. a mandatory doubling or quadrupling of the average fuel efficiency of automobiles during the course of ten or twenty years. See Goldemberg et al. (1987).

For the initial decades of the 21st century, we have assumed values for the AEEI which fall in the middle of these two opposing views: 0.5% annually for the USA and OOECD regions, 0.25 for the Soviet Union and Eastern Europe (SU-EE), 0.0 for ROW and 1.0 for China. The lower values for SU-EE and ROW reflect the fact that these regions will be undergoing further industrialization before moving toward a service-based economy. As a result, they are likely to experience fewer opportunities for reductions in their energy intensiveness than if they were further along toward a postindustrial phase. The higher AEEI value for China (an annual rate of 1.0%) reflects its enormous potential for energy efficiency improvements. This is readily apparent from a five-region comparison of the ratios of total primary commercial energy per unit of GDP. There are serious conceptual difficulties in comparing GDP measurements between these five regions, but nonetheless the basic point is clear. According to Figure 2, China uses about four times as much energy per unit of GDP as any other region. Unless China makes more efficient use of its energy resources, it will become exceedingly difficult to achieve rapid economic growth targets. As the other regions increasingly take on the postindustrial characteristics of the OECD nations, the AEEI differentials between regions are likely to decline. For the second half of the 21st century, we assume an AEEI of 0.5 % annually in all regions.

Supply and Cost Assumptions

Table 1 identifies the alternative sources of *electricity* supply that are included in Global 2100. The first five technologies represent existing sources: hydroelectric and other renewables, gas-, oil-, and coal-fired units, and nuclear power plants. The second group of technologies includes the new electricity generation options that will become available in the future. These differ in terms of their projected costs, carbon emission rates and dates of introduction.

Appendix A (available upon request to the authors) contains a summary of our cost and performance estimates for the electricity supply technologies in each of the five regions. These technologies are intended to be representative of existing and future options. The cost and performance characteristics were adjusted for regional differences using judgmental and first-order adjustment factors considering the wide variety of conditions within the various regions. See Vejtasa and Schulman (1989).

It is expected that new gas-fired capacity for base load electricity will be produced by combustion turbine combined cycle plants. These units have a high thermal efficiency and relatively low costs. If natural gas prices remain at their 1988 levels, this technology would represent an attractive source of electricity. However, as natural gas resources gradually become exhausted, fuel prices will rise. For example, in a baseline projection for the Gas Research Institute, Woods (1988) projects a tripling of U.S. wellhead prices by 2010. With such an increase, gas-fired electricity would lose its competitive advantage over coal. Although there is a broad spectrum of coal technologies under development with attractive cost and performance features, the new coal-fired technology identified in Table 1 is based on a new pulverized coal plant with flue gas desulfurization. These cost estimates are similar to those for Advanced Fluidized-Bed Combustion (AFBC) and Integrated Gasification Combined Cycle (IGCC) plants.

ADV-HC and ADV-LC respectively refer to advanced high- and low-cost carbon-free electricity generating technologies. Any of a number of technologies could be included in these categories – solar, nuclear, biomass, clean coal technologies with CO_2 removal, etc. Given the enormous disagreement which currently exists as to which of these technologies or combination of technologies will ultimately win out in terms of economic attractiveness and public acceptability, we have chosen to refer to them simply as ADV-HC and ADV-LC.

Our cost and performance data are based upon specific designs considered in EPRI's Technical Assessment Guide (1989). The high costs refer to an advanced solar technology with cost and performance characteristics similar to those for concentrator photovoltaic cells. The low costs are representative of an advanced nuclear design with passive safety features. For those who disagree with the cost ranking of these two technologies, nuclear would be identified as ADV-HC and solar as ADV-LC. Alternatively, the low-cost source might be a diversified mix of solar, nuclear and other carbon-free technologies.

Table 2 identifies the eight alternative sources of *non*electric energy that are included within the model. The list is headed by OIL-MX, imports less exports of crude oil. Petroleum is the international "swing fuel," and its price is crucial to any near-term projections of energy supplies and demands. Accordingly, we have explored several alternative crude oil price scenarios, but have reported on only one in this paper. Alternative scenarios would differ in their rate of growth toward the levels needed in order to make "backstop" sources of energy supply become economically viable during the 21st century.

All other fuels are ranked in ascending order of their cost per million BTU of crude oil equivalent. The least expensive domestic source is CLDU – coal employed for direct uses in industries such as iron and steel, cement, etc. Its growth rate is taken to be only 20% that of the GDP. Next in the "merit order" are domestic oil and gas. These exhaustible resources are available at constant marginal costs, but are subject to upper bounds on extraction rates based upon a model of reserves and resource depletion.

For determining the extraction rates of these exhaustible resources, we draw a sharp distinction between current *proven reserves* and the remaining stock of *undiscovered resources*. Because cost estimation is exceedingly hazardous in this area, we do not attempt to provide an explicit economic rationale through continuously rising marginal cost curves. Instead, just two categories of oil and two categories of natural gas are distinguished: low- and high-cost.

Proven reserves of hydrocarbons are depleted by current production, and are augmented by new discoveries out of the remaining stock of undiscovered resources. This is almost but not quite a constant ratio model of exhaustible resource depletion. At any one time, production is a fixed fraction of remaining reserves. New discoveries may not exceed a fixed fraction of the remaining undiscovered resources.

The only element of flexibility lies in the ability to defer reserve additions. With this formulation, Global 2100 is able to incorporate forward-looking resource depletion policies. At the same time, the model is capable of representing an important real-world phenomenon. During a given year, a region may be importing oil – and also engaging in domestic production out of both low- and high-cost resources.

For low-cost resources, our reserve and resource estimates are taken from the 5th percentile point along the probability distributions available from the U.S. Geological Survey work by Masters et al. (1987). High-cost undiscovered resources are arbitrarily assumed to be 50% of those in the low-cost category.

Masters et al. provide a modal (i.e., most likely) estimate of resources along with the 5th and 95th percentile. For practical purposes, the 95th percentile point indicates a lower bound on undiscovered conventional resources, and the 5th percentile indicates an upper bound. That is, according to the USGS, there is only a 5% probability that undiscovered conventional resources will exceed the 5th percentile values. By taking the USGS upper bound on conventional resources, we have attempted to allow for future technological progress, e.g. further reductions in the costs of deep drilling. Had our calculations been based upon the modal or the 95th percentile, the prospects for conventional gas production would be considerably more pessimistic than the case examined here.

By definition, a backstop source is available in unlimited quantities at constant marginal costs. According to Table 2, there are two principal backstop options: SYNF (synthetic fuels based on coal or shale oil) and NE-BAK (e.g., biomass fuels or hydrogen by electrolysis, using a carbon-free source of electricity). NE-BAK emits no carbon, but is likely to be more expensive than synthetic fuels based upon coal or shale oil. One or the other of these high-cost technologies will impose an upper bound upon the future cost of nonelectric energy – depending on whether or not there is a carbon constraint. Note that the carbon emissions coefficient for crude oil is only half of that for synthetic fuels. At a later point, it will be shown that this difference *could* have a significant impact upon international oil prices.

An Unconstrained Carbon Emissions Scenario

Globally, carbon emissions have increased at an average annual rate of 3.2% per year since 1950. The absolute level of emissions rose from 1.6 to 5.7 billion tons of carbon. Over the same period there has been a substantial shift in the pattern of global contributions. For example, in 1950 North America and Western Europe accounted for 68% of total emissions. By 1980 their share had fallen to 43%. In contrast, the portion attributable to China and to other developing countries in Latin America, Southeast Asia, and Africa rose from 7 to 20% over the same period.

A key issue for the present analysis is the size and pattern of future emissions in the absence of measures to slow growth. In this section, we calculate carbon emissions for each of our five geopolitical groupings under an unconstrained "business as usual" energy future.

Figure 5 compares global emissions at two points in time, 1990 and 2100. Although emissions grow considerably over the next century, the average rate of growth slows to 1.4% per year. This is low by historical standards. A good deal of the explanation for this slowdown lies in our assumptions about the slowdown in population growth and in its implications for future GDP. From Figure 1, recall that the growth rates for all regions are projected to taper off significantly.

Energy efficiency improvements also play a role in reducing the growth in carbon emissions. Figure 6 shows average annual GDP and total primary energy (TPE) growth rates for all five regions through 2100. The growth rate in energy demand is considerably lower than that for GDP. Price- and non-price induced energy efficiency improvements have produced a significant decoupling of these two growth rates.

Figure 5 shows a significant shift in the pattern of global contribution in CO_2 emissions. Today, the industrialized nations (USA, OOECD and SU-EE) account for 71% of carbon emissions. By the end of the next century, their contribution is projected to drop to less than one-half of the total. The reason for this shift is apparent from Figure 6. Energy *growth* rates in the developing nations are more than twice as high as in today's industrialized countries.

The Costs of Limiting Carbon Emissions

We now explore the impacts of a carbon emissions limit on each of the five regions. There are many possible ways to define a global agreement on carbon emissions. Although there is considerable disagreement over the appropriate level of reduction, there is agreement that the burden must fall disproportionately on the industrialized nations. If global income inequalities are to diminish, developing countries will experience much faster rates of economic expansion and energy demand growth than their currently industrialized counterparts. The developing nations are unlikely to accept any agreement which fails to provide for some increase in their carbon emissions. For them, the issue will be how far to limit their *growth* in emissions.

There have been a number of calls to reduce global carbon emissions by 20% below current levels. Some of the proposals apply solely to the industrialized countries, while others refer to the world as a whole. Recognizing the need for some emissions growth in the developing countries, we begin by assuming that the 20% reduction is confined to the industrialized nations only. Specifically, we assume that the USA, OOECD and SU-EE agree to stabilize carbon emissions at their 1990 level through the year 2000 and then gradually reduce them by 20% by the year 2020. In return, the developing nations (China and ROW) would limit their emissions to twice their 1990 levels. Although these targets are not as ambitious as those contained in some recent proposals, they nevertheless represent a significant reduction in emissions, especially when compared with a business-as-usual view of the future.

Figure 7 compares carbon emissions with and without the carbon limits. Overall, the proposed limit would lead to a 15% increase in global emissions between 1990 and 2030, but no further increase thereafter. By 2100, this leads to a 75% reduction in the emissions level that would have been reached in the absence of an international agreement.

The feasibility of any scheme to reduce worldwide carbon emissions will depend on the costs to individual nations. Using Global 2100, we may add together the impacts of rising energy costs in each region and calculate the annual losses due to the carbon constraint. Figure 6 shows the annual losses as a percentage of aggregate GDP. For the USA, the effects of a carbon constraint do not begin to have measurable macroeconomic consequences until after 2000. At that point the rise in energy prices begins to have a significant effect upon the share of gross output available for current consumption and investment. By 2030, roughly 3% of the total annual U.S. GDP is lost as a consequence of the carbon constraint. This percentage remains roughly constant for the remainder of the time horizon. For the other OECD countries, measurable macroeconomic consequences do not begin to accrue until 2010, and the annual losses in total consumption are limited to the 1-2% range. The costs of the carbon constraint are somewhat lower than those in the USA. They have a relatively higher proportion of undiscovered oil and gas resources, and their nuclear power industry is larger. Moreover, they have a lower energy/GNP ratio in the base year of 1990. (Recall Figure 2.) The Soviet Union and Eastern Europe will find it more difficult to adjust to a 20% emissions cutback than their western counterparts. By 2030, roughly 4% of total macroeconomic consumption is lost as a consequence of the carbon constraint. The higher costs relative to the OECD follows directly from the assumption that it will be more difficult to decouple GDP and energy growth in this region. Recall that both the elasticity of price-induced substitution and the rate of autonomous energy efficiency improvement are lower for SU-EE than for the USA and OOECD.

Since ROW includes OPEC, Mexico and other potential oil exporters, the costs of a carbon constraint to that region are negligible until 2020. Gradually, however, their oil and natural gas resources also become exhausted. For them too, it becomes increasingly difficult to find low-cost carbon-free energy sources, and the macroeconomic consequences begin to mount. By the end of the 21st century, approximately 5% of GDP is lost as a consequence of the emissions constraint.

Eventually, China would become the region most heavily affected by the international carbon reduction agreement. Figure 6 shows that China's annual GDP losses would exceed 10% by the latter half of the 21st century. Their rapid rate of economic development would place enormous upward pressures on energy demands. Since China's fossil fuel resource base is dominated by coal, it will be costly to accept any constraints on carbon emissions. In a carbon constrained energy future, the principal alternatives for China will be high-cost supply substitutes (NE-BAK) and price-induced conservation. Both of these will bear a high price tag.

According to the present set of calculations, the percentage losses accruing to China would be far greater than those experienced by other regions. The question arises as to how much the carbon limits would need to be relaxed to bring China's losses in line with those of the rest of the world. Figure 7 compares Chinese losses when their emission levels are allowed to increase by factors of two, three, four and five. If international negotiators were to adopt the criterion of equal percentage GDP losses across regions, these results suggest that a quadrupling in emissions (the 4X scenario) will be required for China. This would result in a 37% increase in global emissions between 1990 and 2030, but no further increase thereafter. Even with this exception for China, global emissions in 2100 would be only 30% of the level that would have been reached in the absence of an international agreement.

The Chinese case highlights the difficulty of achieving a 20% reduction in *worldwide* emissions. Clearly, any increases by developing countries would need to be offset by reductions in the industrialized countries. Figure 8, a back-of-envelope calculation, shows global emissions under different international agreements. If China and ROW are permitted to double their emissions, then the industrialized countries would need to reduce theirs by nearly 70% below current levels in order to achieve a worldwide reduction of 20%. If China is permitted to quadruple its emissions, the industrialized countries would need to reduce theirs to zero!

China would not necessarily obtain significantly higher emission rights under other quota rules. For example, Grubb (1989) has suggested that *population* would be an equitable criterion for the allocation of emission rights. For a 20% global reduction, this would require an enormous reduction from the industrialized countries and especially from the USA. The ROW would be entitled to more than double their emissions. China would still not be entitled to double its emissions over the next century—let alone to quadruple them. See Figure 11.

Carbon Taxes

There are a variety of policy instruments available for reducing CO_2 to the desired levels. An economically efficient option would be to impose a tax upon those activities responsible for carbon emissions, and to vary the tax rate according to the carbon content of individual fuels. The purpose would be to discourage those activities with relatively high carbon emissions.

Figure 10 compares the region-by-region time path for the carbon tax that would be required to provide the price signals consistent with the regional carbon limits considered in this paper. It turns out that the *long-run* equilibrium tax level is determined so to make synthetic fuels (SYNF) and nonelectric backstop (NE-BAK) supply technologies equally attractive to the energy consumer. Specifically, the equilibrium tax is determined by the ratio of their cost differential to their carbon coefficient differential:

$$\frac{\text{Cost differential, \$/mmBTU}}{\text{Carbon differential, tons/mmBTU}} = \frac{20\text{-}\$10}{.04 \text{ tons}} = \$250/\text{ton}$$

A carbon tax of $250/ton would imply a five-fold increase in the price of coal—or an increase of $.75/gallon of refined petroleum products. If one region were to sell just 100 million tons of carbon emission rights to another, this tax rate would imply a financial transfer of $25 billion annually!

The specific paths to the equilibrium level depend on several factors. A tax is required earlier in the industrialized regions because of the agreement to reduce carbon emissions by 20% by 2020. In contrast, China and ROW are allowed to increase emissions beyond current levels. Pressures to switch away from carbon intensive fuels do not begin to build until somewhat later.

Although the industrialized countries face identical percentage reductions, the time path for carbon taxes varies considerably. Those which find it less difficult to decouple energy consumption from GDP growth will tend to have an easier and smoother transition. Lacking the demand-side alternatives of its western counterparts, the SU-EE will need to maintain taxes at a higher level to induce consumers to switch away from high-carbon fuels.

A greater abundance of low-cost supply alternatives will smooth the transition to the equilibrium tax. Recall that in comparison with USA, OOECD is better off in terms of supply-side alternatives. It has larger oil and gas resources and a larger stock of nuclear power plants. As a consequence, OOECD will require lower taxes during the transition years.

In the event of an international agreement to limit carbon emissions, region-by-region differences in time paths could be exploited to identify cost-effective strategies for emissions reduction. At a given point in time, regions which find it more difficult to adjust to their emissions limits (those requiring higher taxes) should be willing to purchase emissions rights from regions experiencing less difficulty. For example, based on the carbon taxes reported here, all three industrialized regions would be buyers of emissions rights prior to 2020. Having agreed to reduce emissions by 20% by that year and lacking sufficient supply and demand side alternatives to achieve such reductions without a sizeable tax, countries in these regions should be willing to pay a great deal for the right to emit more carbon. By contrast, the developing regions do not begin to bump up against their limits until 2020. Prior to that year, countries in these regions should be willing to sell some of their emissions rights to the industrialized nations.

Impact of Carbon Limits upon International Oil Prices

With carbon emissions limits as stringent as those considered in this paper, there would be severe restrictions on the role of coal as a "bridge to the future." This in turn *could* affect international oil prices. In the short term, it is true that carbon limits would reduce the demand for oil and for other carbon-based fuels. Over the long term, however, carbon limits are likely to increase the demand for oil.

From Table 2, recall that conventional crude oil has an emissions coefficient only half that of coal and shale-based synthetic fuels. This is why conventional oil would enjoy a premium value in the presence of carbon constraints. In the absence of carbon limits, synthetic fuels impose a relatively low cap on the international oil price, e.g. $10/million BTU – that is, $60/barrel. With carbon emissions limits, there is an increased demand for oil relative to synfuels, and a higher international price is required in order to arrive at a market equilibrium. If a carbon-free backstop costs $20/million BTU ($120/barrel of crude oil equivalent), and the carbon tax is $250/ton, there would be a $30 premium value for the lower carbon coefficient of crude oil. This means that the international value of petroleum would rise to $90/barrel. Carbon limits place a severe penalty on the use of coal and shale-based substitutes for crude oil, and are therefore likely to strengthen the market power of OPEC.

Summary and Conclusions

This paper has sought to improve our understanding of how the costs of limiting carbon emissions are likely to vary among regions. An analysis which encompasses a period exceeding a century is necessarily subject to enormous uncertainty. Nevertheless, we believe that the calculations have provided some useful results.

In the absence of an international agreement to limit growth, carbon emissions are likely to increase considerably, perhaps by a factor of four or more over the next century. During this period, there is apt to be a significant shift in the regional pattern of emissions. In 1990, the industrialized countries accounted for 71% of man-made carbon emissions. By 2100, their share is likely to drop below 50%.

In recent years, there have been numerous proposals to reduce worldwide emissions by 20% or more. Our analysis indicates that this target would be difficult to achieve. If per capita income inequalities are to be significantly reduced, developing countries will have to achieve much higher rates of economic growth than their currently industrialized counterparts. This will place enormous upward pressure on their demands for commercial energy. For developing countries, the issue will not be how far to reduce emissions, but how far to limit their rate of growth.

Against this background, achieving a 20% worldwide reduction would prove difficult. Suppose, for example, that the developing countries were to agree to limit emissions to a 100% increase. The industrialized countries would then need to reduce their emissions by nearly 70% below current levels in order to achieve a 20% reduction in worldwide emissions.

We have explored less ambitious but perhaps more nearly feasible carbon limits: a 20% reduction in emissions by the industrialized countries, a doubling or quadrupling by China and a doubling in the other developing countries. This would result in a 15-37% overall increase between 1990 and 2030, but no further increase thereafter. By 2100, global emissions would be only 25-30% of the level that would have been reached in the absence of an international agreement.

We have also investigated the size of the carbon tax that would be required in each region to induce consumers to reduce their dependence upon carbon-based fuels. Under the assumptions adopted here, it turns out that the long-run equilibrium tax is the same in all regions – $250 per ton of carbon. Such a tax would imply a five-fold increase in the price of coal and an increase of $.75 per gallon of gasoline. If one region were to sell just 100 million tons of carbon emissions rights to another, this tax rate would imply a financial transfer of $25 billion annually.

There are significant regional differences in the time path to the long-run equilibrium carbon tax level. This would point to opportunities for international trade in emissions rights. At a given point in time, those regions which find it more difficult to adjust to their carbon limits should be willing to purchase emissions rights from other regions experiencing less difficulty.

Our analysis has focused exclusively on the *costs* of reducing carbon emissions. We have not attempted to estimate the *benefits* of slowing down the impacts of global climate change. Without such information, it is unclear whether it makes sense to undertake sizeable reductions in emissions. If it does turn out that significant abatement measures are worthwhile, it is important to understand how the costs of these measures might vary between nations. This could be critical to the negotiation of an acceptable international agreement.

Table 1

Identification of Electricity Generation Technologies

Technology Name	Earliest Possible Introduction Date*	Identification
Electric Technologies		
Existing		
HYDRO		Hydroelectric, geothermal and other renewables
GAS-R		Remaining initial gas-fired
OIL-R		Remaining initial oil-fired
COAL-R		Remaining initial coal-fired
NUC-R		Remaining initial nuclear
New		
GAS-N	1995	Advanced combined cycle, gas-fired
COAL-N	1990	New coal-fired
ADV-HC	2010	High cost non-carbon based
ADV-LC	2010	Low cost non-carbon based

* Estimated year when the technology could provide .1 trillion kWh (approximately 20GW of installed capacity at 60% capacity factor).

Table 2

Nonelectric Energy Supplies

Technology Name	Description	Carbon emission coefficient, tons of carbon per million BTU of crude oil equivalent [1]	Unit Cost 1988 $ per million BTU of crude oil equivalent
OIL-MX	Oil imports - exports	.0203	3.30 in 1990, rising to 15 from 2030 onward
CLDU	Coal - direct uses	.0251	2.00 [2]
OIL-LC	Oil - low cost	.0203	2.50 [3]
GAS-LC	Natural gas - low cost	.0145	1.50 [3, 4]
OIL-HC	Oil - high cost	.0203	6.00
GAS-HC	Natural gas - high cost	.0145	5.00 [4]
SYNF	Synthetic fuels	.0408	10.00
NE-BAK	Nonelectric backstop	.0000	20.00

NOTES:

1 Source of carbon emission coefficients: Edmonds and Reilly (1985).

2 CLDU costs are only $1.00 per million BTU in China.

3 OIL-LC costs are only $0.50 per million BTU in ROW. Similarly, GAS-LC costs are only $1.00 per million BTU in ROW.

4 To allow for burner-tip equivalence, an additional $1.25 per million BTU is added to allow for gas distribution costs.

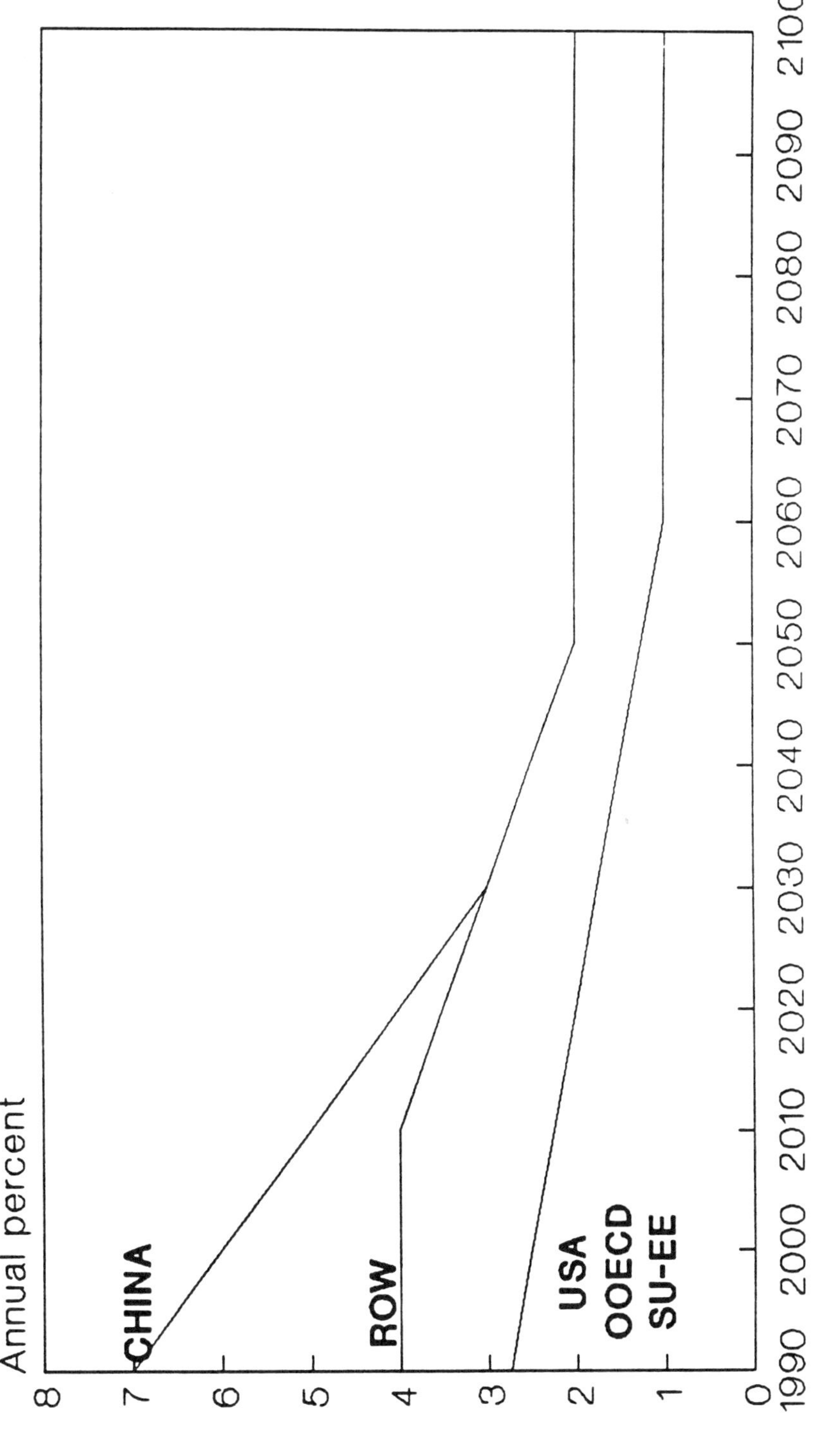

Figure 1 Potential GDP Growth Rates

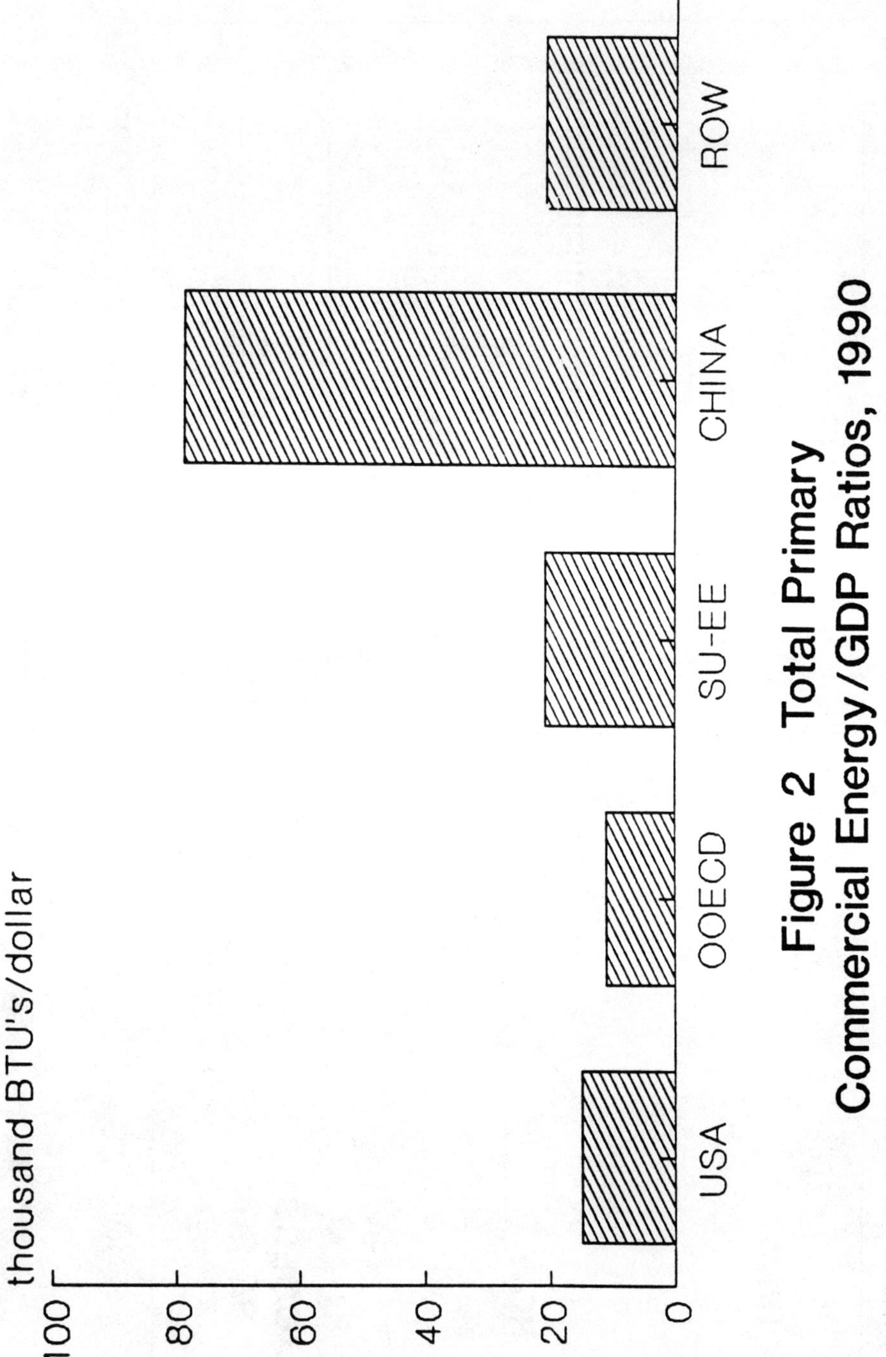

Figure 2 Total Primary Commercial Energy/GDP Ratios, 1990

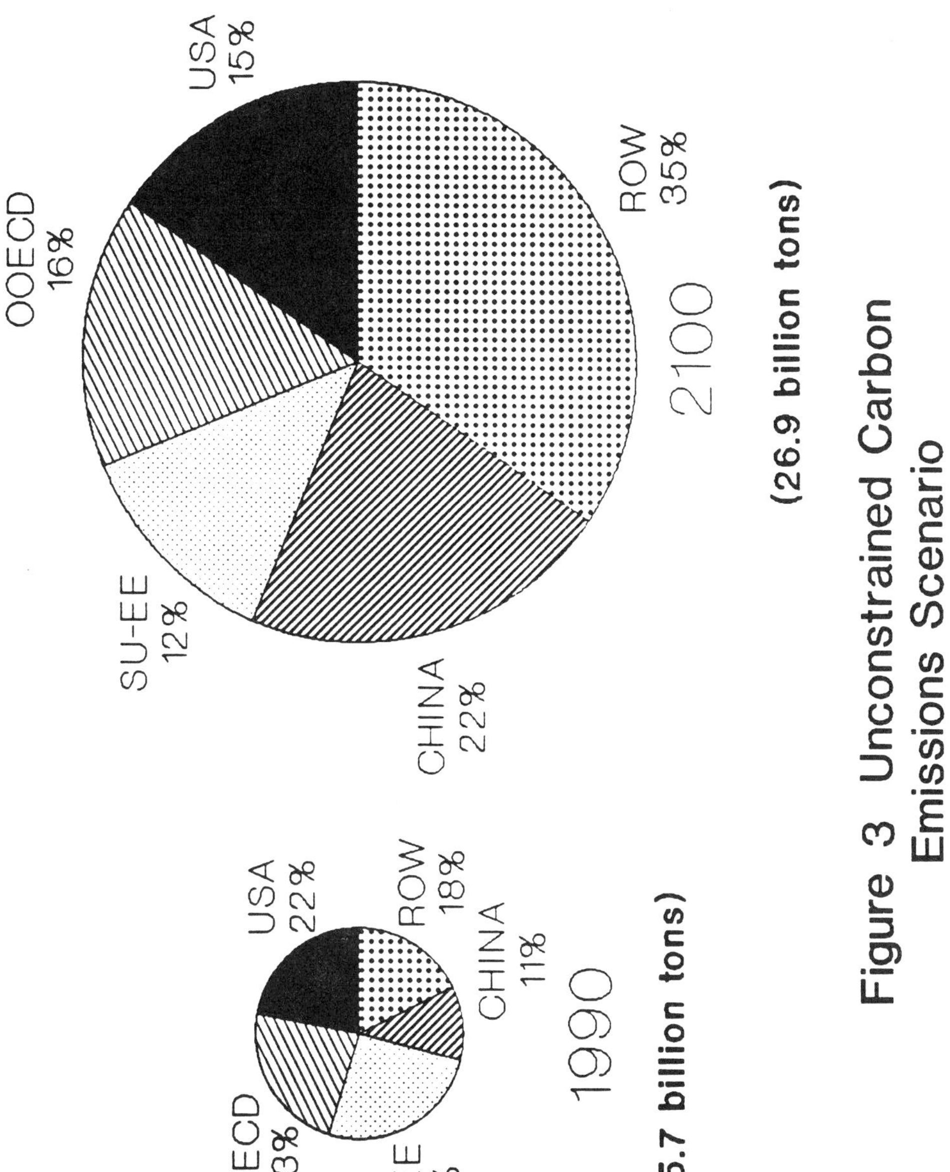

Figure 3 Unconstrained Carbon Emissions Scenario

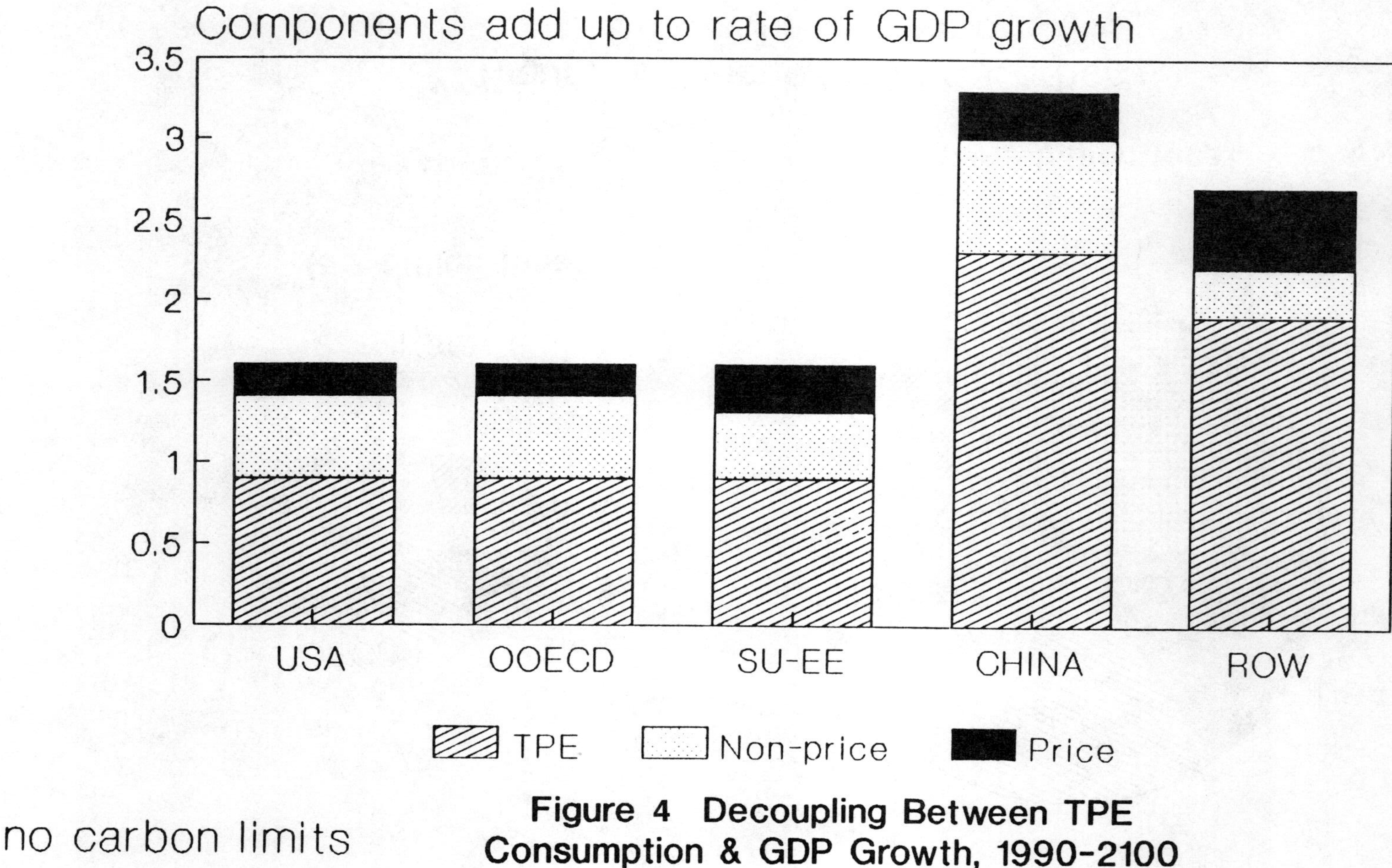

Figure 4 Decoupling Between TPE Consumption & GDP Growth, 1990-2100

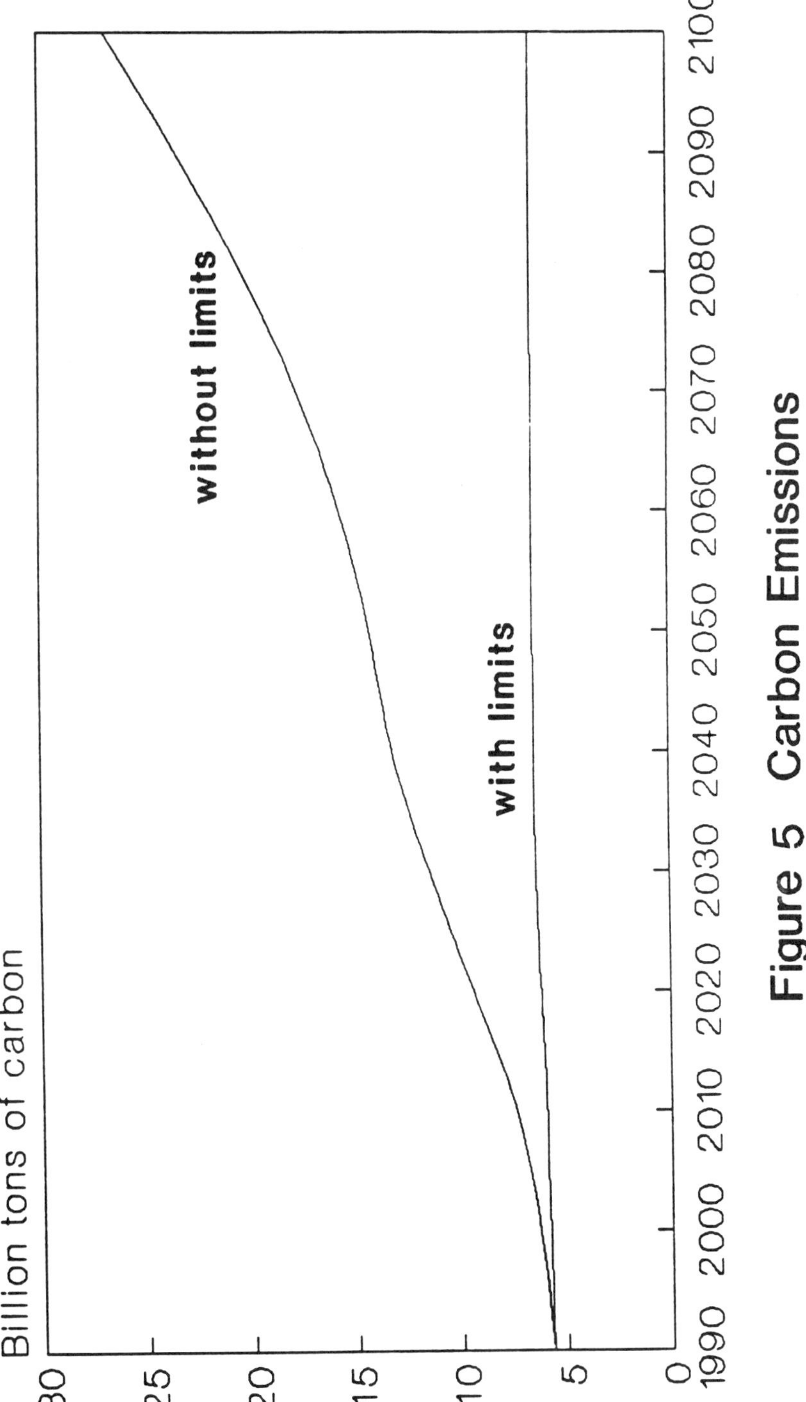

Figure 5 Carbon Emissions
(with and without carbon limits)

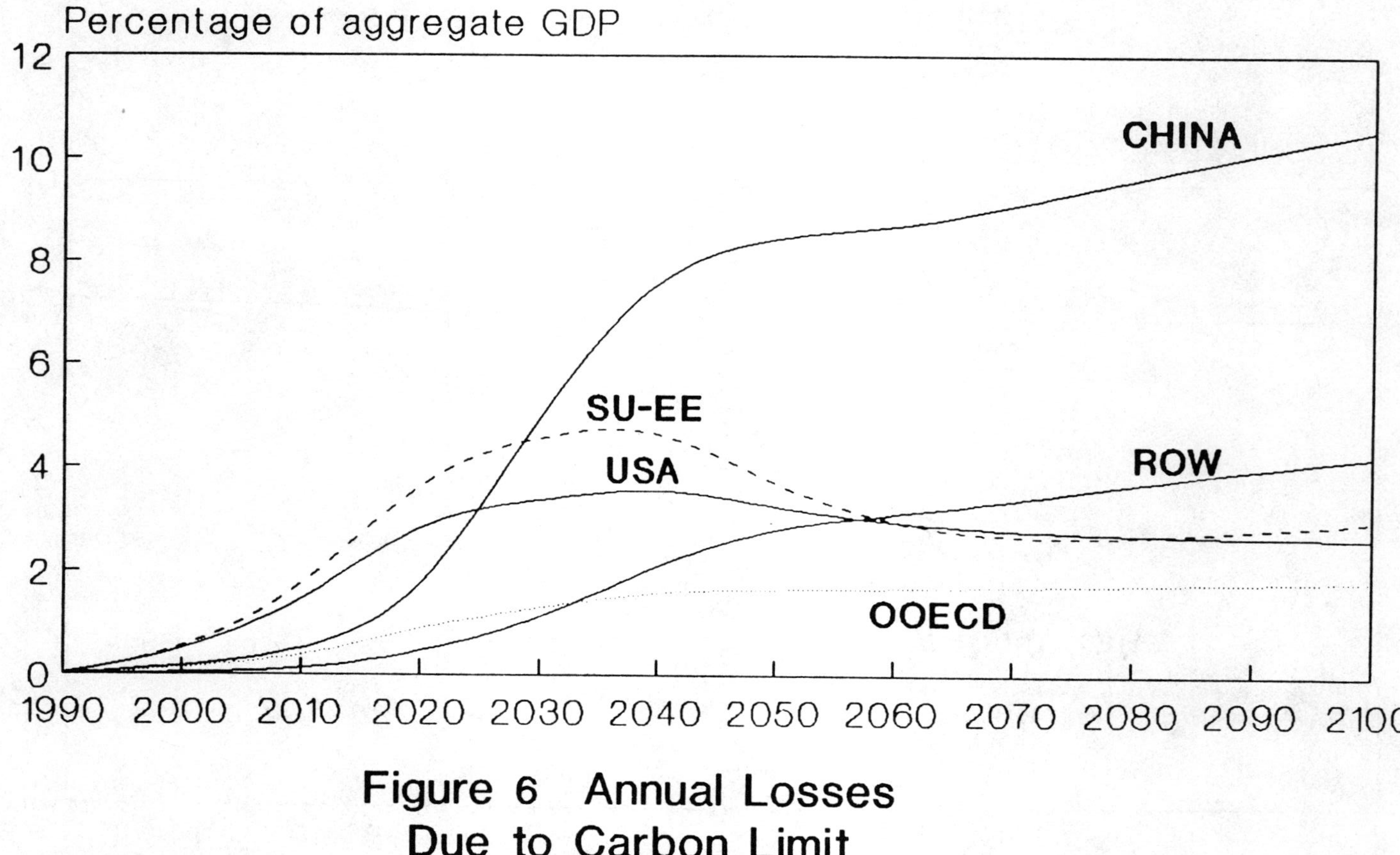

Figure 6 Annual Losses Due to Carbon Limit

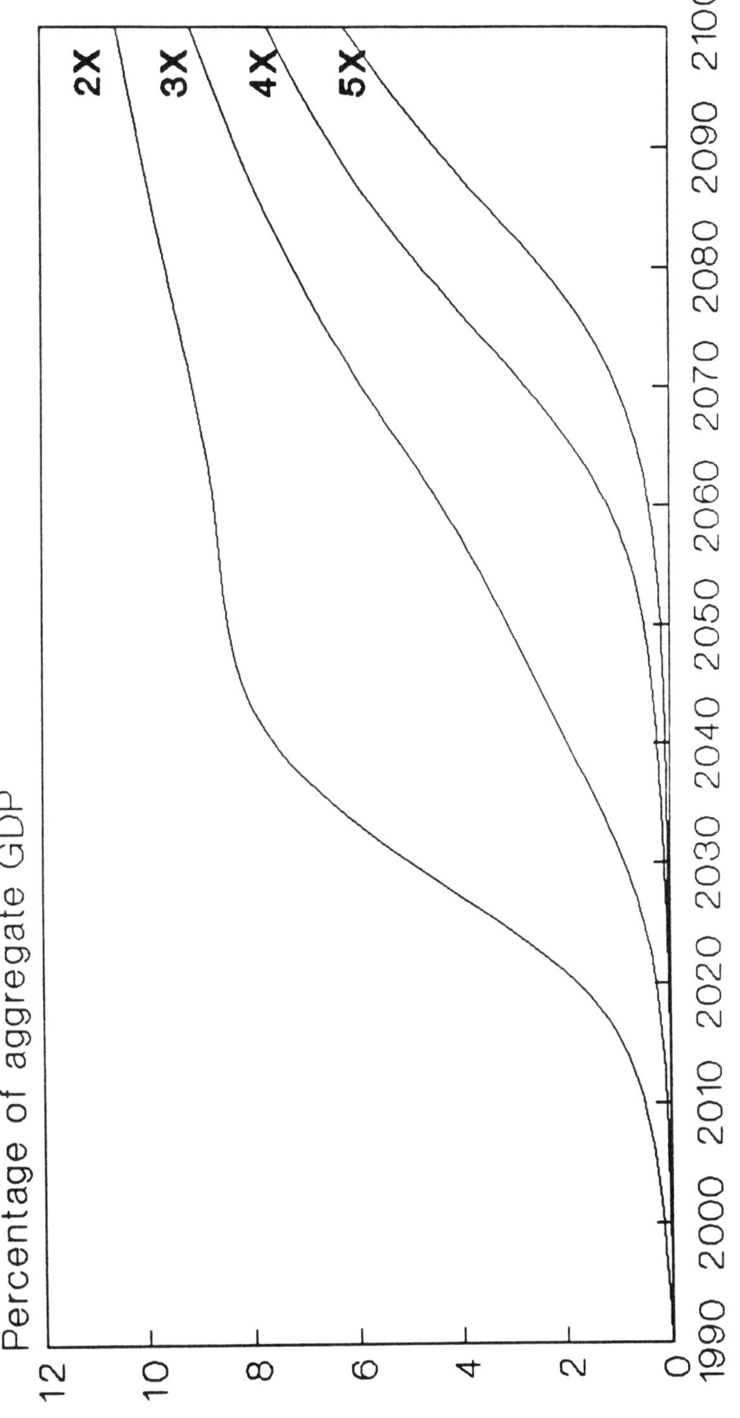

Figure 7 Annual Losses For China Under Different Limits

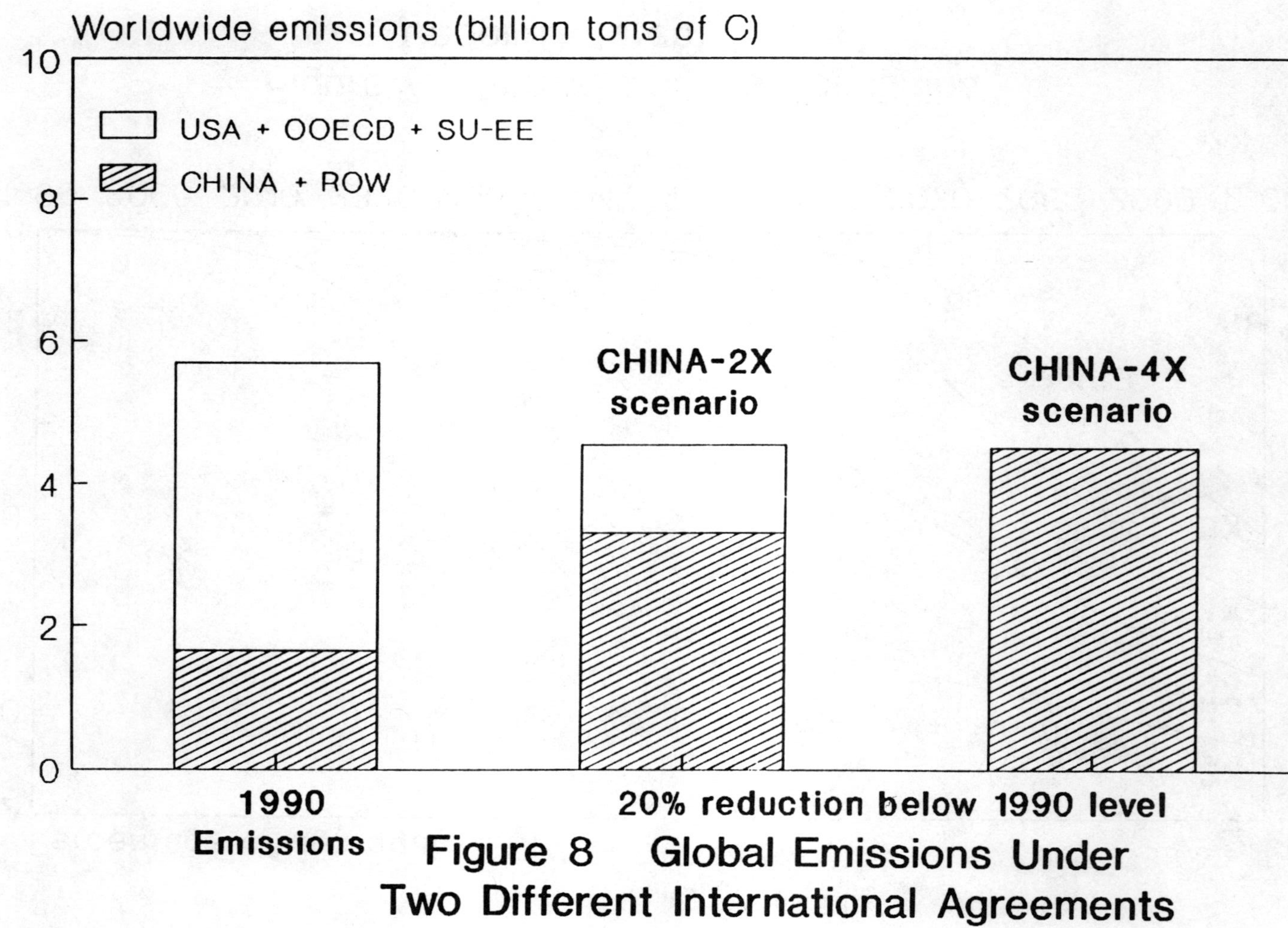

Figure 8 Global Emissions Under Two Different International Agreements

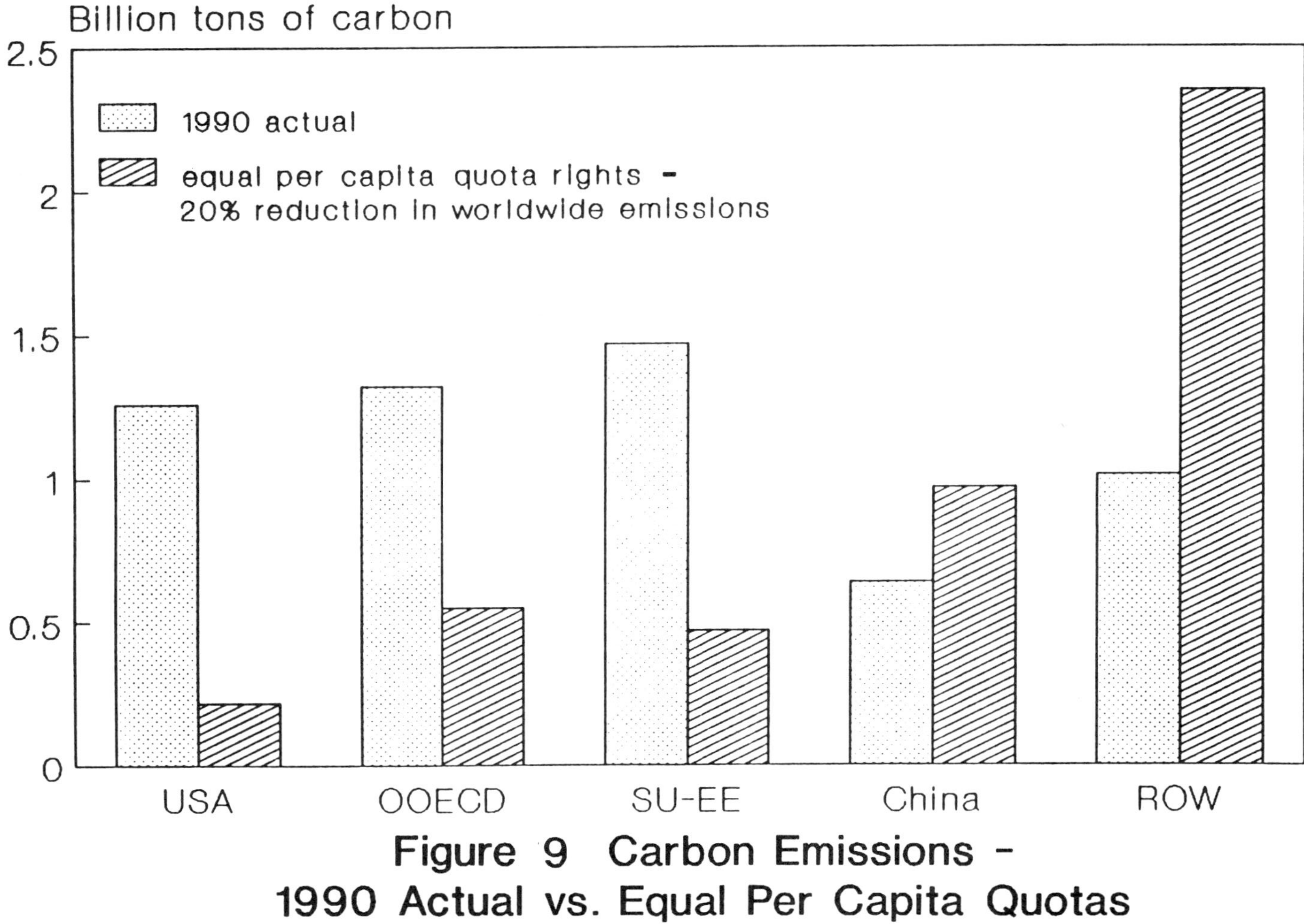

Figure 9 Carbon Emissions -
1990 Actual vs. Equal Per Capita Quotas

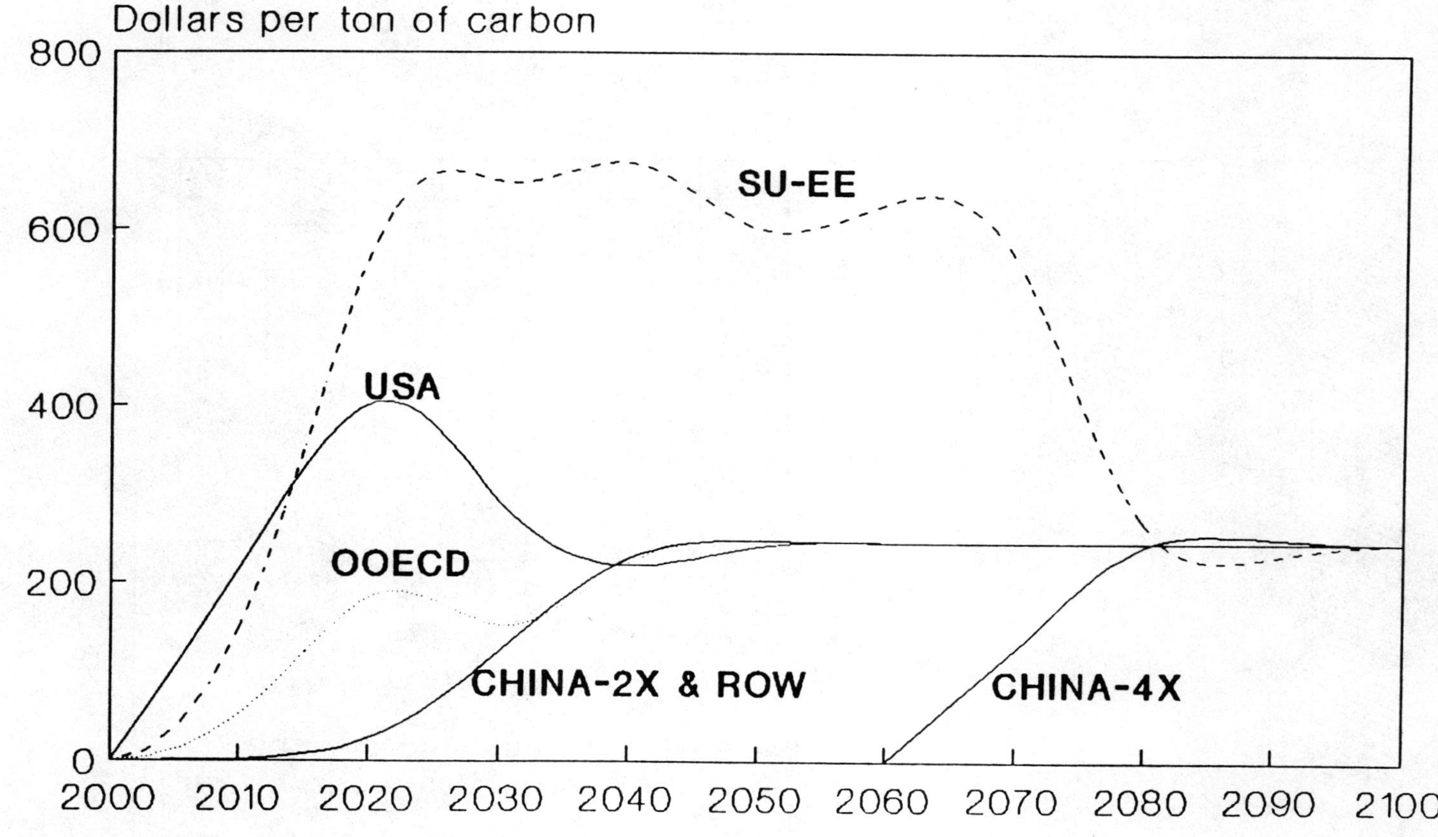

Figure 10 Carbon Taxes

Acknowledgment

The research reported in this paper was funded by the Electric Power Research Institute (EPRI). The views presented here are solely those of the individual authors, and do not necessarily represent the views of EPRI or its members. Earlier versions of this paper have been presented to: the International Association for Energy Economics, New Delhi, January 1990; the International Institute for Applied Systems Analysis, Laxenburg, January 1990; and InterAction Council, Amsterdam, February 1990.

The authors are much indebted to Diane Erdmann and to Lawrence Gallant for research assistance. Helpful comments have been provided by: Robert Dorfman, George Hidy, William Hogan, Henry Lee, Lu Yingzhong, Stephen Peck, Lee Schipper, Stanley Vejtasa, John Weyant and Robert Williams.

References

R.G.D. Allen, 1968. *Macroeconomic Theory*, Macmillan, New York.

A. Brooke, D. Kendrick and A. Meeraus, 1988. *GAMS: A User's Guide*, Scientific Press, Redwood City, California.

S.P.A. Brown and K.R. Phillips, 1989. *An Econometric Analysis of U.S. Oil Demand*, Federal Reserve Bank of Dallas, January.

W.C. Clark, 1985. *On the Practical Implications of the Carbon Dioxide Question*, International Institute for Applied Systems Analysis, Laxenburg, Austria.

J. Edmonds and J.M. Reilly, 1985. *Global Energy–Assessing the Future*, Oxford University Press, New York.

Energy Information Administration (EIA), 1989. *Monthly Energy Review*, U.S. Department of Energy, Washington, D.C., January.

Energy Modeling Forum, 1977. *Report 1, Energy and the Economy*, Stanford University, September.

Electric Power Research Institute (EPRI), 1989 (in preparation). *Technical Assessment Guide*, Electric Power Research Institute, Palo Alto, California.

J. Goldemberg, T.B. Johansson, A.K.N. Reddy and R.H. Williams, 1987. *Energy for a Sustainable World*, World Resources Institute, Washington, D.C.

W. W. Hogan, 1988. *Patterns of Energy Use Revisited*, Harvard University, June.

D.W. Jorgenson and P.J. Wilcoxen, 1989. *Environmental Regulation and U.S. Economic Growth*, Harvard University, July.

A. Manne, 1981. *ETA-MACRO: A User's Guide*, EA-1724, Electric Power Research Institute, Palo Alto, California, February.

A. Manne and L. Schrattenholzer, 1988. "The International Energy Workshop–A Progress Report," *OPEC Review*, Spring.

C.D. Masters, E.D. Attanasi, W.D. Dietzman, R.F. Meyer, R.W. Mitchell and D.H. Root, 1987. "World Resources of Crude Oil, Natural Gas, Natural Bitumen, and Shale Oil," *12th World Petroleum Congress, Proceedings*, vol. 5.

D. Rind, 1989. "A Character Sketch of Greenhouse," *EPA Journal*, vol. 15, no. 1, January-February.

S.A. Vejtasa and B.L. Schulman, 1989. *Technology Data for Carbon Dioxide Emission Model: Global 2100*, SFA Pacific, Inc., Mountain View, California.

T.J. Woods, 1988. *The Long-Term Trends in U.S. Gas Supply and Prices: The 1988 GRI Baseline Projection of U.S. Energy Supply and Demand to 2010*, Gas Research Institute, Washington, D.C., December.

CARBON SEQUESTRATION VERSUS FOSSIL FUEL SUBSTITUTION: ALTERNATIVE ROLES FOR BIOMASS IN COPING WITH GREENHOUSE WARMING

D.O. Hall, H.E. Mynick, and R.H. Williams

Center for Energy and Environmental Studies
Princeton University
Princeton, NJ 08544

Abstract

Displacing fossil fuel with biomass grown sustainably and converted into useful energy with modern conversion technologies would be more effective in decreasing atmospheric CO_2 than sequestering carbon in trees. Some industrial restructuring would be required to bring about a major energy role for biomass. However, the prospect that electricity and liquid fuel from biomass could often be less costly than from coal and petroleum makes this strategy for coping with greenhouse warming inherently easier to implement than many alternatives.

Introduction

Since it was initially proposed,[1] there has been much discussion[2-17] of carbon (C) sequestration by forests as one strategy for offsetting CO_2 emissions to reduce greenhouse warming. While the substitution of biomass for fossil fuel has sometimes been mentioned,[1-9] there has been no systematic comparison of these alternative biomass strategies for coping with greenhouse warming.

In this paper it is shown that, while sequestering C in forests is a relatively low-cost strategy for offsetting CO_2 emissions from fossil fuel combustion, substantially greater benefits can be obtained by displacing fossil fuel with biomass grown sustainably and converted into useful energy using modern energy conversion technologies. Biomass substituted for coal can be as effective as C sequestration, per tonne of biomass, in reducing CO_2 emissions; however, fuel substitution can be carried out indefinitely, while C sequestration can be effective only until the forest reaches maturity. Also, far greater biomass resources can be committed to fossil fuel substitution at any given time than to C sequestration, because (i) producers will tend to seek biomass species with higher an-

Published 1991 by Elsevier Science Publishing Company, Inc.
Global Climate Change: The Economic Costs of Mitigation and Adaptation
James C. White, Editor

nual yields for energy applications, and (ii) biomass for energy can be obtained from sources other than new forests. Thus biomass can play a larger role in reducing greenhouse warming by displacing fossil fuel than by sequestering C. Moreover, biomass energy is potentially less costly than the displaced fossil fuel energy in a wide range of circumstances, so that the net cost of displacing CO_2 emissions would often be negative. Thus bioenergy strategies have "built-in" economic incentives that make them inherently easier to implement than many alternative strategies for coping with greenhouse warming.

Carbon Sequestration

The basic C sequestration proposal calls for planting trees in forest reserves that would be maintained in perpetuity. With this approach, C absorption would continue until the forest matures, which could be some 40 to 100 years, if trees of long rotation are selected. This is not a permanent solution, but it does allow time to develop alternative, zero-CO_2-emitting energy sources. The capacity of growing forests to absorb C from the atmosphere depends on various factors, but 2.7 tonnes of C per hectare per year (tC/ha/yr) is typical[18] of average values assumed in most C-sequestration studies; as biomass, on a dry-weight basis, is about half C, the corresponding biomass productivity would be about twice as large. However, Moulton and Richards have estimated that the *total* forest ecosystem sequestering rate (including roots and soil C) could average 5.3 tC/ha/yr for a U.S. tree-planting program. Such an effort, involving up to 139 million hectares of economically marginal and environmentally sensitive crop lands and pasture lands and understocked forestlands held by private owners other than the forest industry, would have the potential for offsetting up to 56% of present U.S. CO_2 emissions.[12]

Variations on the C sequestration proposal that permit a continuing absorption of C in forests beyond maturation involve cutting down the mature trees, replanting, and either putting the harvested wood into permanent storage ("pickling the trees") or stimulating the market demand for long-lived forest products by offering a "bounty" for harvesting trees for this purpose.[16] The requirements for tree harvesting, transport, and storage will make the "tree pickling" option much more costly than basic reforestation and thus much less interesting, at least until less costly options are exhausted. The market for long-lived forest products is likely to be able to offset only a small fraction of fossil CO_2 emissions; in the period 1985-87 global consumption of sawnwood and wood-based panels averaged only 600 million cubic meters/yr,[19] with a total C content of 0.13 gigatonne/year (Gt/yr). Projected normal demand growth is in the range of 2-3%/yr to the year 2000,[20] and offering a bounty is not likely to change demand growth much. Thus present and prospective sequestering rates in long-lived forest products are small compared to the rate of anthropogenic CO_2 emissions, some 5.9 Gt C/yr in 1985 (Table 1). Sequestering of C in trees will probably be considered primarily in the form of the basic sequestering option, rather than these variations.

The cost of offsetting CO_2 emissions by sequestering C in trees is directly related to the cost of growing biomass. According to Moulton and Richards, average and marginal unit costs for a tree-growing program offsetting 56% of U.S. fossil CO_2 emissions would be \$27/tC and \$48/tC (Figure 1), respectively.[12] The annual cost of such a large-scale U.S. effort, some \$19.5 billion, might be paid for by a carbon tax of \$15/tC on all fossil fuels consumed, the effect of which would be to increase the cost of coal-based electricity generation by 0.4 cents/kwh (a 7% increase) and the cost of gasoline by 1.0 cent/liter, according to our calculation. If the sequestering rate were half the value estimated by Moulton and Richards, the required tax would be twice as large.

These costs are modest relative to the costs presently estimated for recovering and sequestering CO_2 from fossil fuel power plants. Recovering with a chemical absorption process 90% of the CO_2 from the flue gases of coal-fired steam-electric plants and piping the recovered CO_2 to, and sequestering it in, abandoned natural gas wells has been estimated to cost about \$120/tC for the Netherlands.[21] An innovative approach applicable to integrated coal gasifier/combined cycle power plants leads to an estimated cost for CO_2 removal and sequestering of a little more than \$50/tC,[21,22] which is still more than the estimated cost of sequestering C in new forests.[12]

While the cost of offsetting CO_2 emissions by sequestering C in forests is low, it is usually positive, because there are typically no offsetting credits from ancillary benefits. Some alternative strategies for reducing CO_2 emissions have negative net costs because such benefits can exceed the gross costs – e.g. investments in improving energy efficiency that obviate more costly expenditures for energy supply.[23] It has been shown in detailed studies for Sweden[24] and the Netherlands,[25] for example, that major reductions of CO_2 emissions could be achieved in those countries at negative net cost by exploiting cost-effective opportunities for improving energy efficiency. To the extent that there are negative cost opportunities for reducing or offsetting CO_2 emissions, they warrant higher priority than growing trees for C sequestration.

Fossil Fuel Substitution

The major alternative to C sequestration as a strategy for using biomass in coping with greenhouse warming is to grow biomass sustainably for energy markets, with the amount grown equal to that burned in a given period. When biomass is used this way, there is no net atmospheric buildup of CO_2, because the CO_2 released in combustion is compensated for by that extracted from the atmosphere in photosynthesis. The potential for reducing CO_2 emissions through biomass substitution depends on the fossil fuel displaced and on the relative efficiencies of converting biomass and fossil fuel into useful energy.

Suppose first that the conversion efficiencies are equal. Then each GJ of biomass substituted for fossil fuel would reduce emissions by the C content of one GJ of fossil fuel displaced – 0.014 tC, 0.019-0.020 tC, and 0.023-0.025 tC, for natural gas, petroleum, and coal, respectively. Oven-dry biomass, with a heating value of about 20 gigajoules/tonne (GJ/tonne) and a C content of 0.5 tonnes/tonne, can sequester 0.5/20 = 0.025 tC per GJ of heating value. Thus substituting biomass for coal is essentially equivalent to C sequestration, while substituting biomass for petroleum or natural gas would be less effective than C sequestration, in terms of the impact on the atmosphere of producing a tonne of biomass.

In practice the efficiencies of making useful energy will not be the same for biomass and fossil fuels. It is customary to assign much lower efficiencies to biomass. Most biomass used for energy in the world today is in the form of fuelwood, crop residues, or dung for cooking in rural areas in developing countries, at efficiencies of the order of 10% – only about a fifth of the efficiency of typical stoves fueled with natural gas or liquid petroleum gas. Further, compared to the 34-36% efficiencies achieved with modern, large-scale, 400-600 MW coal-fired steam-electric plants, typical biomass-fired steam-electric power plants have efficiencies in the range of 20-25%. The strong scale economies inherent in steam-electric power-generating technology dictate the choice of less costly alloys in boiler construction and thus to the production of lower quality steam and to lower efficiencies at plant scales of tens of megawatts, which are typically needed for biomass applications because of the dispersed nature of the biomass resource. Moreover, if liquid fuels like methanol or ethanol are produced from biomass as alternatives to gasoline in transport applications, conversion losses amount to nearly 50%,[26,27] while refinery losses in making gasoline from petroleum are only about 10%.

This outlook changes, however, if consideration is given to modern conversion technologies and future energy needs. The technologies of choice for producing electricity from biomass at modest scales in the near term are likely to be integrated gasifier/gas turbine cycles, which would offer efficiencies higher than for coal steam-electric power generation, as well as lower capital costs.[28-30] Also, if synthetic liquid fuels from biomass are considered not as alternatives to petroleum-based liquid fuels but as alternatives to synfuels derived from coal[31] – the appropriate comparison for a world faced with the declining availability of secure petroleum supplies – then the conversion efficiencies are comparable for biomass and fossil fuel feedstocks (Table 5).

Thus if biomass is considered primarily as a substitute for coal using modern conversion technologies for producing either electricity or liquid synfuels, the effect on atmospheric CO_2 would be comparable to what could be achieved with C sequestration, per tonne of biomass produced (Figure 1).

Relative Potentials for Reducing Greenhouse Warming

Biomass can play a larger role in reducing global warming when used to displace fossil fuel than when used to sequester C. This is in part because, when biomass is substituted for fossil fuel, the use of a given piece of land is not limited to just the period till the forest matures, as is the case for the basic C sequestration proposal. Additionally, the market for biomass as a substitute for fossil fuel is much larger than that in the variant of the sequestration proposal in which C is stored in long-lived forest products.

Moreover, when biomass is produced for energy markets, producers will seek to maximize the harvestable annual yield of biomass rather than the total amount of C that can be sequestered in a mature forest. This goal shift will probably lead producers to choose short-rotation woody or herbaceous crops instead of long-rotation forests. For long-rotation forests, achievable harvestable yields with present technology are about 4-8 dry tonnes/ha/yr in temperate regions and 10-12 tonnes/ha/yr in tropical areas, compared to yields for short-rotation tree crops of 9-12 tonnes/ha/yr in temperate and 20-30 tonnes/ha/yr in tropical regions.[9,32-35] Moreover, even higher yields are feasible with herbaceous crops. For example, the annual yield of sugar cane, averaged over 17 million hectares of cane harvested globally in 1987, was about 35 dry tonnes/ha/yr of above-ground harvestable plant matter (including the tops and leaves); in some countries (e.g. Ethiopia, Peru, Zimbabwe), the average yield is about twice the global average.[36] Moreover, herbaceous crops can often be grown at relatively high productivity on crop and pasture lands where the soil and climatic conditions are not especially favorable for growing trees. For example, switchgrass (*Panicum virgatum*), a perennial herbaceous crop, has been found to be relatively drought-resistant and to provide good erosion control, while offering good yields on marginal U.S. crop lands (over 10 dry tonnes/ha/yr) with relatively low levels of inputs.[37,38]

Biomass can also play a larger role in coping with greenhouse warming as a fossil fuel substitute than as a store for sequestering C because the land that can be used for energy production is not restricted to new lands for planting forests or alternative crops. In a study carried out for the Oak Ridge National Laboratory (ORNL), it was estimated that comparable contributions to total potential U.S. biomass supplies of 29.3 EJ/yr in the period beyond 2030 would come from those agricultural and forest residues that could be economically recovered in environmentally acceptable ways (8.9 EJ/yr), from growth in existing forests (9.5 EJ/yr), and from biomass energy crops (10.8 EJ/yr) (Table 2).[39]

While some biomass residues are often already being used for energy or other purposes, they could be used much more effectively with modern, energy-efficient conversion technologies. For example, in the cane sugar industry, bagasse (the residue left after crushing the cane to extract the sugar juice) is presently fully used in most parts of the sugar-producing world just to satisfy the steam and electricity requirements of sugar factories. But by employing energy-efficient steam-using equipment in the factory, by using biomass gasifier/gas turbines instead of inefficient steam turbines for electricity generation, and by using for fuel the tops and leaves of the cane plant (now often burned off just before the cane harvest) as well as the bagasse, it is feasible to increase electricity production from cane residues to more than 40-fold on-site needs, while still meeting all on-site steam requirements for sugar processing.[29] Similarly, using residues from kraft pulpmaking for gas turbine-based power generation in energy-efficient pulp mills can result in electricity production that is more than five times on-site needs.[30]

Existing forests can often also provide additional biomass for energy beyond that offered by logging residues. In many temperate zone forests, annual removals are much less than annual growth. For example, a 1980 study by the Office of Technology Assessment of the U.S. Congress estimated that net annual growth in U.S. commercial forests in the 1970s was some 400-800 million tonnes/yr, while annual harvests of "industrial roundwood" for lumber, plywood, pulp, and other forest products were only 180 million tonnes/yr.[40] When harvests are much less than growth, forest yields tend to be lower than what they might otherwise be. Moreover, much of the unharvested stock is often too low in quality for use in traditional forest products markets but is well suited for energy applications. Removal of the low-quality woodstock for energy purposes can simultaneously lead to enhanced yields of high quality wood.[40,41] The increased productivity of high quality wood in regrowth forests managed this way can help ease the pressures to exploit original-growth forests, thereby easing environmental concerns.

Existing forests can also be made more productive by full stocking with trees well suited to the sites. The Office of Technology Assessment estimated that with full stocking net annual growth of biomass on United States commercial forestland could be doubled, to 800-1600 million tonnes/yr, corresponding to an average productivity of 4-8 tonnes/ha/yr.[40]

The potential of using existing forests in the U.S. for bioenergy purposes can be estimated by assuming a biomass productivity of 6 tonnes/ha/yr on the 190 million hectares of commercial timberland (exclusive of the 14 million hectares of timberland in the U.S. that is protected by law from exploitation for environmental and other reasons and the 86 million hectares of other U.S. forest land). Potential biomass production on commercial timberland in excess of current removals (some 200 million tonnes/yr) would be 940 million tonnes/yr or 18.8 EJ/yr – equivalent in energy terms to current coal use in the U.S. Less than the full potential is likely to be exploited. The 1989 ORNL study of the U.S. bioenergy potential targeted recovering for energy about half this amount.[39]

At the global level the potential for utilizing wood from existing forests for energy is quite uncertain, owing to the paucity of data on the total productivity of the world's forests. However, Earl estimated that the annual increment of wood was 17.8×10^9 cubic meters on 3800 million hectares of global forests in 1970.[42] For comparison, the estimated global average annual wood harvests in the period 1985-87 were 3.26×10^9 cubic meters for industrial roundwood, fuelwood, and charcoal.[19] If the productivity of the world's forests today is close to Earl's estimate, some of the unused increment (having an energy content of 125 EJ/yr, equivalent to 1.27 times total world coal consumption in 1988[43]) could be recovered for energy purposes.

In practice the biomass sources used for energy will probably be a diverse mix of residues, increased production from existing forests, and wood or herbaceous crops planted for energy purposes on unforested land or understocked forested land. The appropriate mix will be determined by economics, water and land resources availability, and constraints posed by environmental and soil conservation considerations.

The Costs of Reducing Greenhouse Warming

Producing biomass for energy purposes is more costly than growing trees to sequester C because of the added costs of harvesting, processing, transport, drying, and storage. In the case of short-rotation wood crops, for example, the total cost paid for biomass at an energy conversion facility can be more than three times the cost of growing the biomass (Table 3).[44,45] However, revenues from the sale of energy produced from biomass can be taken as a credit against the cost of providing it. Here the estimated costs of reducing CO_2 emissions are presented for both power generation and liquid fuels production from biomass as alternatives to fossil fuels, using alternative technologies (Figure 1 and Tables 4 and 5).

Electricity produced from biomass in steam-electric power plants would be more costly than from coal, for biomass costing more than about $1/GJ when coal costs about $1.8/GJ, a typical expected lifecycle price for coal power plants that might be ordered in the U.S. today. The corresponding cost of fossil fuel CO_2 displacement by biomass with this technology would be greater than the cost of sequestering C in forests, except in special circumstances where biomass is available at very low cost (e.g. mill residues in the forest products industry).

In contrast, with biomass gasifer/gas turbine technologies, which are expected to be both less capital-intensive than coal steam-electric plants and to have comparable or greater efficiencies, electricity from biomass could be less costly than electricity from coal using biomass priced at more than double the coal price (Table 6). As there are likely to be substantial biomass supplies available at prices less than double the coal price, the corresponding cost of reducing CO_2 emissions would often be negative if biomass gasifier/gas turbine power were substituted for coal steam-electric power (Figure 1 and Table 4).

While the biomass versions of the gas turbine technologies considered here could be commercialized more quickly than the corresponding coal versions (because unproven sulfur removal technology is needed for coal but not for biomass), the latter might be commercialized eventually. If they were to become the norm for coal-based power generation, the biomass versions could still be competitive for biomass prices up to 20% more than the coal price, since the biomass plants would be less capital-intensive (Table 6).

The net costs of reducing CO_2 emissions through biomass substitution for fossil fuels in liquid fuels production with alternative technologies are indicated in Figure 1. Here biomass-derived methanol and ethanol are considered as alternatives to gasoline and coal-derived methanol (Table 11). As for electricity, there appear to be major opportunities for displacing fossil CO_2 emissions with biomass at negative cost. The indicated economics are especially promising for ethanol derived from lignocellulosic feedstocks (e.g. wood) using enzymatic hydrolysis.[27]

As neither the gas turbine technologies nor the alcohol technologies described here are yet commercially available, one cannot assign a high degree of precision to these cost estimates. However, the cost estimates should not be far off, at least for the biomass gasifier/gas turbine power technologies and for the biomass/methanol technologies, since there are no major technological hurdles that must be overcome in commercializing them.

The Potential for Biomass Energy in Coping with Greenhouse Warming

The global CO_2 emissions scenarios advanced by Working Group III of the Intergovernmental Panel on Climate Change (IPCC)[46] provide a useful context in which to examine the global prospects for displacing CO_2 emissions through substituting biomass for fossil fuel. Global emissions levels for three IPCC scenarios through the middle of the next century are presented in Table 1.

For the "business as usual" scenario (Scenario A), the IPCC Working Group I projects that the buildup of greenhouse gases would lead to an increase in the global average temperature at a rate of 0.3°C per decade, to 4°C above the preindustrial level by 2100.[47] For Scenario D, the most ambitious scenario considered by Working Group III for coping with greenhouse warming, CO_2-equivalent greenhouse gas emissions stabilize by 2100 at 560 ppm, double the preindustrial level of CO_2, and the global mean temperature increases 0.1°C per decade or 2°C above the preindustrial level by 2100.[47] This scenario involves a strong emphasis on energy efficiency, a shift to renewables and nuclear energy in the first half of the 21st century, and a reversal of deforestation.

Here we explore the prospects for reducing CO_2 emissions to the Scenario D levels through the use of biomass for energy. For this exercise we construct a new biomass energy-intensive Scenario D' with the same CO_2 emissions levels as Scenario D (Table 1). Our reference scenario is a variant of the IPCC Scenario B, which involves an emphasis on energy efficiency, natural gas as a low-C fossil fuel, a reversal of deforestation, and modest amounts of bioenergy. We choose this as a point of departure because energy efficiency is likely to be the most cost-effective strategy for reducing greenhouse emissions,[24,25] natural gas is widely seen as the fossil fuel of choice in the decades immediately ahead,[48] and a consensus is emerging that deforestation should be curbed, even though it might be difficult to achieve this goal. To avoid double-counting biomass in estimating the potential role of bioenergy, however, we construct for our reference scenario, Scenario B', a variant of Scenario B that involves no biomass for energy. In Scenario B' deforestation is assumed to be halted rather than reversed, and coal is substituted for the biomass used for energy in Scenario B (Table 1). If all the difference in emissions between Scenarios B' and D' were achieved with biomass substituting for coal, fossil CO_2 emissions amounting to 1.7 Gt C/yr by 2025 and 5.4 Gt C/yr by 2050 would have to be displaced (Table 1).

The emissions reduction needed by 2025 could probably be met by using for energy various industrial and agricultural residues, which are prime candidates for initial bioenergy systems. Detailed assessments indicate attractive economics in the sugar cane industries for co-producing electricity plus sugar or alcohol,[29] and in the kraft pulp industry for electricity plus pulp.[30] There are many other residues that could probably also be exploited (Tables 14 and 15).

For 2050, we assume that one-third of the targeted fossil CO_2 emissions reduction is achieved by displacing coal with residues, and two-thirds by displacing coal with biomass crops, both woody and herbaceous, grown on 600 million hectares, at an average productivity of 12 dry tonnes/ha/yr.

While much higher than the productivity of natural forests, the assumed productivity is consistent with what has been achieved to date with experimental trials and demonstrations and with limited commercial plantation experience (see earlier discussion). Considering that the era of modern scientific silvaculture began only around 1970 in both temperate and tropical zones[9] and that the growing of herbaceous crops for energy purposes is even more embryonic, at least this average productivity could plausibly be achieved on a large scale by the second quarter of the next century. For comparison, average productivities of wheat in the U.K. and maize in the U.S. have more than tripled since the mid-1940s. At present maize yields in the U.S. average 7.5 tonnes/ha/yr of grain plus an equal quantity of residues (Table 15). Moreover, the targeted annual productivity corresponds to a 0.4% efficiency for converting solar energy into recoverable biomass energy, while the practical maximum photosynthetic efficiency under field conditions is about 5%,[49] and 2.4% has been attained for Napier grass, under optimal field conditions,[50] suggesting a large potential for long-term gain.

The land area targeted for biomass energy crops in 2050 is equivalent to 15% and 40% of the amount of land now in forests and crop lands, respectively.[19] It is also equivalent to what would be in new forests by 2050 if the ambitious goal for net forest growth of 12 million hectares at the beginning of the next century, agreed to in the November 1989 Nordwijk Declaration,[51] were realized.

Houghton has estimated that 500 million hectares of land in Africa, Asia, and Latin America could be available for reforestation.[3] His criteria for availability were that the land (i) had supported forests in the past, and (ii) was now unused for crop lands or settlements. He estimated that an additional 365 million hectares of land in the fallow cycle of shifting cultivation might also be targeted for reforestation. Independently, Grainger has estimated that some 758 million hectares of degraded lands are available for reforestation.[52] Moreover, some of the world's 1500 million hectares of tropical grasslands might be used for biomass energy crops (e.g. growing perennial grasses). At present about 750 million hectares of these grasslands are burned off each year,[53] and some of this land may be amenable to different management practices if benefits were to accrue to the local populace. While the various estimates of available land are quite uncertain, they suggest that large areas may be available for energy crops in tropical areas.

Considerable land might also be available for energy crops in industrialized countries. In the European Community over 15 million hectares of crop land would have to be taken out of production if agricultural surpluses and Community expenditures on agricultural subsidies were to be brought under control.[34]

In the U.S., 30 million hectares of crop land were idled in 1988 to reduce production or conserve land.[6] The land available for biomass production could be considerably greater than this. About 43 million hectares of crop lands have erosion rates exceeding the maximum rate consistent with sustainable production;[12] shifting this land from annual food crops to various perennial energy crops could greatly reduce erosion. An additional 43 million hectares of crop land have "wetness" problems – poor drainage, high water tables, or flooding; when used for ordinary agriculture these lands could potentially contribute to surface- and groundwater pollution[12] – problems that could be eased with the production of some types of energy crops as alternatives. Moreover, the amount of idle crop land might increase substantially. A 1987 report of the New Farm and Forest Products Task Force estimated that over the next quarter century new crops will be needed for some 60 million hectares of existing crop land.[54] There are also 60 million hectares now in pasture, range, and forest considered capable of supporting biomass production for energy.[27]

The contribution of biomass from energy crops could be reduced either by greater use of biomass residues or by the extraction, with improved management, of additional biomass from existing forests. If the global emissions reduction of Scenario D' in the middle of the next century were achieved with equal shares from residues, energy crops, and existing forests (like the ORNL estimate of potential U.S. biomass supplies[39]), existing forests would contribute for energy an amount of biomass equivalent to about half of the annual increment[42] in excess of current removals,[19] and thus the assumed contribution from energy crops would be half as large. However, because of the uncertainties in forest statistics worldwide, we have not included in Scenario D' a contribution from wood from existing forests.

We conclude that the CO_2 emissions levels of Scenario D could plausibly be achieved without exploiting low-C energy supplies other than natural gas and biomass. It might be feasible to reduce emissions further by exploiting other renewable energy technologies for which the prospects are auspicious,[27,55] as recognized implicitly by Working Group III in formulating Scenario D.

Toward Sustainable Biomass Production

If biomass is to play a major role in the energy economy, strategies for sustaining high yields over large areas and long periods are needed. The experience of sustaining high sugar cane yields over centuries in the Caribbean and in countries like Brazil suggests that this will be feasible, but good management practices and new research are required to achieve this wider goal.

Achieving sustainable production and maintaining biological diversity may require polycultural strategies (e.g. mixed species in various alternative planting configurations) for biomass production in many areas. Biomass energy systems can usually accommodate a variety of feedstocks. At present, however, monocultures are favored for energy crops, in large part because management techniques in use today tend to be adapted from monocultural systems for agriculture. Polycultural management techniques warrant high priority in energy crop research and development.

While net biomass energy yields for short rotation tree crops are typically twelve times energy inputs,[56] it is desirable, both economically and environmentally, to try to reduce energy inputs. For example, the nutrient status of afforested lands might be maintained by recycling nutrients and by choosing suitable mixed species and clones.[57,58] The promise of such strategies is suggested by 10-year trials in Hawaii, where yields of 25 dry tonnes/ha/yr have been achieved without N-fertilizer when *Eucalyptus* is interplanted with N_2-fixing *Albizia* trees.[59]

Research can lead not only to improvements in present techniques for producing energy crops but also to new approaches. For example, long-term experiments in Sweden have shown that: (i) in most forests trees grow at rates far below their natural potential, (ii) nutrient availability is usually the most important limiting factor, and (iii) optimizing nutrient availability can result in four- to six-fold increases in yield. Under nutrient-optimized conditions all tree species investigated have behaved similarly to C_3 crop plants, with about the same total biomass yield per unit of light intercepted by the leaves during the growing season.[60] Growing trees under nutrient-optimized conditions thus could make it possible to achieve high yields with existing species and clones, thus facilitating the incorporation of pest resistance and other desirable characteristics, and the maintenance of a diverse landscape mosaic. To the extent that crop lands and wastelands would be converted to energy crops this way, it may be feasible not only to maintain but to improve biological diversity. An additional advantage of pursuing non-nutrient-limited production strategies is that the trees thus produced shift a percentage of their increased overall yield from roots to above-ground production—again similarly to the experience with agricultural crops.[35]

Nutrient-induced yield increases can be achieved without nutrient leaching when good forest management is practiced. But achieving sustainable high yields this way requires implementing techniques being developed for matching nutrient applications to the time-varying need for nutrients.[60,61]

Achieving high levels of biological diversity will also require maintaining some of the land in biomass-producing regions in a "natural" condition. For example, some bird species require for survival dead wood and the associated insect populations. Experience in Swedish forests suggests that maintaining a relatively modest fraction of forest area in such natural reserves is adequate to maintain a high level of bird species diversity.[62] Research is needed to understand how best to achieve desirable levels of biological diversity under the wide range of conditions under which biomass might be grown for energy.

While major expansions are needed for research efforts relating to sustainable biomass production, there is time for the needed research and extensive trials, because major bioenergy industries can be launched in the decades immediately ahead using as feedstocks primarily residues from the agricultural and forest products industries.

Developments Needed in Biomass Energy Conversion Technology

Research and development (R&D) are needed on converting biomass efficiently and cost-effectively into modern energy carriers, if biomass is to play a major role in the global energy economy.

While there has been relatively little R&D on biomass energy conversion, there has been considerable effort aimed at "modernizing coal" through thermochemical conversion, for both electricity and fluid fuels applications. Some of this coal conversion technology can be adapted to biomass.

For the near term the prospects are auspicious for commercializing biomass gasifer/gas turbine power-generating technologies designed originally for coal. While commercially ready coal gasifer/gas turbine technologies cannot provide electricity at lower cost than existing coal steam-electric power systems, simplified versions under development offer the potential for substantially lower cost.[63,64] Such simplified technologies could probably be commercialized more quickly for biomass than for coal, because biomass contains negligible sulfur, the cost-effective removal of which is the major technological hurdle that must be overcome before these technologies can be commercialized for coal.[28-30] Recently, a Finnish/Swedish consortium announced plans to build a demonstration plant in Sweden with such technology and have it running in two to three years.[65]

For the longer term, power generation R&D should focus on technologies well matched to the characteristics of biomass. Gasifiers should be designed to exploit the fact that biomass is much more reactive and thus easier to gasify than coal. Power-generating technologies other than gas turbines should also be developed – e.g. advanced fuel cells for applications at smaller scales than the 5-100 MW scales for which gas turbines are well suited.

Methanol can be derived from biomass using thermochemical conversion technology like that used for coal. While methanol is likely to be less costly from biomass than from coal in small-scale plants,[66] methanol can be produced from coal in plants of much larger capacity, giving rise to scale economies that cannot practically be exploited with biomass, owing to the dispersed nature of the biomass resource. Alternative liquid fuel technologies designed to exploit the unique characteristics of biomass – e.g. technologies based on biological processes – might be able to compensate for this scale disadvantage.

Fuel ethanol is produced from sugar cane via fermentation on a large scale in Brazil. Though with present technology this ethanol is not competitive at the pre-August 1990 world oil price, the co-production of electricity from cane residues using gasifier/gas turbine power generating technologies at alcohol distilleries could make the ethanol competitive even at this low oil price.[29] For temperate climates, the production of ethanol from low-cost lignocellulosic feedstocks (e.g. wood) via enzymatic hydrolysis techniques is promising. Analyses carried out at the U.S. Solar Energy Research Institute (SERI) suggest that, with emphasis on R&D, ethanol produced this way could be competitive with gasoline from petroleum by the turn of the century for biomass costing less than $3/GJ (Table 11 and Figure 1).[27]

Finally, R&D on the growing, harvesting, and preparation of biomass feedstocks should be coordinated with the R&D on biomass conversion.[67] It may often be possible to substantially reduce costs for costly items (e.g. biomass drying), as well as overall costs, by taking a systems approach to development.

Industrial Infrastructure Issues

Fully exploiting the biomass energy potential will probably require evolving industries quite different from those that now provide energy, because biomass energy systems would be different from the energy systems now in place—they would be rural-based, relatively labor-intensive, variable from region to region, and more decentralized. Structurally, these industries would have characteristics of today's agricultural and forest products industries, as well as of today's energy industries. Public policy changes may well be needed to facilitate their orderly development.

While articulation of the needed policies is beyond the scope of the present analysis, these changes could probably be brought about by creatively using familiar policy instruments. For example, general policies promoting co-generation and power from renewable energy sources, like the 1978 Public Utility Regulatory Policies Act (PURPA) in the U.S., could be helpful in nurturing a biomass-based power industry. The expansion of biomass-based power generation in the U.S., from about 250 MW in 1980[68] to some 9,000 MW in 1990[27] was due in large part to the influence of this Act. Likewise, policies aimed at removing agricultural subsidies and simultaneously providing interim incentives to farmers to shift production to biomass for energy[34] could be quite helpful in nurturing bioenergy industrial development.

Conclusion

Biomass strategies are attracting considerable attention as options for coping with greenhouse warming. While, to date, emphasis has been on planting trees to sequester carbon, the growing of biomass for energy provided by modern energy conversion systems would enable biomass to play much wider roles. Though C-sequestering strategies will be important where the produced biomass cannot be practically harvested for energy (e.g. in areas remote from energy markets or on steep slopes) or where the creation of new forest reserves is deemed desirable for environmental or economic reasons, biomass energy strategies will usually be preferred. Moreover, since biomass energy will often be less costly than fossil fuel energy, biomass energy strategies will be inherently easier to implement than many other proposed strategies for coping with greenhouse warming.

The techniques and technologies for growing biomass and converting it into modern energy carriers must be more fully developed, and new industrial infrastructures must be evolved in order to realize the full potential for bioenergy. Despite such challenges, bioenergy industries could be launched in the decades immediately ahead, starting off using residues from agriculture and forest product industries. Initially, biomass could be converted into modern energy carriers using technologies developed for coal that could be adapted to biomass with little incremental effort. If at the same time the R&D needed on the sustainable production and conversion of biomass is given high priority, and if policies are adopted to nurture the development of bioenergy industries, these industries will be able to innovate and diversify as they grow and mature.

Methodology for the Calculations Presented in the Figures and Tables

The calculations presented in the following figures and tables were carried out on a self-consistent basis. All costs are presented in 1989 dollars. Where costs were originally presented in the dollars of other years, they were converted to 1989$ using the US GNP deflator. Fuel energy is presented in terms of the higher heating value (HHV).

For electricity production, the costs are evaluated assuming a 6.1% real discount rate [the value recommeneded by the Electric Power Research Institute (EPRI) for evaluating utility investments], an insurance rate of 0.5% per year, and a 30-year system life. In Tables 7-9, *two values are shown* (in the form A/B) for the fixed capital charges and busbar costs: A, with a property tax rate of 1.5% of the initial capital cost per year and a 38% corporate income tax rate, and existing tax preferences [corresponding to an annual capital charge rate of 0.1030 for fossil fuel systems and 0.1007 for renewable and nuclear systems (EPRI, Technical Assessment Guide, Palo Alto, CA, 1986)]; and B, with zero corporate income and property taxes [corresponding to an annual capital charge rate of 0.0784]. The latter capital charge rate is used when evaluating the cost of CO_2 emissions offsets.

The schedule of fixed capital expenditures during construction of power plants is assumed either to reflect average experience, or, if relevant experience is not available, equal annual payments are asumed for an idealized plant construction period, as recommended by the EPRI. For the latter case, interest charges during construction, as a fraction of the fixed overnight construction cost is given by:

$$IDC = [(1 + i)^g/g]/CRF(i,g) - 1,$$

where

$$i = \text{discount rate},$$
$$g = \text{idealized construction period, in years},$$
$$CRF(i,g) = i/[1 - (1 + i)^{-g}].$$

Biomass fuel costs were evaluated using a 5% real discount rate, while a 10% real discount rate was used for evaluating the costs of liquid synthetic fuels production.

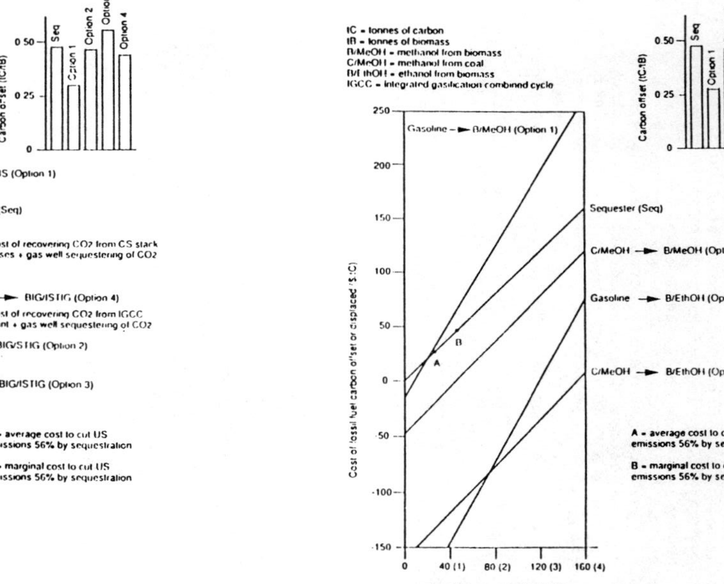

Figure 1. Alternative ways to use biomass to cope with greenhouse warming. Sequestering carbon in forests is compared to alternative strategies for substituting biomass for coal in power generation (left) and to alternative strategies for substituting biomass for fossil fuels in liquid fuels production (right).

The top histograms show the CO_2 emissions reduction (in tonnes C/dry tonne of produced biomass) for sequestering and for four alternative options in which biomass is substituted for fossil fuels. The lower graphs show the cost of fossil CO_2 emissions offset or displaced (in $/tonne C) versus the cost of biomass for these alternative strategies. On the lines labeled "sequester," A = the average cost and B = the marginal cost of offsetting US CO_2 emissions 56% through tree-planting on 139 million hectares, at an average total forest ecosystem sequestering rate of 5.3 tonnes C/ha/yr, as estimated by Moulton and Richards (see text). The cost of fossil fuel displaced is $(C_b - C_f)/E_f$, where C_b (C_f) is the unit cost of the energy output from the biomass (fossil fuel) system, and E_f is the CO_2 emission rate of the fossil fuel displaced (in tonnes of C per unit of energy output). The carbon dioxide emissions avoided are assumed to be 23 kg C/GJ of input coal (HHV basis) and 0.76 kg C/liter of gasoline displaced.

On the left (see Tables 4, and 6-9 for cost and performance estimates of alternative technologies):
- Option 1: Substitute biomass-fired steam-electric power for coal-fired steam-electric power.
- Option 2: Substitute biomass-integrated gasifier/STIG power for coal-fired steam-electric power.
- Option 3: Substitute biomass-integrated gasifier/ISTIG power for coal-fired steam-electric power.
- Option 4: Substitute biomass-integrated gasifier/ISTIG power for coal-integrated gasifier/ISTIG power.

On the right (see Tables 5 and 10-11 for cost and performance estimates of alternative technologies):
- Option 1: Substitute methanol from biomass for gasoline @ 25 cents/liter (wholesale price projected for 2000 in the US).
- Option 2: Substitute methanol from biomass for methanol from coal.
- Option 3: Substitute ethanol from biomass for gasoline @ 25 cents/liter.
- Option 4: Substitute ethanol from biomass for methanol from coal.

Table 1. Alternative Global CO_2 Emissions Scenarios[a-f] (10^9 tonnes of C/year)

	Commercial Energy					Deforestation[g]			Cement		Total			
Year	A	B	B′	D	D′	A	B=D	B′=D′	A	B=B′=D=D′	A	B	B′	D=D′
1985	5.1	5.1	5.1	5.1	5.1	0.7	0.7	0.7	0.1	0.1	5.9	5.9	5.9	5.9
2000	6.5	5.6	5.6	5.7	5.4	1.0	-0.2	0.0	0.2	0.2	7.7	5.5	5.8	5.6
2025	9.9	6.6	6.6	5.4	4.9	1.4	-0.5	0.0	0.2	0.2	11.5	6.4	6.8	5.1
2050	13.5	7.6	8.1	3.0	2.7	1.4	-0.3	0.0	0.3	0.2	15.2	7.5	8.3	2.9

[a] Scenarios A, B, and D, developed by Working Group III of the Intergovernmental Panel on Climate Change (Table 8, Appendix, in Intergovernmental Panel on Climate Change, "Formulation of Response Strategies," Report prepared for IPCC by Working Group III, June 1990) are for the averages of the high and low economic growth variants of the scenarios developed by this Working Group. Due to rounding, totals do not always equal the sums of the components.

[b] Scenario A, the "Business as Usual" scenario: the energy supply is coal-intensive; only modest increases in energy efficiency are achieved; deforestation continues until the tropical forests are depleted.

[c] Scenario B: the supply mix shifts toward low-C fuels, notably natural gas; there are large increases in energy efficiency; deforestation is reversed.

[d] Scenario D: the measures of Scenario B are complemented by a shift to renewables and nuclear power in the first half of the next century, to the extent that emissions remain stable near the 2.9 Gt C/yr level after 2050.

[e] Scenario B′, developed by the authors (see text): like Scenario B, except deforestation is halted, not reversed, and, in 2050, 23.3 EJ/yr of coal, with a CO_2 emission rate of 0.5 Gt C/yr, is substituted for the 23.3 EJ/yr of biomass energy in Scenario B.

[f] Scenario D′, developed by the authors (see text): the same total emissions as Scenario D; the difference in emissions between Scenarios B′ and D′ (1.7 Gt C/yr in 2025 and 5.4 Gt C/yr in 2050) is achieved entirely by substituting biomass for fossil fuel (Table 1b).

[g] The contribution of deforestation to global emissions in 1985 assumed by Working Group III (Intergovernmental Panel on Climate Change, "Formulation of Response Strategies," Report prepared for IPCC by Working Group III, June 1990) in the construction of its scenarios is lower than many other estimates. In its report assessing the scientific aspects of greenhouse warming, Working Group I assigned to deforestation a value of 1.6 ± 1.0 Gt C/yr for the 1980s (Chapter i, in Climate Change: the IPCC Scientific Assessment, J.T. Houghton, G.J. Jenkins, and J.J. Ephraums, eds., Cambridge University Press, Cambridge, 1990).

Table 2. Potential Biomass Supplies for Energy in the US, as Estimated by the Oak Ridge National Laboratory[a]

Feedstock	Net Raw Biomass Resource[b,c] (EJ/year)	Cost[c] ($/GJ) Current	Target
Residues			
Logging Residues	0.8	> 3	< 2
Urban Wood Wastes and Land Clearing	1.2	2	2
Forest Manufacturing Residues	2.1	1	<1
Environmentally Collectible Agricultural Residues	2.0	1-2	1
Municipal Solid Waste and Industrial Food Waste	2.4	2-3	< 1.5
Animal Wastes	0.5	< 4	3.5
Subtotal	8.9		
Biomass from Existing Forest			
Commercial Forest Wood	4.5	< 2	< 2
Improved Forest Management	4.5		< 2
Shift 25% of Wood Industry to Energy	0.5	2	2
Subtotal	9.5		
Biomass from Energy Crops			
Agricultural Oil Seed	0.3		
Wood Energy Crops	3.2	3	2
Herbaceous Energy Crops			
Lignocellulosics	5.5	4	2
New Energy Oil Seed	0.4		
Aquatic Energy Crops			
Micro-Algae	0.3		
Macro-Algae	1.1	3.5	2
Subtotal	10.8		
Total	29.3[b]		

[a] Source: Table 2.4-3, page 85, in W. Fulkerson et al., Energy Technology R&D: What Could Make a Difference? A Study by the Staff of the Oak Ridge National Laboratory, vol. 2, Supply Technology, ORNL-6541/V2/P2, December 1989.

[b] These are biomass supplies net of estimated losses in production and handling, before conversion to fluid fuels or electricity.

Table 3. Delivered Cost of Wood Chips from Populus Plantation Systems ($/ODT)

Production Cost[a,b,c]	
Establishment[d]	5.27
Land rent[e]	6.43
Maintenance[f]	
Insecticides/Fungicides	0.93
Fertilizer	1.07
Management	2.64
Land Taxes	0.96
SUBTOTAL	17.30
Harvesting[g,h]	
Harvester, Tractor	4.58
Baler	3.87
SUBTOTAL	8.45
Transport[g]	
Loader/Unloader	4.46
Tractor/Trailer[i]	5.15
SUBTOTAL	9.61
Chipper/Conveyor[g]	3.15
Storage/Drying[g]	
Storage[j]	6.77
Drying[k]	11.08
SUBTOTAL	17.85
TOTAL	56.36 ($2.90/GJ[l])

[a] *For short-rotation populus on good-quality agricultural land. Based on the use of a production model incorporating findings from the US DOE Short-Rotation Woody-Crop Program (C.H. Strauss and L.L. Wright, "Woody Biomass Production Costs in the United States: An Economic Summary of Commercial Populus Plantation Systems,"* Solar Energy, *45(2), pp. 105-110, 1990).*

[b] *The levelized production cost is given by:*

$$[CRF(i,N)*E + i*L + M]/\{i*Y_t/[(1+i)^t - 1]\},$$

where

i = discount rate = 0.05
N = plantation life = 12 years (two rotations)
$CRF(i,N)$ = capital recovery factor = $i/[1 - (1+i)^{-N}]$ = 0.1128
t = rotation period = 6 years
L = land price = $1800/ha
E = plantation establishment cost = $654/ha
M = annualized maintenance cost = $78.5/ha/yr
Y_t = yield at harvest = 95 ODT/ha

[c] *While the average annual yield is 95/6 = 15.8 t/ha/yr, the levelized yield used in the economic analysis is:*

$$i*Y_t/[(1+i)^t - 1] = 0.1470*95 = 14.0 \text{ tonnes/ha/yr}.$$

Notes to Table 3, cont.

[d] *The establishment cost includes mowing/brushing, plowing, herbicides, liming, fertilization, planting.*

[e] *The land rent (i*L) is for a land price of $1800/ha (typical for a good corn production site).*

[f] *The maintenance costs include (i) insecticides, fungicides applied every other year beginning in year 2 @ a cost of $26/ha/application, corresponding to an annual levelized cost of $13/ha/yr; (ii) fertilizers applied every other year beginning in year 3 @ a cost of $37/application, corresponding to an annual levelized cost of $15/yr); (iii) management @ $37/ha/yr, and (iv) land taxes @ 0.75% of the land price per year or $13.5/ha/yr.*

[g] *Source: C.H. Strauss, S.C. Grado, P.R. Blankenhorn, and T.W. Bowersox, "Economic Valuations of Multiple Rotation SRIC Biomass Plantations," Solar Energy, 41(2), pp. 207-214 (1988).*

[h] *For a harvesting strategy in which trees are cut, crushed, field-dried, and baled before loading and transport to the storage/conversion site. [It has been found that for bolts of crushed wood averaging 10 cm in diameter, moisture contents (wet basis) have dropped from 50% to 20-30% after 6 days in the field (P.E. Barnett, "Evaluation of Roll Splitting as an Alternative to Chipping Woody Biomass," in Biomass Energy Research Conference, University of Florida, Gainsville, March 12-14, 1985. Crushing tree-length stems with diameters up to 18 cm at a rate of 14 m/minute requires only modest amounts of energy--some 0.88 kWh/tonne (C. Ashmore, "Preliminary Analysis of Roll Crushing of Hybrid Poplar Using the FERIC Roll Crusher," unpublished, 1985).]*

[i] *Round-trip truck transport costs for a conversion facility located 40 km from the harvesting site.*

[j] *For 6 months of storage, with the wood covered by heavy polyethylene film.*

[k] *Drying with unheated, forced-air system, based on a study by Frea (W.J. Frea, "Economic Analysis of Systems to Pre-Dry Forest Residues for Industrial Boiler Fuel," Energy from Biomass and Wastes VIII, D.L. Klass, ed., Institute of Gas Technology, 1984).*

[l] *Poplar has a heating value of 19.38 GJ/tonne (HHV basis).*

Table 4. When 1 Tonne of Wood[a] Displaces Coal[b] in Power Generation:[c]

Technology Shift:	Coal Energy Displaced (GJ)	CO_2 Emissions Displaced (tonnes C/tonne wood)	Cost of Displaced CO_2 ($ per tonne C)
Option 1 CS-->BS	13.31	0.306	- $61.44 + 63.33*$P_b$
Option 2 CS-->BIG/STIG	20.61	0.474	- 124.88 + 40.89*P_b
Option 3 CS-->BIG/ISTIG	24.37	0.560	- 141.89 + 34.61*P_b
Option 4 CIG/ISTIG-->BIG/ISTIG	19.75	0.454	- 94.13 + 42.69*P_b

Biomass-Based Electricity Production with Alternative Technologies[a]

	Heat Rate (MJ/kWh)	Busbar Cost (cents/kWh)
BS = 27.6 MW Steam-Electric Plant[d]	15.36	3.60 + 1.536*P_b
BIG/STIG = 2 x 51.5 MW BIG/STIG Plant[d]	9.92	2.06 + 0.992*P_b
BIG/ISTIG = 111 MW BIG/ISTIG Plant[d]	8.39	1.65 + 0.839*P_b

Coal-Based Electricity Production with Alternative Technologies[b]

	Heat Rate (MJ/kWh)	Busbar Cost (cents/kWh)
CS = 2 x 500 MW Steam-Electric Plant w/AFBC[d]	10.55	5.09
CIG/ISTIG = 109 MW CIG/ISTIG Plant[d]	8.55	3.50

[a] Here biomass is poplar with HHV (LHV) = 19.38 (18.17) GJ/dry tonne, containing 25 kg C/GJ (HHV basis). P_b is the wood price, in $/GJ.

[b] For Illinois #6 coal with HHV (LHV) = 29.6 (28.5) GJ/dry tonne, a C content of 23 kg/GJ (HHV basis), and for delivered coal costing $1.83/GJ (West North Central Region, US).

[c] See Figure 1 for graphical presentation.

[d] See Table 6.

Table 5. When 1 Tonne of Wood[a] Displaces Fossil Fuel-Based Liquid Fuels:[b]

Technology Shift:	Fossil Fuel Displaced (GJ)	CO_2 Emissions Displaced (tonnes C/tonne wood)	Cost of Displaced CO_2 ($ per tonne C)
Option 1 G-->B/MeOH	14.0	0.278	- $17.7 + 69.8*$P_b$
Option 2 C/MeOH-->B/MeOH	20.1	0.462	- $47.7 + 42.0*$P_b$
Option 3 G-->B/EthOH	13.2	0.264	- $214.1 + 72.1*$P_b$
Option 4 C/MeOH-->B/EthOH	19.5	0.448	- $161.8 + 42.5*$P_b$

Alcohol from Wood with Alternative Technologies[a]	Efficiency (%, HHV)	Production Cost (cts/l, gasoline-equiv.)[c]
B/MeOH = MeOH from biomass[d] (IGT fluidized bed gasifier)	57.7	23.85 + 5.32*P_b
B/EthOH = EthOH from biomass[e] (enzymatic hydrolysis of wood)	53.5	8.90 + 5.49*P_b
Fossil Fuel-Based Liquid Fuels		
G = Gasoline from petroleum, 2000[f]	90.0	25.2
C/MeOH = MeOH from coal,[g] (Texaco, entrained-flow gasifier)	55.7	29.9

[a] *Here biomass is poplar with HHV (LHV) = 19.38 (18.17) GJ/dry tonne, containing 25 kg C/GJ (HHV basis). P_b is the wood price, in $/GJ.*

[b] *See Figure 2 for graphical presentation.*

[c] *Assuming 1 GJ of alcohol is equivalent to 1.2 GJ of gasoline, so that 1 liter of MeOH (EthOH) is worth 0.59 liters (0.80 liters) of gasoline.*

[d] *From Table 11 the cost is (14.07 + 3.14*P_b)/0.59 cents/liter.*

[e] *From Table 11 the cost is (7.12 + 4.39*P_b)/0.80 cents/liter.*

[f] *The US wholesale gasoline price, as projected for 2000 by the US Dept. of Energy [Energy Information Administration, Annual Energy Outlook 1990 with Projections to 2010, DOE/EIA-0383(90)].*

[g] *For coal costing $1.58/GJ, the cost per liter is (12.48 + 3.25*1.58)/0.59 (Table 11).*

Table 6. Busbar Costs for Alternative Power Technologies[a] (in 1989 cents/kWh)

	CS[d]	BS[e]	CIG/STIG[f]	BIG/STIG[g]	CIG/ISTIG[f]	BIG/ISTIG[g]
Fuel[a]	$1.055*P_c$	$1.536*P_b$	$1.011*P_c$	$0.992*P_b$	$0.855*P_c$	$0.839*P_b$
Variable O&M	0.72	0.50	-0.16	0.10	- 0.13	0.09
Fixed O&M	0.32	0.80	0.86	0.62	0.73	0.52
Capital	2.12	2.30	1.68	1.34	1.34	1.04
Total	3.16 + $1.055*P_c$	3.60 + $1.536*P_b$	2.38 + $1.011*P_c$	2.06 + $0.992*P_b$	1.94 + $0.855*P_c$	1.65 + $0.839*P_b$
Example:						
P_c = \$1.8/GJ[b]	5.1		4.2		3.5	
P_b = \$2.9/GJ[c]		8.1		4.9		4.1

[a] P_c = biomass price, and P_b = biomass price, in \$/GJ (HHV basis); O&M = operation and maintenance cost. The capital charges presented here are for the case with zero corporate income and property taxes.

[b] The levelized price of coal, 2000-2030, delivered to utilities in the West/North Central United States, as projected by the US Dept. of Energy.

[c] The delivered cost of wood chips from short rotation populus tree crops, including the costs of 40 km transport, drying, and 6-months storage (Table 3).

[d] CS = a subcritical, coal-fired steam-electric plant (two 500 MW units) with atmospheric fluidized bed combustors, a 10.55 MJ/kWh heat rate, an installed capital cost of \$1610/kW, and a 68% capacity factor. See Table 7.

[e] BS = a 27.6 MW biomass-fired steam-electric plant, having a 15.36 MJ/kWh heat rate, an installed capital cost of \$1925/kW, and a 75% capacity factor. See Table 8.

[f] CIG/STIG = a coal-integrated gasifier/steam-injected gas turbine and CIG/ISTIG = a coal-integrated gasifier/intercooled steam-injected gas turbine. Both systems use an air-blown, pressurized, fixed-bed gasifier with hot-gas cleanup. The CIG/STIG plant consists of two 50.5 MW units; its heat rate is 10.11 MJ/kWh; its installed capital cost, \$1410/kW. The CIG/ISTIG plant consists of one 109.1 MW unit; its heat rate is 8.55 MJ/kWh; its installed capital cost, \$1120/kW. The capacity factor is assumed to be 75%. See Tables 9 and 10.

[g] BIG/STIG = a biomass-integrated gasifier/steam-injected gas turbine and BIG/ISTIG = a biomass-integrated gasifier/intercooled steam-injected gas turbine. The cost/performance characteristics of these systems are based on the corresponding coal designs (note f), without the hot-gas sulfur removal technology, which is not needed for biomass. A BIG/STIG plant consists of two 51.5 MW units; its heat rate is 9.92 MJ/kWh; its installed cost, \$1120/kW. A BIG/ISTIG plant consists of one 111.2 MW unit; its heat rate is 8.39 MJ/kWh; its installed cost, \$875/kW. The capacity factor is assumed to be 75%. See Tables 9 and 10.

Table 7. Electricity Cost for New US Coal-Fired Steam-Electric Plants w/AFBC (cents/kWh)

Coal[b]	1.93
Variable O&M	0.72
Fixed O&M	0.32
Capital[c]	2.79/2.12
Total	5.76/5.09

[a] Assuming EPRI estimates for heat rate (10.55 MJ/kWh), overnight construction cost ($1169/kW), other capital ($81/kW), and O&M costs. East or West Central US siting, for a subcritical plant (two 500 MW units) with atmospheric fluidized bed combustors (Electric Power Research Institute, Technical Assessment Guide, Palo Alto, CA, 1986).

[b] For a 30-year levelized coal price of $1.83/GJ, appropriate for the West/North Central region of the US, 2000-2030, as projected by the US Dept. of Energy.

[c] Following a US Department of Energy analysis of steam-electric power plants [Energy Information Administration, Annual Outlook for US Electric Power: Projections Through 2010, DOE/EIA-0474(90), June 14, 1990], it is assumed that the average capacity factor is 68% and that construction profile for these plants is:

Year before plant begins operating	Annual expenditure as % of overnight construction cost
8	4.0
7	14.0
6	33.0
5	34.0
4	11.0
3	3.0
2	1.0

so that interest during during construction adds 31% to the overnight construction cost.

Table 8. Cost of Electricity from Biomass-Fired Steam-Electric Plant[a] *(cents/kWh)*

Fuel[b]	$1.536 * P_b$
Variable O&M	0.50
Fixed O&M	0.80
Capital	2.95/2.30
Busbar Cost	$1.536 * P_b$ + (4.25/3.60)

[a] *Based on an EPRI design for a 24 MW condensing/extraction cogeneration plant producing 20,430 kg/hour (45,000 lb/hour) of steam at 11.2 bar (165 psia) for process (Electric Power Research Institute, Technical Assessment Guide, Palo Alto, CA, 1986). Here it is assumed that this steam is instead condensed, thus producing an additional 3.6 MW of electric power.*

[b] *For a heat rate of 15..6 MJ/kWh [corresponding to steam conditions of 86 bar (1265 psia) and 510 °C (950 °F) at the turbine inlet and a turbine efficiency of 80%]. Here P_b is the price of biomass, in $ per GJ.*

[c] *Assuming EPRI values for the overnight construction cost ($1693/kW), other capital ($127/kW), and idealized construction period (3 years).*

[d] *Assuming a 75% capacity factor and equal annual capital expenditures during the construction period.*

Table 9. Estimated Busbar Cost for IG/STIG and IG/ISTIG Power Plants Fueled with Coal and Biomass (in cents/kWh)

	CIG/STIG[a]	BIG/STIG[b]	CIG/ISTIG[a]	BIG/ISTIG[b]
Fuel[a]	$1.011*P_c$	$0.992*P_b$	$0.855*P_c$	$0.839*P_b$
Operating Labor[b]	0.30	0.20	0.28	0.19
Maintenance[c]	0.42	0.32	0.33	0.24
Administrative costs[d]	0.14	0.10	0.12	0.09
Water requirements[e]	0.028	0.028	0.026	0.026
Catalysts and binder[f]	0.018	-	0.016	-
Solids disposal[g]	0.071	0.069	0.060	0.059
H_2SO_4 byproduct credit[h]	- 0.273	-	- 0.231	-
Capital[i]	2.21/1.68	1.72/1.34	1.76/1.34	1.34/1.04
Totals	2.91/2.38 + $1.011*P_c$	2.44/2.06 + $0.992*P_b$	2.36/1.94 + $0.855*P_c$	1.95/1.65 + $0.839*P_b$

[a] Here P_c and P_b are the prices for delivered coal and biomass feedstocsk, respectively, in \$/GJ. Heat rates for CIG/STIG (@ 101.0 MW_e) and CIG/ISTIG (@ 109.1 MW_e) are 10.11 MJ/kWh and 8.55 MJ/kWh, respectively (J.C. Corman, "System Analysis of Simplified IGCC Plants," General Electric Company, Schenectady, NY, Report on Department of Energy Contract No. DE-AC21-80ET14928, September 1986). The output and performance of the biomass versions of these systems are estimated by starting with the coal systems and modifying them to account for the major differences arising from operation on biomass. The biomass gasification efficiency is assumed to be the same as the coal gasification efficiency. One difference is that only about 40% as much high pressure steam is needed to gasify a GJ of biomass as a GJ of coal, and the steam not needed for gasification can be injected into the turbine. However, this is not likely to have a significant effect on overall performance, since the injection of high-pressure steam into the combustor gives rise to approximately the same mass flow through the turbine and would require the same steam heating in the combustor as injection into the gasifier. An important difference, however, is that some low-pressure steam needed for the sulfur recovery unit with coal is not needed in the biomass systems. Here it is assumed that this low-pressure steam is injected into the turbine to increase power output and efficiency. As a result, the output and heat rate of the BIG/STIG are 103.0 MW_e and 9.92 MJ/kWh, while the corresponding quantities for BIG/ISTIG are 111.2 MW_e and 8.39 MJ/kWh, respectively.

[b] The coal-based systems required 3 operators for the gasification system, 4 for the hot-gas cleanup, and 3 for the power plant. At \$22.55 per hour, operating labor costs for the coal systems are \$1.977 million per year. Because hot-gas desulfurization is not needed for the biomass systems, it is assumed that 7 operators are needed for the biomass systems--four less because hot gas desulfurization is not needed and one more because of increased fuel handling requirements. Thus annual operating labor costs would be \$1.384 million.

[c] Annual maintenance costs (40% labor and 60% materials) are estimated to be

Notes for Table 9, cont.

$2.812 million for CIG/STIG (including $0.634 million for chemical hot-gas cleanup) and $2.342 million for CIG/ISTIG (including $0.591 million for chemical hot-gas cleanup). The corresponding values for BIG/STIG and BIG/ISTIG, without chemical hot gas cleanup, are $2.178 million and $1.751 million, respectively.

[d] Annual administrative costs, assumed to be 30% of O&M labor, are $0.930 million for CIG/STIG, $0.874 million for CIG/ISTIG, $0.677 million for BIG/STIG, and $0.625 million for BIG/ISTIG.

[e] Raw water costs are $0.189 million per year for all systems.

[f] Annual catalysts and binder costs $0.121 million ($0.113 million) for CIG/STIG (CIG/ISTIG) and zero for BIG/GT systems.

[g] Annual costs for solids disposal are $0.469 million ($0.428 million) for CIG/STIG (CIG/ISTIG) and are assumed to be the same for the corresponding BIG/GT systems.

[h] Annual H_2SO_4 byproduct credits are $1.815 million for CIG/STIG, $1.659 million for CIG/ISTIG, and zero for BIG/GT systems.

[i] For the unit capital costs given in Table 10 and an assumed 75% capacity factor.

Table 10. Estimated Installed Capital Cost (in $/kW) for IG/STIG and IG/ISTIG Power Plants Fueled with Coal and Biomass

	CIG/STIG[a]	BIG/STIG[b]	CIG/ISTIG[a]	BIG/ISTIG[b]
I. Process Capital Cost				
Fuel Handling	44.4	44.4	41.2	41.2
Blast Air System	15.1	15.1	10.8	10.8
Gasification Plant	180.5	180.5	93.3	93.3
Raw Gas Physical Clean-up	9.9	9.9	8.6	8.6
Raw Gas Chemical Clean-up	197.4	0.0	169.3	0.0
Gas turbine/HRSG	330.4	330.4	287.7	287.7
Balance of Plant				
Mechanical	45.1	45.1	37.0	37.0
Electrical	72.9	72.9	54.3	54.3
Civil	73.5	73.5	68.1	68.1
SUBTOTAL	969.2	771.8	770.3	601.0
II. Total Plant Cost				
Process Plant Cost	969.2	771.8	770.3	601.0
Engineering Home Office (10%)	96.9	77.2	77.0	60.1
Process Contingency (6.2%)	60.1	47.9	47.8	37.3
Project Contingency (17.4%)	168.6	134.3	134.0	104.6
SUBTOTAL	1294.8	1031.2	1029.1	803.0
III. Total Plant Investment				
Total Plant Cost	1294.8	1031.2	1029.1	803.0
AFDC (3.05%, 2 yr construction)	39.5	31.5	31.4	24.5
SUBTOTAL	1334.3	1062.7	1060.5	827.5
IV. Total Capital Requirement				
Total Plant Investment	1334.4	1062.7	1060.5	827.5
Preproduction Costs (2.8%)	36.3	28.9	28.8	22.5
Inventory Capital (2.8%)	36.3	28.9	28.8	22.5
Initial Chemicals, Catalysts	2.8	0.0	2.6	0.0
Land	1.5	1.5	1.5	1.5
TOTAL	1411	1122	1122	874

[a] The CIG/STIG plant consists of two 50.5 MW_e STIG units, each coupled to a Lurgi Mark IV dry-ash, air-blown, fixed bed gasifier. The CIG/ISTIG plant consists of a single 109.1 MW_e ISTIG unit, coupled to a Lurgi Mark IV single dry-ash, air-blown, Lurgi Mark IV fixed bed gasifier. Costs were estimated according to the rules set forth in the EPRI Technical Assessment Guide [J.C. Corman, "System Analysis of Simplified IGCC Plants," General Electric Company, Schenectady, NY, Report on Department of Energy Contract No. DE-AC21-80ET14928, September 1986].

[b] The biomass versions of these plants have outputs of 103.0 MW_e and 111.2 MW_e for BIG/STIG and BIG/ISTIG, respectively (see note a, Table 9). It is assumed that BIG/STIG (BIG/ISTIG) costs are the same as CIG/STIG (CIG/ISTIG) costs, except that the raw gas chemical clean-up phase required for coal would not be needed for biomass, because of its negligible sulfur content.

Table 11. Costs for Alcohol Production from Coal and Biomass Feedstocks[a]

	C/MeOH[b]	B/MeOH[c]	B/EthOH[d]
Annual Production (10^9 liters)	2.103	0.384	0.261
Onstream Time (hours/yr)	8000	8000	8000
Fixed Capital Investment ($\$10^9$)	1.436	0.265	0.098
Working Capital ($\$10^6$)	53.1	13.0	-[e]
Production Cost (cents/liter)			
Fixed Investment	9.321	9.420	5.13
Working Capital	0.253	0.338	-[e]
Wood	-	$3.14*P_b$	$4.39*P_b$
Coal	$3.25*P_c$	-	-
O&M	2.905	4.309	1.99
Total	$12.48 + 3.25*P_c$	$14.07 + 3.14*P_b$	$7.12 + 4.39*P_b$
Total Cost (cents/liter, gasoline-equiv.)[f]	$21.15 + 5.51*P_c$	$23.85 + 5.32*P_b$	$8.90 + 5.49*P_b$
Example: P_c = \$1.6/GJ[g], P_b = \$2.3/GJ[h]	30.0	36.1	21.5

[a] For an annual capital charge rate on fixed (working) capital of 0.1365 (0.10), based on a 10% real discount rate, a 15-year plant life, and an insurance cost of 0.5% of the fixed capital cost per year.

[b] C/MeOH = methanol from coal, with a Texaco pressurized, entrained-flow, oxygen-blown coal gasifier plus methanol synthesis plant. The conversion efficiency, coal-to-methanol, is 55.7% (HHV basis). See Table 12.

[c] B/MeOH = methanol from biomass, with a pressurized, steam/oxygen-blown, fluidized bed biomass gasifier being developed by the Institute of Gas Technology plus methanol synthesis plant. The conversion efficiency, biomass-to-methanol, is 57.7% (HHV basis). See Table 12.

[d] B/EthOH = ethanol from biomass. Performance and cost projections are US Dept. of Energy estimates of what could be achieved by 2000 with an intensive research, development, and demonstration effort targetting enzymatic hydrolysis technology applied to lignocellulosic feedstocks. The conversion efficiency, wood-to-ethanol, is 53.5% (HHV basis). See Table 13.

[e] Included with the fixed capital cost.

[f] Assuming that in gasoline engines modified for alcohol use, 1 GJ of alcohol is worth 1.2 GJ of gasoline (LHV basis), so that 1 liter of MeOH (EthOH) is worth 0.59 (0.80) liters of gasoline.

[g] For a plant in the Midwest US burning Illinois No. 6 coal.

[h] As in Table 6, except that for alcohol production biomass drying is not necessary.

Table 12. Costs for Methanol Production Using Coal and Biomass Feedstocks

	Methanol from Coal[a,c]	Methanol from Biomass[b,d]
Annual Production (billion liters)	2.103	0.384
Onstream Time (hours/year)	8000	8000
Fxd Invstmnt[e] (billion $)	1.436	0.265
Working Capital (million $)	53.1	13.0
Production Cost (cents/liter)		
Fixed Investment[f]	9.321	9.420
Working Capital[f]	0.253	0.338
Wood[g]	-	$3.14*P_b$
Coal[h]	$3.25*P_c$	-
Slfr Byprdt Crdt	- 0.451[c]	-
Slag Disposal	0.115	-
Ctlsts & Chmcls	0.563	0.479
Electricity	-	0.817
Steam	-	- 0.366
Water	0.084	0.113
Fuel	-	0.282
Operating Labor	0.220	0.253
Maintenance	1.292	1.492
Direct Overhead	0.099	0.113
Gnrl Plnt Ovrhd.	0.983	1.126
TOTAL	12.48 + $3.25*P_c$	14.07 + $3.14*P_b$
CO_2 Emission Rate[i] (tonnes C/GJ)	0.0412	-

[a] US Dept. of Energy, "Assessment of Costs and Benefits of Flexible and Alternative Fuel Use in the US Transportation Sector. Technical Report Three: Methanol Production and Transportation Cost," August 1989, based on a study prepared for the DOE's Office of Policy, Planning, and Analysis, by Chem Systems, Inc. Cost estimates are drawn from reports published by the Electric Power Research Institute and other sources. It is assumed that the methanol plant is located at the coal mine mouth in Illinois.

[b] Chem Systems, Inc., "Assessment of Cost of Production of Methanol from Biomass," report to the Solar Energy Research Institute, December 1989.

Notes for Table 12, cont.

[c] Based on the use of a Texaco pressurized, entrained-flow, oxygen-blown coal gasifier, producing a gas consisting primarily of hydrogen and carbon monoxide.

[d] Based on the use of a pressurized, steam/oxygen-blown, fluidized bed biomass gasifier being developed by the Institute of Gas Technology.

[e] The overnight construction cost is $1,290 million for the MeOH-from-coal plant and $238 million for the MeOH-from-biomass plant. With a 3-year construction program with 30% of the cost paid at the end of the 1st yr, 50% at the end of the 2nd, and 20% at startup, the total installed cost becomes $1,436 million for the MeOH-from-coal plant and $265 million for the MeOH-from-biomass plant, assuming a 10% discount rate.

[f] For an annual capital charge rate on fixed (working) capital of 0.1365 (0.10), based on a 10% real discount rate, a 15-year plant life, and an insurance cost of 0.5% of the fixed capital cost per year.

[g] Here P_b is the price of biomass, in $/GJ, and the wood conversion efficiency is 57.7% (HHV basis).

[h] Here P_c is the price of coal, in $/GJ, and the coal conversion efficiency is 55.7% (HHV basis).

[i] Dry Illinois #6 coal (68% C) has a heating value of 29.6 GJ/tonne and a CO2 emission rate of 0.0229 tonnes C/GJ.

Table 13. Projected Cost for Ethanol Production in 2000 from Lignocellulosic Feedstocks[a]

Annual Production[b] (billion liters)	0.261
Onstream Time (hours/year)	8000
Installed Capital Cost (billion $)	0.098
Production Cost (cents/liter)	
Capital[c]	5.13
Wood[d]	$4.39*P_b$
O&M	1.99
Total	$7.12 + 4.39*P_b$

[a] Performance and cost projections for 2000 are what is estimated could be achieved with an intensive research, development, and demonstration effort targetting enzymatic hydrolysis technology and lignocellulosic feedstocks, according to a 1990 Department of Energy Interlaboratory White Paper (Office of Policy Planning and Analysis, US Dept. of Energy, "The Potential of Renewable Energy," SERI/TP-260-3674, March 1990).

[b] For a wood handling capacity of 2110 dry tonnes/day, a 91% average capacity factory, and an ethanol yield of 450 liters of ethanol (@ 23.5 MJ/liter, higher heating value) per tonne of dry wood feedstock. For wood with a higher heating value of 19.75 GJ/tonne, this corresponds to a conversion efficiency of 53.5%.

[c] Assuming a 10% real discount rate and a 15-year plant life, the capital recovery factor is 0.1315. Including an insurance cost of 0.5% of the initial capital cost per year brings the total annual capital charge rate (neglecting taxes) to 0.1365.

[d] Here P_b is the price of biomass, in $/GJ.

Table 14. Scenario for CO_2 Emissions Reduction via Biomass Energy Use[a] (Gt C/yr)

2025	Electricity and alcohol from sugar cane[b]	0.7
	Electricity from kraft pulp industry residues[c]	0.2
	Energy from other residues[d]	0.8
	Total	1.7
2050	Electricity and alcohol from sugar cane	0.7
	Electricity from kraft pulp industry residues	0.2
	Energy from other residues	0.9
	Energy from biomass energy crops[e]	3.6
	Total	5.4

[a] A scenario for reducing global CO_2 emissions from the Scenario B' level to the Scenario D' level (Table 1) through bioenergy use only.

[b] Assuming that sugar cane production grows at the historical rate of 3%/year, from 968 million tonnes of cane (tc) in 1987 to 2976 million tonnes in 2025 and that electricity is coproduced in excess of onsite needs @ 885 kWh/tc with BIG/ISTIG technology or the equivalent (using for both plant energy and excess electricity 2.85 GJ of bagasse and 5.0 GJ of the cane tops and leaves per tc). Assuming this displaces electricity that would otherwise be produced from coal, CO_2 emissions would be reduced 0.640 Gt C in 2025. Also assuming that in 2025 45% of the cane is used to produce ethanol, at a rate of 70 liters/tc, and that this alcohol displaces gasoline, CO_2 emissions would be further reduced by 0.058 Gt C/yr in 2025 (J.M. Ogden, R.H. Williams, and M.E. Fulmer, "Cogeneration Applications of Biomass Gasifier/Gas Turbine Technologies in the Cane Sugar and Alcohol Industries: Getting Started with Bioenergy Strategies for Reducing Greenhouse Gas Emissions," Proceedings of the Conference on Energy and Environment in the 21st Century, MIT Press, Cambridge, MA, 1990)

[c] Assuming that chemical pulp production grows to 2025 at the rates projected to 2000 by the Food and Agricultural Organization (FAO), so that global production increases at an average rate of 3.1%/yr, from 105 million tonnes in 1988 to 330 million tonnes in 2025. It is further assumed that electricity is coproduced at a rate of 2544 kWh/tonne of pulp (tp) in excess of onsite needs with BIG/ISTIG technology or the equivalent (using for both plant energy and excess electricity 7.0, 25.3, and 8.4 GJ/tp of hog fuel, black liquor, and forest residues, respectively). Assuming the produced electricity displaces electricity that would otherwise be produced from coal, CO_2 emissions in 2025 would be reduced by 0.204 Gt C (E.D. Larson, "Biomass-Gasifier/Gas Turbine Applications in the Pulp and Paper Industry: an Initial Strategy for Reducing Electric Utility CO_2 Emissions," Proceedings of the Ninth EPRI Conference on Coal Gasification Power Plants, Palo Alto, CA, 17-19 October, 1990).

[d] Since residues from other major forest product and agricular industries are large compared to those from the sugar cane and kraft pulp industries (Table 15), it is assumed that comparable emissions reductions could be achieved through use of some of these residues for energy.

[e] Assuming that biomass is produced on 600 million hectares at an average productivity of 12 dry tonnes/ha/yr and that the produced biomass displaces coal and thus CO_2 emissions at an average rate of 3.6 Gt C/yr.

Table 15. Selected Global Residue Production Rates (EJ/year)

Forest Product Industries[a]	
Kraft Pulp[b]	
Hogfuel	0.7
Black Liquor	2.7
Forest Residues	0.8
Subtotal	4.2
Sawnwood and Wood Panels[c]	
Mill Residues	3.6
Forest Residues	6.2
Subtotal	9.8
Agricultural Industries[a]	
Sugar Cane[f]	7.6
Wheat[g]	12.9
Rice[g]	10.6
Maize[g]	7.3
Barley[g]	3.8
Subtotal	42.2
Total	56.2

[a] *Assuming higher heating values of 20 GJ and 15 GJ per dry tonne of woody and agricultural residues, respectively.*

[b] *Assuming hog fuel, black liquor, and logging residues (which excludes roots, stumps, branches, needles, and leaves) of 7.0 GJ, 25.3 GJ, and 8.0 GJ per tonne of pulp, respectively (characteristic of the kraft pulp industry in the US Southeast), for the 1988 global chemical pulpwood production of 105 million tonnes (E.D. Larson, "Biomass-Gasifier/Gas Turbine Applications in the Pulp and Paper Industry: an Initial Strategy for Reducing Electric Utility CO_2 Emissions," Proceedings of the Ninth EPRI Conference on Coal Gasification Power Plants, Palo Alto, CA, 17-19 October, 1990).*

[c] *Assuming mill (note d) and forest (note e) residues of 0.30 tonnes and 0.52 tonnes per cubic meter of sawnwood/wood panel products, respectively (characteric of the US forest products industry in 1976), for the 1985-87 world sawnwood/wood panels production rate of 600 million cubic meters (World Resources Institute, World Resources 1990-91, Oxford University Press, New York, 1990).*

[d] *Primary and secondary mill residues of the US forest products industry not used by the pulp industry in 1976 amounted to 34.7 million dry tonnes (Office of Technology Assessment, Energy from Biological Processes, vol. III, Appendices, Part A: Energy from Wood, September 1980), while US sawnwood and wood panels production amounted to 115.4 3 million cubic meters (FAO, 1978 Yearbook of Forest Products, United Nations, Rome, 1980). Thus 34.7/115.4 = 0.30 tonnes of mill residues were produced for each cubic meter of sawnwood and woodpanels produced.*

Notes to Table 15, cont.

[e] *US forest residues totalled 76.4 million tonnes in 1976 (Office of Technology Assessment, Energy from Biological Processes, vol.III--Appendices, Part A: Energy from Wood, September 1980). Assuming each of the 40 million tonnes of pulp produced in the US in 1976 (FAO, 1978 Yearbook of Forest Products, United Nations, Rome, 1980) was associated with 0.42 tonnes of forest residues (E.D. Larson, "Biomass-Gasifier/Gas Turbine Applications in the Pulp and Paper Industry: an Initial Strategy for Reducing Electric Utility CO_2 Emissions," Proceedings of the Ninth EPRI Conference on Coal Gasification Power Plants, Palo Alto, CA, 17-19 October, 1990), the residues associated with sawnwood/woodpanels production in 1976 amounted to 59.6 million tonnes. Thus some 59.6/115.3 = 0.52 tonnes of forest residues were associated with each cubic meter of sawnwood and wood panels production.*

[f] *Assuming bagasse amounting to 2.8 GJ and recoverable cane tops and leaves amounting to 5.0 GJ per (wet) tonne of harvested stem (J.M. Ogden, R.H. Williams, and M.E. Fulmer, "Cogeneration Applications of Biomass Gasifier/Gas Turbine Technologies in the Cane Sugar and Alcohol Industries: Getting Started with Bioenergy Strategies for Reducing Greenhouse Gas Emissions," Proceedings of the Conference on Energy and Environment in the 21st Century, MIT Press, Cambridge, MA, 1990), for the 1987 cane production rate of 968 million tonnes worldwide (Food and Agriculture Organization of the United Nations, FAO Production Yearbook, vol. 41, 1987).*

[g] *Global grain production rates, 1986 (US Dept. of Commerce, Statistical Abstract of the United States 1990, US Government Printing Office, Washington, DC, 1990) and associated residue production rates, assuming residue production coefficients characteristic of US grain production in the period 1975-77 (note h) were:*

Grain	1986 Production (million tonnes)	Residue Coefficient	Residue Production (million tonnes)
Wheat	538	1.6	861
Rice	473	1.5	710
Maize	485	1.0	485
Barley	182	1.4	255

[h] *Selected US grain production rates, 1975-77 (US Dept. of Agriculture, Agricultural Statistics 1978, US GPO, Washington, DC, 1978) and grain residue production rates (Office of Technology Assessment, Energy from Biological Processes, vol. II, Technical and Environmental Analyses, September 1980), along with the corresponding residue coefficients, were:*

Grain	Annual Production (ave., 1975-77) (million tonnes)	Residue Production (ave., 1975-77) (million tonnes)	Residue Coeficient
Wheat	57.2	90.7	1.6
Rice	5.2	7.8	1.5
Maize	155.6	155.3	1.0
Barley	8.5	12.1	1.4

Acknowledgments

This research was supported by the U.S. Environmental Protection Agency, the William and Flora Hewlett Foundation, the John Merck Fund, the New Land Foundation, and the Winrock International Institute for Agricultural Development.

References

1. F.J. Dyson, "Can We Control the Carbon Dioxide in the Atmosphere?" *Energy*, 2, pp. 287-291, 1977.

2. G. Marland, *The Prospect of Solving the CO_2 Problem through Global Reforestation*, U.S.-DOE/OER Report DOE/NBB-0082, 1988.

3. R.A. Houghton, "The Future Role of Tropical Forests in Affecting the Carbon Dioxide Concentration of the Atmosphere," *Ambio*, 19(4), pp. 204-209, 1990.

4. Office of Policy, Planning and Evaluation, U.S. Environmental Protection Agency, *Policy Options for Stabilizing the Global Climate*, D.A. Lashoff, and D.A. Tirpak, eds., draft report to Congress, February 1989.

5. D.O. Hall and F. Rosillo-Calle, "CO_2 Cycling by Biomass: Global Bioproductivity and Problems of Devegetation and Afforestation," Chapter 9 in *Balances in the Atmosphere and the Energy Problem*, E.W.A. Lingemon, European Physical Society Symposium, Geneva, 1990.

6. M.C. Trexler, *Reforesting the United States to Combat Global Warming*, World Resources Institute, Washington, D.C., July 1990.

7. L.L. Wright, R.L. Graham, A.F. Turhollow, "Short-Rotation Woody Crop Opportunities to Mitigate Carbon Dioxide Buildup," presented at the *North American Conference on Forestry Responses to Climate Change*, Washington, D.C., May 1990.

8. K.F. Wiersum and P. Ketner, "Reforestation: a Feasible Contribution to Reducing the Atmospheric Carbon Dioxide Content?" in *Climate and Energy*, P.A. Okken, R. Swart, and S. Zwerver, eds., pp. 107-124, Kluwer Academic Publishers, Dordrecht, 1989.

9. J.L. Kulp, *The Phytosystem as a Sink for Carbon Dioxide*, EPRI Report EN-6786, May 1990.

10. N. Myers, "The Greenhouse Effect: A Tropical Forestry Response," *Biomass*, 18, pp. 73-78, 1989.

11. S. Postel and L. Heise, *Reforesting the Earth*, Worldwatch Paper No. 83, Washington, D.C., 1988.

12. R.J. Moulton and K.R. Richards, *Costs of Sequestering Carbon through Tree Planting and Forest Management in the United States*, USDA Forest Service General Technical Report, December 1990.

13. ICF, Inc., *Preliminary Technology Cost Estimates of Measures Available to Reduce U.S. Greenhouse Gas Emissions by 2010*, study prepared for the U.S. Environmental Protection Agency, August, 1990.

14. D.B. Botkin, "Can We Plant Enough Trees to Absorb all the Greenhouse Gases?" presented at the *University of California Global Climate Change Workshop No. 2: Energy Policies to Address Global Climate Change*, University of California at Davis, September 6-8, 1989.

15. D.O. Hall, "Carbon Flows in the Biosphere: Present and Future," *Journal of the Geological Society*, 146, pp. 175-181, 1990.

16. W.D. Nordhaus, "The Cost of Slowing Climatic Change: A Survey," *The Energy Journal*, 12(1), 1991.

17. D.J. Dudek and A. LeBlanc, "Offsetting New CO_2 Emissions: A Rational First Greenhouse Policy Step," *Contemporary Policy Issues*, 8, pp. 29-42, 1990.

18. J.T. Houghton, G.J. Jenkins, and J.J. Ephraums, eds., *Climate Change: the IPCC Scientific Assessment*, Cambridge University Press, Cambridge, 1990, Chapter 10.

19. World Resources Institute, *World Resources 1990-91*, Oxford University Press, New York, 1990.

20. K. Kuusela, "Forest Products—World Situation," *Ambio*, 16(2,3), pp. 80-85, 1987.

21. C.A. Hendriks, K. Blok, and W.C. Turkenburg, "The Recovery of Carbon Dioxide from Power Plants," in *Climate and Energy: The Feasibility of Controlling CO_2 Emissions*, P.A. Okken, R.J. Swart, and S. Zwerver, eds., Kluwer Academic Publishers, Dordrecht, 1989.

22. C. A. Hendriks, K. Blok, and W.C. Turkenburg, "Technology and Cost of Recovery and Storage of Carbon Dioxide from an Integrated Gasifier Combined Cycle Plant," unpublished manuscript, 1990.

23. R.H. Williams, "Innovative Approaches to Marketing Electric Efficiency," in *Electricity: Efficient End Use and New Generation Technologies and Their Planning Implications*, T.B. Johansson, B. Bodlund, and R.H. Williams, eds., Lund University Press, Lund, 1989.

24. E. Mills, D. Wilson, and T.B. Johansson, "Beginning to Reduce Greenhouse-Gas Emissions Need Not Be Expensive," invited theme paper for the *Second World Climate Conference*, Geneva, 29 October-7 November, 1990.

25. K. Blok, E. Worrell, R.A.W. Albers, R.F.A. Cuelenaere, and W.C. Turkenburg, "The Cost-Effectiveness of Energy Conservation from the Point of View of Carbon Dioxide Reduction," paper prepared for the *13th World Energy Engineering Congress*, Atlanta, 10-12 October, 1990.

26. Chem Systems, Inc., *Assessment of Cost of Production of Methanol from Biomass*, report to the Solar Energy Research Institute, December 1989.

27. Office of Policy, Planning and Analysis, U.S. Dept. of Energy, *The Potential of Renewable Energy*, An Interlaboratory White Paper, SERI/TP-260-3674, 1990.

28. E.D. Larson and R.H. Williams, "Biomass-Gasifier Steam-Injected Gas Turbine Cogeneration," *Journal of Engineering for Gas Turbines and Power*, 112, pp. 157-163, April 1990.

29. J.M. Ogden, R.H. Williams, and M.E. Fulmer, "Cogeneration Applications of Biomass Gasifier/Gas Turbine Technologies in the Cane Sugar and Alcohol Industries: Getting Started with Bioenergy Strategies for Reducing Greenhouse Gas Emissions," *Proceedings of the Conference on Energy and Environment in the 21st Century*, MIT Press, Cambridge, MA, 1990.

30. E.D. Larson, "Biomass-Gasifier/Gas Turbine Applications in the Pulp and Paper Industry: An Initial Strategy for Reducing Electric Utility CO_2 Emissions," *Proceedings of the Ninth EPRI Conference on Coal Gasification Power Plants*, Palo Alto, CA, 17-19 October, 1990.

31. U.S. Dept. of Energy, *Assessment of Costs and Benefits of Flexible and Alternative Fuel Use in the U.S. Transportation Sector. Technical Report Three: Methanol Production and Transportation Cost*, August 1989, based on a study prepared for the DOE's Office of Policy, Planning and Analysis, by Chem Systems, Inc.

32. W.T. Gladstone and F.T. Ledig, "Reducing Pressure on Natural Forests Through High-Yield Forestry," in *Forest Ecology and Management*, 35, pp. 69-78, 1990.

33. G. Siren, L. Sennerby-Forsse, and S. Ledin, "Energy Plantations – Short Rotation Forestry in Sweden," in *Biomass: Regenerable Energy*, D.O. Hall and R.P. Overend, eds., John Wiley and Sons, 1987.

34. F.C. Hummel, W. Palz, and G. Grassi, eds., "Main Conclusions and Proposals," in *Biomass Forestry in Europe: A Strategy for the Future*, Elsevier Applied Science, London, 1988.

35. M.G.R. Cannell, "Physiological Basis of Wood Production," *Scandinavian Journal of Forest Research*, 4, pp. 459-490, 1989.

36. Food and Agriculture Organization of the United Nations, *FAO Production Yearbook*, vol. 41, 1987.

37. D.J. Parrish, D.D. Wolf, W.L. Daniels, J.S. Cundiff, and D.H. Vaughan, "A Five-Year Screening Study of Herbaceous Energy Crops," *Energy from Biomass and Wastes XIV*, D.L. Klass, ed., Institute of Gas Technology, Chicago, IL, 1990.

38. J.H. Cushman and A.F. Turhollow, "Selecting Herbaceous Energy Crops for the Southeast and Midwest/Lake States," *Energy From Biomass and Wastes XIV*, D.L. Klass, ed., Institute of Gas Technology, Chicago, IL, 1990.

39. W. Fulkerson et al., *Energy Technology R&D: What Could Make a Difference? A Study by the Staff of the Oak Ridge National Laboratory*, vol. 2, Supply Technology, ORNL-6541/V2/P2, December 1989.

40. Office of Technology Assessment, *Energy from Biological Processes*, Congress of the United States, Washington, D.C., 1980.

41. Green Mountain National Forest, *Developing Markets for Low Quality Wood: The Wood Supply in Bennington County and Southwestern Vermont*, Summary Report, October 5, 1990.

42. D.E. Earl, *Forest Energy and Economic Development*, Clarendon Press, Oxford, 1975.

43. Energy Information Administration, *International Energy Annual 1988*, DOE/EIA-0219(88), United States Department of Energy, Washington, D.C., 1989.

44. C.H. Strauss and L.L. Wright, "Woody Biomass Production Costs in the United States: An Economic Summary of Commercial *Populus* Plantation Systems," *Solar Energy*, 45(2), pp. 105-110, 1990.

45. C.H. Strauss, S.C. Grado, P.R. Blankenhorn, and T.W. Bowersox, "Economic Valuations of Multiple Rotation SRIC Biomass Plantations," *Solar Energy*, 41(2), pp. 207-214, 1988.

46. Intergovernmental Panel on Climate Change, *Formulation of Response Strategies*, report prepared for IPCC by Working Group III, June 1990.

47. J.T. Houghton, G.J. Jenkins, and J.J. Ephraums, eds., *Climate Change: The IPCC Scientific Assessment*, Cambridge University Press, Cambridge, 1990, Policymakers Summary.

48. W. Vergara, N.E. Hay, and C.W. Hall, *Natural Gas: Its Role and Potential in Economic Development*, Westview Special Studies in Natural Resource and Energy Management, Westview Press, Boulder, CO, 1989.

49. J.R. Bolton and D.O. Hall, "Photochemical Conversion and Storage of Solar Energy," *Annual Review of Energy*, 4, pp. 353-401, 1979.

50. J. Coombs, D.O. Hall, and P. Chartier, *Plants as Solar Collectors: Optimizing Productivity for Energy: An Assessment Study*, Solar Energy R&D in the European Community, Series E, Volume 4: Energy from Biomass, D. Reidel Publishing Company, Dordrecht, 1983.

51. "The Noordwijk Declaration on Atmospheric Pollution and Climatic Change," *Ministerial Conference on Atmospheric and Climatic Change*, Noordwijk, the Netherlands, November 1989.

52. A. Grainger, "Estimating Areas of Degraded Tropical Lands Requiring Replenishment of Forest Cover," *International Treecrops Journal*, 5, pp. 1-2, 1988.

53. W.M. Hao, M.H. Liu, and P.J. Crutzen, "Estimates of Annual and Regional Releases of CO_2 and Other Trace Gases from Fires in the Tropics," in *Proceedings of the III International Symposium on Fire Ecology*, J.J. Goldhammer, ed., Springer, Berlin, in press.

54. New Farm and Forest Products Task Force, *New Farm and Forest Products—Responsive to the Challenges and Opportunities Facing American Agriculture*, Report to the U.S. Secretary of Agriculture, 1987.

55. C.J. Weinberg and R.H. Williams, "Energy from the Sun," *Scientific American*, 262(9), September 1990.

56. F.T. Ledig, "Improvement of Eucalyptus for Fuel and Fiber in California," in *Biomass Production by Fast-Growing Trees*, J.S. Pereira and J.J. Landsberg, eds., Kluwer Academic Publishers, Dordrecht, 1989.

57. I. Stjernquist, "Modern Wood Fuels," in *Bioenergy and the Environment*, J. Pasztor and L.A. Kristoferson, eds., Westview Press, Boulder, CO, 1990.

58. C.P. Mitchell, "Nutrients and Growth Relations in Short-Rotation Forestry," *Biomass*, 22, pp. 91-105, 1990.

59. D.S. DeBell, C.D. Whitesell, and T.H. Schubert, "Using N_2-Fixing *Albizia* to Increase Growth of *Eucalyptus* Plantations in Hawaii," *Forest Science*, 35(1), pp. 64-75, 1989.

60. S. Linder, "Nutritional Control of Forest Yield," in *Nutrition of Trees*, Marcus Wallenberg Foundation Symposia, 6, Falun, Sweden, pp. 62-87, 1989.

61. J.P. Kimmins, "Modelling the Sustainability of Forest Production and Yield for a Changing and Uncertain Future," *The Forest Chronicle*, 6, pp. 271-280, June 1990.

62. S.G. Nilsson, "Density and Species Richness of Some Forest Bird Communities in South Sweden," *Oikos*, p. 392-401, 1979.

63. J.C. Corman, *Systems Analysis of Simplified IGCC Plants*, report prepared for the U.S. Department of Energy by General Electric Corporate Research and Development, Schenectady, N.Y., ET-14928-13, September 1986.

64. A.A. Pitrilo and L.E. Graham, "DOE Activities Supporting IGCC Technologies," in *Proceedings of a Conference on Integrated Gasification Combined Cycle Plants for Canadian Utility Applications*, Ch. 1A, Canadian Electrical Association, Montreal, 1990.

65. A. Ahlstrom Corporation, "Sydkraft AB, Sweden and A. Ahlstrom, Finland Goes Ahead with Unique Gasification Process," press release, November 1990.

66. R.H. Williams, "Low-Cost Strategies for Coping with CO_2 Emission Limits," *The Energy Journal*, 11(3), 1990.

67. D.O. Hall and R.P. Overend, eds., *Biomass: Regenerable Energy*, Wiley, New York, 1987.

68. D.C. Rinebolt, "Biomass Energy: A Maturing Technology," *Biologue*, 6(5), pp. 20-23, 1990.

INTRODUCTION TO RESPONDENTS PANEL ON CURRENT ANALYSES OF ECONOMIC COSTS

Rosina Bierbaum

Oceans and Environment Program
Office of Technology Assessment
United States Congress
Washington, DC 20510

As I look out over the audience, I'm struck by how many of us were here in a previous life. I guess there *is* life after acid rain. The last time I spoke to this group was five years ago on the costs and benefits of acid rain control. At that time, I'm pretty sure we thought no more contentious environmental issue could confront the Congress. We were wrong. What we're discussing today – climate change – is much bigger. A lot of our earlier analytic and economic analyses were really just warming up – if you'll pardon the pun – for this all-encompassing issue.

This session is meant to be a series of responses to the papers on the costs of greenhouse gas emissions reductions. I'm sure everybody has a lot of comments. We've had time to sort out, cogitate and ruminate on these papers. To get us focused, let me just recap for a few minutes what each author has said. I'll give you a really brief summary and overview, somewhat tongue-in-cheek.

Lester Lave started us off by reminding us of the inherent bias in any "well-documented analysis" – and that's in quotes – and that we tend to use real data points, which may cause us to ignore behavioral change, technology change or cost reduction. And he suggested that many analyses, therefore, might be *over-estimates* of costs.

We heard then from the later speakers a lot of "good news" and "bad news." Alex Cristofaro from EPA told us that, if we considered all the greenhouse gases and their greenhouse warming potential (GWP), the United States could actually freeze its total greenhouse warming potential for the next ten years. This freeze in GWP would be primarily due to the comparatively big hits gotten by reducing CFCs, as well as reductions we may get under the recent Clean Air Act by controlling landfill methane, from the NO_x provisions, and some possible reduction in energy use. He did point out, however, that CO_2 would rise over this period, and if we wish to freeze CO_2, additional measures would be needed. He cited a recent analysis by Dale Jorgenson suggesting that achieving a freeze in CO_2 might require a $17 per ton carbon tax.

Copyright 1991 by Elsevier Science Publishing Company, Inc.
Global Climate Change: The Economic Costs of Mitigation and Adaptation
James C. White, Editor

Howard Geller, the next speaker, then took us through historic cost savings for cars and refrigerators and showed us that energy and CO_2 savings could be gotten at a negative cost. He argued that Corporate Average Fuel Efficiency (CAFE) standards for cars saved us two million barrels of oil per day at minus $200 per ton of carbon. He cited a benefit-cost ratio of eight to one for refrigerator efficiency improvements. Further, based on a New York utility study that he was part of, Mr. Geller said that a 30-percent cost-effective conservation potential existed at least for that New York utility.

Having just heard that the United States could get emission reductions for free, Rich Richels sobered us up. If China quadruples its CO_2 emissions (as we expect) and the less developed countries or LDCs merely double their CO_2 emissions, the sum of those two groups would almost double world emissions. A world goal of a 20 percent *drop* from today's emissions is going to be tough. He said it will be difficult to decouple energy and GDP growth in eastern Europe and the Soviet Union – important concerns – if we consider them to be part of a 20-percent reduction. He commented that the rest of the OECD may have cheaper CO_2 emissions reductions available than the United States, and suggested a 20-percent reduction could cut 1 to 2 percent of their GNP as opposed to 1 to 5 percent of ours.

After that depressing news, Bob Williams took us through some very optimistic calculations for the potential of biomass to offset coal use in the United States and suggested that 56 percent of U.S. carbon emissions could be offset by planting 139 million hectares of biomass crops.

The next papers are meant to take a hard look at the analyses we heard about yesterday. We will look at them from several points of view, certainly from the economic point of view and also from the political, social and pragmatic perspectives.

We have a well-versed and diverse group of speakers here. They will discuss the pros and cons of these analyses that I quickly summarized as well as the ramifications of all economic analyses of CO_2 emission reductions.

A CRITICAL ANALYSIS OF CLIMATE CHANGE POLICY RESEARCH

Dale S. Rothman and Duane Chapman*

*Resource Economics
Department of Agricultural Economics
Cornell University
Ithaca, NY 14853.

Abstract

After more than a decade of warning from scientists, the policy community has begun to take up the challenge of global climate change. This paper considers some of the recent efforts to analyze policymaking in this area. Several shortcomings are seen in the present policy research, including: inconsistencies in the data and methods used in some of the analyses, a myopic vision of options available, an overly anthropocentric focus on estimating costs and benefits, inadequate treatment of uncertainty and irreversibility, and the lack of recognition of differential motives of developing and developed countries. We also have concerns with how results of the analyses have been presented and interpreted, and the limited amount of peer review that has taken place.

Introduction

After more than a decade of warning from scientists, the policy community has begun to take up the challenge of global climate change. The last few years have seen an explosion in policy research related to this issue.[1] This paper considers some of the recent efforts, focusing on those which attempt to estimate the costs and/or benefits of various policies. These represent the research that has received the most attention within the policy community and that has been at the center of much of the policy debate in the U.S. The paper is not intended to be a summary of work to date;[2] rather its purpose is to raise several concerns about the questions that are being addressed, how the research is being performed, and how the results are being presented to the policy community and the public in general.

Much of the research reviewed here reflects preliminary effort to address issues in this area and includes several works in progress. We recognize that we have not evaluated all of the work that has been, or is being done. Nevertheless, it is our opinion that it is the appropriate time to raise these sorts of questions and concerns. The reader should,

Copyright 1991 by Elsevier Science Publishing Company, Inc.
Global Climate Change: The Economic Costs of Mitigation and Adaptation
James C. White, Editor

for the most part, interpret our conclusions as general critiques and not as criticisms of the specific works reviewed.

We begin by laying out the basic concepts that several authors have suggested should be considered in climate change policy research. This is followed by a catalog of the concerns that we have with the present body of research. The paper concludes with a summary of our major conclusions and our perception of desirable future research in this area.

Climate Change Policy Research: What Should be Stressed?[3]

> One approach would be to take a 'certainty equivalent' or 'best guess' approach, ignore uncertainty and the costs of decisionmaking, and plunge ahead....It is appropriate as long as the risks are symmetrical and the uncertainties are unlikely to be resolved in the foreseeable future. Unfortunately, neither of these conditions is likely to be satisfied for the greenhouse effect. (Nordhaus 1990c)

> These features – cost-benefit thinking, appropriate behavior under uncertainty and incentives to cooperate – define what we call 'thinking economically about climate change.' (Barbier and Pearce, 1990)

At a very fundamental level, the evaluation of policies to address potential global climate change must include a careful weighing of costs and benefits. (It would be imprudent after all to pursue policies for which the costs of the policy clearly outweigh the benefits.) Climate change presents major challenges, however, calling for innovative analyses. This section considers some of the difficulties, principally the wide range of costs and benefits that must be considered, the importance of uncertainty, risk, and irreversibility, and finally the global and intertemporal aspects of climate change policy.

In analyzing policies to address global climate change, a full account must be made of the costs and benefits. Indirect costs, i.e. impacts on international trade patterns and other aspects of the macroeconomy, should be included as well as direct costs, i.e. tax rates on fuels, or subsidies for conservation or reforestation. Estimates of the benefits of climate change policy need to incorporate joint benefits, such as benefits from reductions in levels of other pollutants, and improved levels of energy efficiency. Researchers must be careful to recognize both costs and benefits that are not easily monetized, as these are likely to be significant for any policy related to global climate change.

Global climate change policy research must face the issues of uncertainty, risk, and irreversibility. Uncertainty refers here to a lack of knowledge that can be improved with further information. Currently, there remain significant gaps in our understanding of the climate system and how it may be affected by increasing concentrations of greenhouse gases. Similarly, our understanding of the natural and socio-economic systems which both determine and bear the effects of global climate change is incomplete; so even if we were to completely decipher the climate system, we would be left with uncertainties concerning how many of the forces driving it might change and what might be the impacts of this change. Some of the uncertainty is expected to be reduced with further study, but this may take a long time and some of it may not be reducible. The concept of risk encompasses this irreducible uncertainty, but it also includes the idea of inherent variability. An example of this would be the interannual variability of rainfall and temperature in a particular region. With better knowledge it may be possible to determine how these might change with global climate change, but the underlying variability will remain. A simpler example would be rolling a fair die. Even if it were determined that the die is fair, the number showing on the die when rolled would still remain an uncertain event. Irreversibility is important because many of the effects that continuing emissions of greenhouse gases imply for natural and social systems cannot be undone once they occur, at least not in the time scales which are considered for climate change policy. This introduces an important asymmetry into the analysis of uncertain events.

Global climate change is not limited to a single nation or region of the world, nor to a single generation. Actions taken today in one part of the world have repercussions that will be felt throughout the globe and for many years to come. This causes severe difficulties in evaluating future costs and benefits, regional differences in costs and benefits, and international implications of climate change policies. It also raises important questions concerning the effectiveness of unilateral action, and of intergenerational and regional equity which cannot be ignored.

To address Barbier and Pearce's third criteria – incentives to cooperate – policy analyses must, in particular, recognize the differential motivations of the developed and developing countries. The following quote, although representative of only a single opinion, gives a flavor of these differences:

> The developed countries have already attained most reasonable goals of development and can afford to substitute environmental protection for further growth of material output. On the other hand, the developing countries can be expected to participate in the global effort only to the extent that this participation is fully consistent with and complementary to their immediate economic and social development objectives. (Munasinghe, 1990)

Approaches to Estimating Costs: Top-Down vs. Bottom-Up

We begin our categorization of concerns with a review of the two major strategies that have been employed in climate change research. Estimates of the costs of reducing or offsetting greenhouse gas emissions are influenced strongly by the approach taken. These can be broadly categorized into macroeconomic, or "top-down," and microeconomic, or "bottom-up" approaches. The latter provide lower estimates of costs, in general, but there are significant deviations among the estimates within each category.

Top-down approaches are characterized by aggregate economy-wide macro models, with particular attention paid to the energy sector. Manne and Richels (1990a, 1990b), Edmonds and Barnes (1990), Morris et al. (1990), CBO (1990), and Jorgensen and Wilcoxen (1990) provide examples of this approach. The models attempt to look at the wider repercussions of taxes or other policies, and tend to estimate only the costs imposed on the economy. The more flexible that the energy system is assumed to be in the model, the lower will be the estimates of the costs for reducing greenhouse gas emissions. This is brought out clearly in the CBO (1990) study which compares estimates of the response of U.S. CO_2 to a carbon tax using models developed by the Department of Energy, Data Resources Incorporated, and Jorgensen. They find that the same tax schedule of $10/ton of carbon in 1990 rising to $100/ton of carbon in 2000 produces changes in the levels of U.S. CO_2 emissions from 1988-2000 of +5%, -6%, and -27% for the EIA, DRI, and Jorgensen models, respectively. One of the main reasons given for the difference in the estimates is the degree of adjustment assumed to occur in response to the taxes. Similarly, Morris et al. (1990) and Jorgensen and Wilcoxen (1990), using more disaggregated models, find much lower estimates of the marginal costs for achieving similar reductions in CO_2 emissions in the United States than do Manne and Richels (1990a).[4]

The bottom-up analyses take a more micro perspective. Individual technologies or specific actions that either reduce or offset greenhouse gas emissions (i.e. by planting trees) are analyzed to make an estimate of the costs. Drennen and Chapman (1990), Dudek and LeBlanc (1990), and Geller (1990) provide examples of this approach. Wider effects on the economy, and feedbacks such as changing relative prices as the technology enters into the market on a large scale are generally ignored. Also, bottom-up analyses normally include estimates of benefits as well as costs. For these reasons, bottom-up analyses tend to provide lower estimates of costs, sometimes even showing a negative net cost of reducing greenhouse gas emissions. For example, Drennen and Chapman (1990) estimate the cost of reducing CO_2 emissions by using compact fluorescent light bulbs to be -$56 per ton of CO_2 equivalent, and Geller (1990) has calculated that increased efficiency standards on automobiles and refrigerators in the past fifteen years have produced net costs of carbon avoidance of -$60 and -$90 per ton of CO_2 equivalent, respectively. Geller (1990) further argues that similar initiatives could reduce carbon emissions in the U.S. by 10% from 1988 levels by the year 2000, at a net savings of $75 billion per year. Specific policies may be found to be quite expensive, however; for example, Drennen and Chapman (1990) estimate that the reduction of greenhouse gas emissions by altering the diet of cattle has a cost of over $350 ton of CO_2 equivalent.

Future research should consider trying to link these two approaches, or at least to narrow the gap. More detailed models, as employed by Morris et al. (1990) and Jorgensen and Wilcoxen (1990) may or may not be the proper direction to take, but they need to be considered if large-scale modeling is to be used. For the moment, as Zimmerman (1990a) and Wuebbles and Edmonds (forthcoming) point out, one must be careful when comparing costs across studies to recognize the differences in the two approaches.

The Importance of Assumptions: Carbon Taxes and GNP Loss

Too often the results of policy analyses, especially those that involve the use of computer models, are taken at face value without a consideration of the assumptions that underlie the analyses. The following two examples illustrate how a few assumptions can directly determine key results.

The first example focuses upon the long-run equilibrium carbon tax estimated by Manne and Richels as necessary to maintain a level of emissions that is 20% below current levels in the United States, other OECD countries and the Soviet Union and Eastern Europe; elsewhere the tax limits the growth in emissions to 100% above current levels. The equilibrium value that they present is $250/ton of carbon (Manne and Richels 1990a,1990b). This number is determined by three assumptions: the cost of synthetic fuels – $10/million British Thermal Units (mmBTU), the cost of a non-electric backstop technology with no carbon emissions – $20/mmBTU, and the carbon emission rate for synthetic fuels – 0.04 tons carbon/mmBTU:

Cost Differential/Carbon Differential = ($20-$10)/(.04 tons) = $250/ton

Manne and Richels do recognize the dependence of this number on these assumptions, particularly the cost differential, but do not present evidence to indicate that they have done extensive tests or sensitivity analysis in this area. This number has also been cited by others, with little or no acknowledgment of the importance of the underlying assumptions.[5]

The second example involves Edmonds and Barnes' (1990) estimates of Gross National Product losses that may occur as a result of policies that aim to reduce emissions of greenhouse gases. They estimate the global "GNP penalty" to be 3% for a global effort to stabilize CO_2 emissions at current levels in the year 2025, and 8% for a global effort to reduce emissions by 50% in the same year. If only Organization for Economic Cooperation and Development (OECD) countries act, the GNP penalty for these countries is a staggering 17% to simply stabilize global emissions at current levels.

Before we get too excited by these results though, we must recognize how they are derived. Edmonds and Barnes use a "GNP feedback elasticity parameter which reduces actual GNP as the cost of energy services rise" (Edmonds and Barnes, 1990). The authors do not say what the value of this parameter is, but in an earlier piece by Reilly, Edmonds, Gardner, and Brenkert (1987) the mean values provided are -0.14 for OECD countries, 0.05 for Mideast countries, and -0.20 for other countries. A value of -0.14 implies that a 1% increase in the costs of energy services will result in a 0.14% decline in GNP. Edmonds and Barnes recognize that this simplistic estimation of costs is one of the major limitations of their present analysis:

> The importance of the GNP feedback results should not be overemphasized. These results are based on a single equation relationship whose key parameter values reflect the results of analysis on the 1970s period. The period

of analysis was too brief to allow a full analysis of long-term adjustment responses and **small changes in this highly uncertain parameter can lead to large changes in dollar value of GNP, even if percentage changes are small.** (Edmonds and Barnes, 1990; our emphasis)

It would seem wise to heed the caution expressed here. Care must be taken to acknowledge the limitations of analysis in the presentation of results. This leads us to our next question: given that much of our knowledge concerning the physical and social science aspects of global climate change are uncertain at best, do the analyses try to address these uncertainties?

Measuring Costs from the Wrong Baseline[6]

A common measure of part of the costs of policies to address global climate change is lost economic output.[7] This is generally expressed as a reduction in Gross National or Gross Domestic Product (GNP or GDP). The procedure employed is to run the economic model (most of these studies are top-down) without policies to project a baseline path of economic growth; the model is then run with the climate change policies. Since policies represent an added constraint on the economy, the resulting path of economic growth is expected to be lower than the baseline. The difference between the two paths provides an estimate for the costs of the policies.

The flaw in this method lies with the baseline path. It is estimated under the assumption of no climate change, and therefore no impact on economic growth from changing climate. If this impact is not zero (it is generally assumed to be negative), then the baseline estimates of economic growth are incorrect, and the resulting cost estimates are measured inappropriately.

Ignoring The Costs of Delaying Action

Delaying the policy response to the greenhouse gas buildup would substantially increase the commitment to global warming. (EPA, 1989)

[T]he risks of delaying action appear to be large, and the costs of reducing emissions are likely to increase as the time allowed for these reductions is shortened. (EPA, 1989)

The longer emissions continue to increase at present-day rates, the greater reductions would have to be for concentrations to stabilize at a given level. (IPCC, 1990)

As the above quotes illustrate, it is well recognized that there may be significant costs associated with delaying action to address global climate change. These costs can take more than one form. The longer that action to prevent or reduce global climate change is postponed, the larger will be the eventual climate change. Also, delaying either mitigative or adaptive action today may require more extreme action in the future that will need to be accomplished over a shorter period of time, thereby implying larger costs. A perverse result, described by Cline (1989b) and Oppenheimer and Boyle (1990), is that delaying action now may lead to costs in the future that surpass the ability of society to take action at that time.

The cost of delaying action has escaped the purview of studies which set out to estimate the costs and benefits of climate change policy, however. A single example of research addressing this cost has been found: Manne and Richels in "Buying Greenhouse Insurance" (1990c). In this paper, the authors present a decision-analysis framework for determining the optimal level of near-term emissions reductions in the face of scientific uncertainty, in their words – "optimal hedging strategy." Although, Manne and Richels are primarily interested in seeing how the optimal hedging strategy varies with: "(a)the probability of eventual carbon limitations, (b)the accuracy and timing of the climate research program, and (c)the prospects for new supply and conservation technologies," their tentative results do underscore the value of pursuing some level of emissions reductions in the face of uncertainty, and may provide some estimates of the costs of delaying action (Manne and Richels, 1990c). Hopefully, more research will be pursued addressing this issue.

De-Emphasizing Benefits in Cost-Benefit Analysis

Very few of the studies to date have addressed the benefits side of the cost-benefit calculus. Some authors have pointed to the large degree of uncertainty in current knowledge concerning the social and economic impacts of global climate change as a justification for focusing only on the costs of policies.[8] Should this be considered acceptable? By emphasizing only the costs of policy do we not introduce a bias into our decision-making? The emphasis of this conference, as reflected in the title, *Global Climate Change: The Economic Costs of Mitigation and Adaptation*, reflects this bias.

Many of the studies that have tried to explicitly estimate benefits from global climate change policy have been summarized by William Nordhaus (1990c).[9] His procedure is to identify sectors of the U.S. economy which are highly or moderately sensitive to climate, using the 1981 sectoral breakdown of the U.S. economy. Based upon detailed studies done by the EPA (1988), the Coolfront Workshop (1989), and Binkley (1988), he then estimates the total damage to the U.S. economy that would result from a 3°C increase in the global mean surface temperature. His final estimate is that the net economic damage of such a warming would be on the order of 0.25% of national income for the U.S. Recognizing that a number of factors are missing in his analysis, he estimates that this number may really be as high as 1%, with his "hunch...that the overall impact upon human activity is unlikely to be larger than 2 percent of total output" (Nordhaus, 1990c). Later, he states that "climate change will lead to a combination of gains and losses with the likelihood of but a small impact on the overall economy and with no strong presumption that modest and gradual global greenhouse warming will on balance be harmful" (Nordhaus, 1990c).

William R. Cline (1990b) provides several reasons why Nordhaus' estimates of the benefits from global climate change policy are likely to be too low. He argues that: 1) the net impact on agriculture will be negative, especially in developing countries, not zero as Nordhaus assumes for his central estimate; 2) the land loss resulting from rising sea levels will be much higher than that estimated for the U.S.; 3) costs associated with species and forest loss, human disutility, and storm intensity have been ignored;[10] and 4) the estimates do not consider the effects beyond a doubling of CO_2 (Cline, 1990b).

Neglect of Joint Benefits

Even considering Cline's critique, we have other concerns with analyses regarding the benefits of global climate change policy. The first of these is that many of the benefits that may occur as a result of global climate change policies have been omitted. Whereas many of the studies address the indirect negative impacts that policies may have, i.e. reduction in the growth of GNP, very few try to account for the indirect positive impacts that may also occur. These include health and other benefits associated with reduced levels of other atmospheric pollutants due to lowered consumption of greenhouse gases and economic gains from increased expenditures on renewable technologies. Metz (1990) states that a primary reason that the European Community is pursuing policies related to global climate change is that they provide significant benefits with respect to acid rain, smog, and other environmental problems; including these ancillary benefits results in only slight negative or positive effects on the economy.

Anthropocentric Nature of Benefit Estimates

> Fair assessment of the trade-offs among policy options will therefore require a thorough analysis of both monetized and non-monetized costs and benefits. (Barbier and Pearce 1990)

Most of the analysis of benefits to date has been overly anthropocentric. It is recognized that many natural ecosystems are likely to suffer severe damage as a result of the pace and extent of the expected global climate change in the next century and beyond (EPA, 1989). Recognition of the benefits from preventing species loss and forest destruction, even when present, tends to focus only on the advantages to human society— "[s]pecies diversity enlarges the options open to future society for medicines and agricultural varieties" (Cline 1990b).

Is it proper to include in our analyses only those items that can be measured in terms of their direct impact on the economy, specifically only those aspects that are valued by humans? Recently the U.S. EPA has been requested to give greater emphasis to the health of natural ecosystems affected by climate change as distinct from the human consequences of that change (EPA, 1990; "E.P.A. Acts to Reshuffle," 1991). We should encourage this development in ourselves and other researchers.

Incorrectly Comparing Marginal Costs and Benefits: Focusing on an Equivalent Doubling of CO_2

> Accumulation of trace gases and consequential global warming is a continuous process that would not stop with the mere doubling of CO_2 but, in the absence of policy action, would persist into the indefinite future...[P]olicy analysis of global warming will tend erroneously toward inaction if it remains focused on the damages that might be expected from a doubling of carbon dioxide, and considers only the next fifty years or so. Consideration of a longer time horizon and much greater global warming can reverse the benefit-cost calculus toward an outcome favorable to preventative action. (Cline 1990d)

The principal analyses of the benefits associated with global climate change policy have concentrated on the potential impacts of an equivalent doubling of CO_2 (Nordhaus 1990c, EPA 1988, Coolfront Workshop 1989, Binkley 1988). Cline (1990d) argues, however, that rather than considering a doubling of CO_2 levels with a warming around 3°C, policy analyses should really take a longer-term view and consider a more than seven-fold increase in CO_2 levels with a warming of over 10°C. Here, we would like to point out a more fundamental concern in any comparison of marginal costs and benefits.

Nordhaus (1990c) has developed curves showing rising marginal costs and constant marginal benefits of reducing emissions; the latter is calculated assuming an equivalent doubling of the CO_2 concentration in the atmosphere. He then uses these estimates to determine an efficient level of greenhouse gas reductions of 17% by finding the point at which these curves cross.[11] This methodology is fundamentally flawed. First, a 17% reduction in greenhouse gas emissions will not result in a stabilization of atmospheric concentrations at the level of an equivalent doubling of CO_2. For example, the Intergovernmental Panel on Climate Change (IPCC) estimates that CO_2 emission reductions of 50% below 1985 quantities would be required to stabilize atmospheric concentrations of CO_2 at twice their preindustrial levels. Secondly, the social and economic damage resulting from global climate change is likely to increase at an accelerating pace as greenhouse gas concentrations increase beyond an equivalent doubling, so the marginal benefit from reducing emissions is not constant. Therefore, with emission reductions of only 17%, the marginal benefit will be much higher than those estimated using an assumed equivalent doubling of CO_2. A more proper comparison of marginal costs and benefits would accordingly lead to a higher level of reductions.

Inadequate Treatment of Uncertainty/ Lack of Sensitivity Analysis

A common element in our knowledge concerning all aspects of global climate change, from the scientific to socio-political-economic, is some degree of uncertainty. The further out into the future that we look, the more that this uncertainty influences the results of the research. Earlier, a few examples were given of the importance of assumptions in many of the models employed. Little of the climate change policy research on estimating costs and/or benefits seen to date has included extensive sensitivity analyses or attempts to incorporate uncertainty, however.

Nordhaus and Yohe (1983) and Edmonds et al. (1986) do provide us with examples of using uncertainty analysis in the forecasts of future emissions of greenhouse gases, and Nordhaus (1990a) uses probabilistic scenario analysis to estimate possible likely paths of future warming. Edmonds et al. (1986) use a Monte-Carlo analysis to generate 400 model scenarios by drawing independent samples from the distributions of 79 important input parameters. Nordhaus and Yohe (1983) use a similar technique. In both of these studies, the results are summarized as subjective probability distributions for future emissions, GNP, etc. Nordhaus (1990a) uses statistical techniques to combine the distributions of the important parameters in his model to directly arrive at a probability distribution for future global warming.

Very little of this sort of analysis has been done in the area of estimating the costs of adaptation or mitigation, however. Scenario analysis is employed by a few studies, but these are limited to only a few scenarios and consider only different policy options and do not specifically test model assumptions (Edmonds and Barnes 1990; Manne and Richels 1990a, 1990b; Morris et al. 1990). Studies which do include sensitivity analyses of input assumptions test only one or two assumptions. Still, the results show that small changes in input parameters can have large effects (Manne and Richels 1990a, 1990b; Morris et al. 1990; Edmonds and Barnes 1990). These conclusions should make us conscious that more effort needs to be expended in this area.

Manne and Richels (1990c) have updated their Global 2100 modeling system to be probabilistic. Hopefully, this will allow them to incorporate more uncertainty into their analysis in the future. The decision-analysis framework which they present also looks to be a step in the right direction with respect to incorporating risk and irreversibility directly into their analysis.

Peer Review

The policy research on climate change should be subject to the same level of scrutiny and peer review as is the scientific research. Since most of the research reviewed here is recent and/or ongoing, it is understandable that much of the work is yet to be published. However, this does not lessen the need for the underlying research used to formulate policy recommendations to be made available, and in sufficient detail to allow for adequate review. Although most researchers are willing to disseminate details and drafts of unpublished research, or work in progress,[12] very little of the analysis has been presented in sufficient detail to allow for careful checks on assumptions, calculations, etc.

Conclusions

This paper has presented a set of concerns that we have with current climate change policy research, particularly that research which has focused on estimating costs and benefits of climate change policy. We have cited the need for more effort in certain areas: combining top-down and bottom-up approaches, benefits estimation, incorporation of indirect costs and benefits, sensitivity analysis, testing of assumptions, inclusion of uncertainty, risk, and irreversibility, and peer review. We have also noted cases where some of the current analysis is logically flawed: specifically measuring costs from the wrong baseline and improperly comparing marginal costs and benefits. It is our hope that this effort will stimulate further discussion on these topics and result in better and more appropriate research in the future.

Notes

1. This is not to deny the existence of earlier policy analysis related to global climate change. See in particular Nordhaus (1979), Kellogg and Schware (1981), and *Changing Climate* (1983).

2. For more general surveys of estimates of the costs and benefits associated with global climate change policy, the reader is referred to Nordhaus (1990d), DOE (1990), Hoeller and Nicolaisen (forthcoming), and Edmonds and Wuebbles (forthcoming).

3. For a further consideration of some of the issues expressed in this section, the reader may wish to refer to Barbier and Pearce (1990) and Nordhaus (1990c).

4. Morris et al.(1990), in their base case, estimate the marginal cost for the U.S. to reduce its CO_2 emissions to 20% below 1988 levels by the year 2010 to be under $40/ton carbon. Jorgensen and Wilcoxen (1990), in a remarkably similar result, determine that a tax of just over $40/ton carbon is sufficient for the U.S. to achieve a reduction of 20% from 1990 levels by the year 2005. Manne and Richels (1990a), however, estimate that for the U.S. to achieve reductions of 20% from 1990 levels by the year 2020, and holding steady thereafter, would require a tax that would rise to nearly $600/ton carbon before settling at a level of $250/ton in the long run.

5. See Nordhaus (1990d), CBO (1990), *N.Y. Times* (19 November 1989), Wuebbles and Edmonds (forthcoming), and Hoeller and Nicolaisen (forthcoming). Of these, only Nordhaus (1990d) notes that the value of $250/ton "is purely a product of the assumption about the cost of carbon-free technologies."

Richels has indicated that he and Manne have come across model scenarios where the long-run tax is not determined by the non-electric sector in certain regions of the world. Also, he emphasizes that the key assumption is the cost differential between the carbon-based and noncarbon-based technologies, rather than the absolute value of either one. (Richard Richels, personal communication, Nov. 1990)

6. This subsection reflects a concern raised by Mary Beth Zimmerman (1990b) of the Alliance to Save Energy at the conference *Global Climate Change: The Economic Costs of Mitigation and Adaptation*, Washington, D.C., December 4-5, 1990. Her argument was disputed by David Montgomery of the Congressional Budget Office and principal author of the CBO (1990) study referenced in this paper, who expressed the opinion that it is proper to separate costs and benefits in the manner presently done in most studies.

7. For just a few examples see Manne and Richels (1990a, 1990b), Edmonds and Barnes (1990), Jorgensen and Wilcoxen (1990), CBO (1990), and Nordhaus (1990e).

8. Nordhaus likes to refer to our knowledge of potential climate change as *terra infirma*, implying unstable terrain, and our knowledge of the social and economic impacts of such change as *terra incognita*, or unknown terrain to indicatc how much greater

are the uncertainties in this aspect of climate change research (Nordhaus 1990c, 1990e, 1990f, 1990g).

9. These include EPA (1988) and the Coolfront Workshop (1989).

10. See Cline (1990c) for a discussion of the costs of increased hurricane damage due to global warming. He provides a conservative estimate of $750-1,000 million annually with a doubling of atmospheric CO_2.

11. This is his estimate for a medium damage scenario. Using low and high damage scenarios, the optimal levels of reduction are 10% and almost 50%, respectively (Nordhaus, 1990c). In this paper, Nordhaus does describe a greenhouse damage function that increases with greenhouse gas concentration, but he indicates that he has little confidence in this assumption, and does not use it in further analysis.

More recently, Nordhaus (1990g) has sketched out an intertemporal general-equilibrium model of global climate change. This model does include an economic damage function which rises at an increasing rate with global climate change, and is used to determine optimal levels of emissions controls. Only preliminary results have been presented at this time.

12. One notable exception that the authors have encountered is with the McKinsey and Co. study prepared for the November 1989 *Ministerial Conference on Atmosphere Pollution and Climatic Change*, held in Noordwijk, the Netherlands. Goldemberg (1990) cites the results of this study. He reports costs of controlling greenhouse gas emissions are significantly lower than estimated in other studies; a tax of just $1 per barrel of oil-equivalent, or $6 per ton of coal-equivalent used to fund CFC phase-out, forest management, and energy conservation would reduce global emissions by 20%. When the authors contacted McKinsey and Co., they were told that the report could not be made available.

References

Allen, Myles R. (1990). "Proposed Index of Global Warming Effect Puts CO_2's Share to over 70%," *Energy Policy* 18(3), pp. 485-6.

Barbier, E.B. and D.W. Pearce (1990). "Thinking Economically about Climate Change," *Energy Policy* 18(1), p. 11.

Binkley, C.S. (1988). "A Case Study of the Effects of CO_2-Induced Climatic Warming on Forest Growth and the Forest Sector: B. Economic Effects on the World's Forest Sector," in M.L. Parry, T.R. Carter, and N.T. Konijn, eds., *The Impacts of Climatic Variations on Agriculture*, Dordrecht, Netherlands, Kluwer Academic Publishers, 1988, pp. 197-218.

Brown, Peter (1990). Comments at *Global Climate Change: The Economic Costs of Mitigation and Adaptation*, Washington, D.C., December 4-5, 1990.

CBO (1990). *Carbon Charges as a Response to Global Warming: The Effects of Taxing Fossil Fuels*, Congress of the United States, Congressional Budget Office, August.

Chapman, Duane and Thomas Drennen (1990). "Equity and Effectiveness of Possible CO_2 Treaty Proposals," *Contemporary Policy Issues* VIII(3), pp. 16-28.

Cline, William R. (1989a). "Political Economy of the Greenhouse Effect," unpublished paper, August.

———. (1989b). "Policy Decision Under Uncertainty," unpublished paper,November.

———. (1990a). "Greenhouse Warming Abatement: Lower Benefits and Higher Costs?" unpublished paper, February.

———. (1990b). "Economics of the Greenhouse Effect: Knowns and Unknowns," unpublished paper, April.

———. (1990c). "Global Warming and Costs of Hurricane Damage," unpublished paper, July.

———. (1990d). "Economic Stakes of Global Warming in the Very Long Term," unpublished paper, November.

Coolfront Workshop (1989). *Climate Impact Response Functions: Report of a Workshop Held at Coolfront, West Virginia*, September 11-14, 1989, National Climate Program Office, Washington, D.C.

Cristofaro, Alex (1990). "The Cost of Reducing Greenhouse Gas Emissions in the United States," presented at *Global Climate Change: The Economic Costs of Mitigation and Adaptation*, Washington, D.C., December 4-5, 1990.

DOE (1990). U.S. Department of Energy, *The Economics of Long-Term Global Climate Change: A Preliminary Assessment*, Report of an Interagency Task Force, September.

Drennen, Thomas and Duane Chapman (1990). "Biological Emissions and North-South Politics," presented at the *Conference on Global Change: Economic Issues in Agriculture, Forestry, and Natural Resources*, Washington, D.C. Nov. 19-21.

Dudek, Daniel J. and Alice LeBlanc (1990). "Offsetting New CO_2 Emissions: A Rational First Greenhouse Policy Step," *Contemporary Policy Issues* VIII(3), pp. 29-42.

Edmonds and Barnes (1990). *Estimating the Marginal Cost of Reducing Global Fossil Fuel CO_2 Emissions*, PNL-SA-18361, Pacific Northwest Laboratory, Washington, D.C., August 1.

Edmonds, J.A., J.M. Reilly, R.H. Gardner, and A. Brenkert (1986). *Uncertainty in Future Global Energy Use and Fossil Fuel CO_2 Emissions 1975 to 2075*, TRO36, DO3/NBB-0081, National Technical Information Service, U.S. Department of Commerce, Springfield, Virginia 22161.

EPA (1988). *The Potential Effects of Global Climate Change on the United States*, U.S. Environmental Protection Agency, Draft Report to Congress, October.

EPA (1989). *Policy Options for Stabilizing Global Climate*, U.S. Environmental Protection Agency, Draft Report to Congress, February.

EPA (1990). *Reducing Risk: Setting Priorities and Strategies for Environmental Protection*, U.S. Environmental Protection Agency Science Advisory Board, September.

Flavin, Christopher and Nicolas Lenssen (1990). "Saving the Climate Saves Money," *World Watch* 3(6), November-December, pp. 26-33.

Geller, Howard (1990). "Saving Money and Reducing the Risk of Climate Change Through Greater Energy Efficiency," presented at *Global Climate Change: The Economic Costs of Mitigation and Adaptation*, Washington, D.C., December 4-5, 1990.

Goldemberg, Jose (1990). "How to Stop Global Warming," *Technology Review* 93(8), pp. 24-31.

Hoeller and Nicolaisen (1990). "A Survey of Studies of the Costs of Reducing Greenhouse Gas Emissions," General Economics Division, Economics and Statistics Division, OECD, Paris. Draft manuscript presented at a *Workshop on Economic/Energy/ Environmental Modeling for Climate Policy Analysis*, Washington, D.C., October 22-23, 1990.

Hogan, William W. and Dale W. Jorgensen (1990). *Productivity Trends and the Cost of Reducing CO_2 Emissions*, Harvard University Global Environmental Policy Project Discussion Paper.

IPCC (1990). *First Assessment Report: Overview*, Intergovernmental Panel on Climate Change, World Meteorological Organization and United Nations Environment Programme, August 31, 1990.

Jorgensen, Dale W. and Peter J. Wilcoxen (1990). *The Cost of Controlling U.S. Carbon Dioxide Emissions*, Harvard University Working Paper.

Kellogg, William W. and Robert Schware (1981). *Climate Change and Society: Consequences of Increasing Atmospheric Carbon Dioxide*, Boulder, Colorado: Westview Press.

Lashof, Daniel A, and Dilip R. Ahula (1990). "Relative Contributions of Greenhouse Gas Emissions to Global Warming," *Nature* vol. 344, pp. 529-31.

Manne, Alan S. (1990). "Global 2100: An Almost Consistent Model of CO_2 Emission Limits," for presentation at *Applied Equilibrium Modeling Conference*, Bern, March 1990.

Manne, A.S. and R.G. Richels (1990a). "CO_2 Emission Limits: An Economic Cost Analysis for the USA," *The Energy Journal* 11(2), April 1990, p. 51.

———. (1990b). "Global CO_2 Emission Reductions – The Impacts of Rising Energy Costs," forthcoming in *The Energy Journal*.

———. (1990c). "Buying Greenhouse Insurance," in forthcoming monograph *Global 2100: the Economic Costs of CO_2 Emission Limits*.

Metz, Bert (1990). "Climate Policy in the European Community and Its Economic Aspects," presented at *Global Climate Change: The Economic Costs of Mitigation and Adaptation*, Washington, D.C., December 4-5, 1990.

Mintzer, Irving M. (1987). *A Matter of Degrees: The Potential for Controlling the Greenhouse Effect*, World Resources Institute, Research Report #5, April.

Morris, et al. (1990). *A Least Cost Energy Analysis of U.S. CO_2 Reduction Options*.

Munasinghe, Mohan (1990). "Energy-Environmental Issues and Policy Options for Developing Countries," paper prepared for the *AEI Bellagio Conference on Energy and the Environment in Developing Countries*, November 26-30, 1990.

Nature (1990). "Greenhouse Numbers," opinion page, vol. 347, p. 410.

New York Times (1991). "E.P.A. Acts to Reshuffle Environmental Priorities," 26 January, p. A11.

New York Times (1989). "Staggering Cost is Foreseen to Curb Warming of Earth," 19 November, p. A1.

New York Times (1988). "Major 'Greenhouse' Impact is Unavoidable, Experts Say," 19 July, p. C1.

Nordhaus, W.D. (1979). *The Efficient Use of Energy Resources*, Yale University Press, New Haven, CT, 1979.

Nordhaus, William D. (1990a). "Uncertainty about Future Climate Change: Estimates of Probable Likely Paths," unpublished paper, January 1990.

———. (1990b). "Contribution of Different Gases to Global Warming: A New Technique for Measuring Impact," unpublished paper, February 1990.

———. (1990c). "To Slow or Not to Slow: The Economics of the Greenhouse Effect," unpublished paper, February 1990.

———. (1990d). "A Survey of Estimates of the Cost of Reduction of Greenhouse Gas Emissions," unpublished paper, February 1990.

Nordhaus, William D. (1990e). "Economic Policy in the Face of Global Warming," unpublished paper, March 1990.

———. (1990f). "Global Warming: Slowing the Greenhouse Express," forthcoming in Henry Aaron, ed., *Setting National Priorities*, 1990.

———. (1990g). "Greenhouse Economics: Count before You Leap," *The Economist*, 7 July 1990, p. 21.

———. (1990h). "An Intertemporal General-Equilibrium Model of Economic Growth and Climate Change," paper presented at a *Workshop on Economic/Energy/Environmental Modeling for Climate Policy Analysis*, Washington, D.C., October 22-23, 1990.

Nordhaus, William D. and G.W. Yohe (1983). "Future Paths of Energy and Carbon Dioxide Emissions," in *Changing Climate: Report of the Carbon Dioxide Assessment Committee*, National Research Council, National Academy Press, Washington, D.C.

Oppenheimer, Michael and Peter Boyle (1990). *Dead Heat: The Race Against the Greenhouse Effect*. New York: Basic Books, Inc.

Perman, R. (1990). "Greenhouse Effect Economics," letter to *Nature* vol. 347, p. 10.

Reilly, J.M., J.A. Edmonds, R.H. Gardner, and A.L. Brenkert (1987). "Uncertainty Analysis of the IEA/ORAU CO_2 Emissions Model," *The Energy Journal* 8(3), p. 1.

Rodhe, Henning (1990). "A Comparison of the Contribution of Various Gases to the Greenhouse Effect," *Science* vol. 248, pp. 1217-1219.

Schneider, Stephen (1989). *Global Warming: Are We Entering the Greenhouse Century*. San Francisco: Sierra Club.

Victor, David G. (1990). "Calculating Greenhouse Budgets," letter to *Nature* vol. 347, p. 431.

Yamaji K., Matsuhashi R., Nagata Y., and Kaya Y. (1990). "An Integrated System for CO_2/Energy/GNP Analysis: Case Studies on Economic Measures for CO_2 Reduction in Japan," paper presented at a *Workshop on Economic/Energy/Environmental Modeling for Climate Policy Analysis*, Washington, D.C., October 22-23, 1990.

Zimmerman, Mary Beth (1990a). *Assessing the Costs of Climate Change Policies: The Uses and Limitations of Models*, The Alliance to Save Energy.

———. (1990b). "The Benefits of Mitigation and Adaptation," comments at *Global Climate Change: The Economic Costs of Mitigation and Adaptation*, Washington, D.C., December 4-5, 1990.

BENEFIT-COST ANALYSIS

W. David Montgomery

Charles River Associates
John Hancock Tower
200 Clarendon Street, T-43
Boston, MA 02116

I'd like to concentrate on three topics related to global climate change. Previous presentations have contained very important lessons on the timescales involved in global warming and the value of investing in additional information before taking action. I would like to draw out and emphasize these lessons.

Other speakers have made statements on two other topics: 1) the applicability of benefit-cost analysis to the problem of global climate change, and 2) the notion that we can accomplish our goals for free or better. I find myself profoundly disagreeing with some of the statements.

I will begin with benefit-cost analysis, and will attempt to illustrate a particular approach involving the comparison of the benefits of actions taken to defer global climate change and the benefits of those actions. In doing so I will use some of the results from our Congressional Budget Office study.[1]

Benefit-Cost Analysis

I'd like to sketch out how a relatively conventional but technically competent economist such as myself might think about the benefits and costs of policy actions and environmental changes, illustrate how this kind of analysis could be used in a sound manner, and discuss what kinds of questions it can and cannot answer about issues like global warming. To perform this kind of analysis, I start with a baseline[2] that requires assumptions about population growth and projections of economic activity and energy use over the relevant time period—let's say at least to the year 2100. With this baseline, one can begin to construct a measure of economic welfare and a projection of what carbon dioxide emissions will be in the absence of any government policy.

I think we would all admit that trying to construct measures of economic welfare is a difficult enterprise, subject to all sorts of criticism, and that Gross National Product (GNP) is not the best measure of economic welfare. It's also not the measure that most of us use when we do a benefit-cost analysis. Economists have long worked to figure out

Copyright 1991 by Elsevier Science Publishing Company, Inc.
Global Climate Change: The Economic Costs of Mitigation and Adaptation
James C. White, Editor

how to value amenities (if this is even possible), nonmarketed commodities, or environmental improvements. To the extent possible, a good, fleshed-out benefit-cost analysis will include those things.

For simplicity, we sometimes express measures in terms of GNP. Most studies I know of that do so take a fairly extended view of what the GNP is and, in fact, represent it as, for instance, the value of the consumption and investment activities that take place in the economy and which are a source of economic welfare and individual satisfaction.

So we start with a baseline that gives us a measure of economic activity as well as a measure of emissions in the absence of any controls.

Costs

Costs in this discussion represent a reduction in economic welfare resulting from a particular program to reduce emissions, and are tied to a particular quantity of emissions reduction. Benefits come fairly far down the road, realized when reduction in emissions reduces concentrations of CO_2 in the atmosphere. We believe that such a reduction in concentrations will delay rises in temperature, which will delay a variety of adverse effects on our physical and biological environment, thus delaying certain losses in economic welfare. We will arrive at this (perhaps indefinite) delay in economic welfare loss after a lot of deductive reasoning and as the result of a particular set of policy actions that reduce emissions by a certain amount.

There are many misconceptions regarding the dividing line between costs and benefits.[3] One is that we fail to account for the damage likely to occur as a result of global warming, thus not measuring our baseline correctly. I think that for the most part those who make this claim simply disagree with how we have defined costs and benefits. In the context of this discussion, "costs" are the reductions in economic activity resulting from emissions reduction. "Benefits" are the improvements in the economy resulting from a delay in global warming.

Some think that the costs for any policy action taken should always be expressed as the net of the benefits it provides. I find it easier for exposition to keep those things separate, but it really doesn't matter. The point is that nothing is missed; it's all counted somewhere. It is simply a question of where it's classified.

If we take into account the effects of global warming that may occur in the absence of policies, we may find that our projections of economic activity are excessively optimistic. That is, global warming is going to cause some reduction in economic activity because of the damage that it causes to our economies.

The highest estimate I've heard at this conference was that a ten-degree rise in world-wide temperature would result in perhaps a five-percent world-wide GNP loss after a couple of centuries. Even if we correct our baseline accordingly, it means that maybe we are off by five percent (at worst) in our projection of emissions. That won't make much difference in any policy analysis. It really doesn't matter whether we take the benefits into account in setting the baseline.

I see the crux of this problem as when, not whether, global warming occurs. There are two key related factors. One, carbon dioxide builds up cumulatively in the atmosphere. Once we have put it there, it stays there for a long, long time unless we set more policies that eliminate it. Two, it is the stock of CO_2 in the atmosphere which, with some lags, produces rises in temperature.

Let me try to make these points more concrete. We examined a number of time periods on the baseline we began with at the Congressional Budget Office (CBO).[4] I'm going to use the longest time period because it is the most interesting one, given the other discussions at this conference.

In this baseline we see a substantial rise in carbon dioxide emissions world-wide (see Figure 1). But there's no reason for self-flagellation; the United States is not responsible for this rise. These conclusions about the relative share of the United States and other countries agrees with those that Rich Richels discussed yesterday:[5] that growth in population and income in the developing countries will lead to a large percentage of this CO_2 increase. We can draw our own implications about what this means for "eco-imperialism."

Moving to the nature of controls, we took this baseline and against it imposed both carbon charges adopted unilaterally by the United States and multilaterally by nations of the world (see Figure 2). We took a carbon charge that began at $100 per ton in the year 2000 and rose to $300 per ton of carbon by the year 2100. This is also similar to what Richels was talking about. A tax of $300 per ton of carbon is $200 per ton of coal, perhaps a tenfold increase over today's coal price. That would result in a stabilization of U.S. emissions at a level perhaps 20 percent or so below the current level. That involves a reduction of about 50 percent from the baseline emissions for the United States.

Again, the level to which we reduced emissions looks very much like the constraint imposed by Manne and Richels,[6] except that their baseline was higher. So, it took about a 75-percent reduction from the baseline to achieve this stabilization at 20 percent below current levels.

We conducted an experiment with our model at CBO, levying a charge that was high enough to achieve a 75-percent reduction from baseline. The cost to the United States was a loss of about one percent of GNP for a charge of $100 per ton, and a loss of about two percent of GNP by the time this charge rose to about $300 per ton. We found the cost of a charge large enough to achieve a 75-percent reduction in emissions from the baseline to be about three percent of GNP. So, for the United States we are looking at a cost of approximately two to three percent of GNP to stabilize emissions at either their current levels, or in the range of 20 percent below current levels.

Benefits

Now, how might we think about the benefit side? What effect does a reduction in emissions have on, for instance, global carbon dioxide emissions? What we found here was somewhat less of a reduction than in the results of Manne and Richels,[7] though still in the same ballpark: a reduction from baseline that is substantial, that doesn't quite get us to stabilization, yet reduces world-wide emissions by at least 50 percent (if not more). Such a reduction in total world-wide emissions would be realized by universal adoption of a $300-per-ton charge.

That brings us to concentrations in the atmosphere. I have a feeling that in this regard Bill Cline made a fundamental error yesterday, because stabilization of emissions at current levels or at 20 percent below current levels will not achieve stable temperatures. Indeed, even a cessation of emissions will not achieve stable temperatures, because we have not yet achieved equilibrium with the concentrations already in the atmosphere. Some temperature increase is built in even if we stabilize emissions at current levels. Concentrations in the atmosphere will continue to grow because we are putting CO_2 in a lot faster than it can be taken out.

The most ambitious policy we examined at CBO is one that would delay when we would get to serious concentrations in the atmosphere (Figure 3) from the middle of the 21st century to the century beyond. But to understand whether the cost of two to three percent of GNP throughout the next century is worth the saving we must attempt to attach an economic value to that delay.

Again, we are delaying the time at which we reach certain levels of concentration. Suppose a change in this time were to reduce the temperature that the world reaches in 2050 by one degree. We are not going to accomplish a reduction of ten degrees, such as Bill Cline was talking about, because much of that increase is already built into the kind of atmospheric forcing coming from our current emissions rates. Maybe we'll get a one-degree reduction from the kind of emission reduction I was illustrating.

You can pick any number you like between one degree and the ten degrees cited by Bill Cline and then calculate how much of the five percent of GNP we might save world-wide by avoiding that ten-degree increase in temperature. I am going to pick the number one because it is easy to deal with.

Suppose that through policy we avoid a one-percent loss in GNP due to this change. A policy of adopting carbon charges that costs us something like two to three percent of GNP for 100 years, and accomplishes a benefit in some later period of time of saving perhaps one percent of GNP, may not be worthwhile. What should we do then?

The responsible continuation of benefit-cost analysis suggests that we try again. Suppose we did not take such stringent measures? Suppose we had a lesser degree of control? We would realize a lesser reduction in emissions, but a reduction that still may bring about a worthwhile change in temperature and whose benefits are more commensurate with costs. We just have to keep trying until we see where things balance out.

Bill Nordhaus tries to achieve this balance in his work,[8] which has not been discussed at this conference. He concludes that relatively modest carbon charges and only modest levels of reduction in CO_2 emissions can be justified when one balances these very crude estimates of benefits against the crude estimates of the costs.

I do not want to argue, nor do I really have an opinion, about where the appropriate dividing line is, or whether the benefits of stabilizing emissions are worth the costs. The point is that we have to be very careful in comparing these margins. We have to look at the costs associated with each particular policy option, and be sure that the benefits that we compare to that cost in fact are the benefits of the actual predicted change in temperature and other consequences resulting from that policy option. In particular, we should not make the error of comparing the costs of a policy option that does not achieve zero increase in temperature with the benefits of achieving that zero increase. We must keep the margins tracking with each other.

The Value of Information

I'd like to comment on the value of information. Perhaps this discussion of benefit-cost analysis is irrelevant because we do not need to decide now how much to reduce emissions. In fact, we have a choice of whether to impose controls now, whether to invest in accumulating new information, or whether to invest in new research and development activities. Some interesting work on these subjects has been done by Manne and Richels,[9] as well as by Steve Peck and Tom Teisberg at the Electric Power Research Institute.[10] I'm drawing on things that I have heard and read in their work.

First, we do not really know the full scope of the global warming problem. Bill Clark pointed out the difference between the problem definition in Germany and what he characterizes as being the U.S. view.[11] The view in Germany is that emissions, concentrations of CO_2, and global temperatures are going to rise very rapidly if we do not do something. The view in the United States is that almost nothing is going to happen. Both countries have established their policy positions along these lines.

Whether or not either view is correct, they demonstrate that our belief in the seriousness of the problem has a lot to do with how many controls we believe should be imposed. We do not know the answer right now. We might feel we have to hedge by undertaking a lot of controls. But suppose we invest in a decade or so of atmospheric and climatological research to get a better idea of the size of the problem. We could then commit ourselves with much greater confidence to a policy of more or less stringent control. It might be worth spending a lot of money over the next ten years to obtain such information.

Development of New Technologies

Likewise, waiting for development of new energy technologies could afford us the lowest-cost solution. We might decide to forego some control now if we know more stringent control will cost only a fourth as much 20 years from now. What matters is the total amount of carbon that gets into the atmosphere over that long period of time. It does not matter when we put it in or when we take it out. So, if it is cheaper to take it out in the future, that might make more sense. We have to, of course, balance such a decision with the possibility of falling off the cliff before we get to the point where those new technologies are available; that is what worries many who see a looming potential catastrophe.

Alex Cristofaro's idea about controlling all pollutants and taking credit for the way in which we have controlled pollutants other than carbon dioxide is a very interesting one.[12] Even if we do wipe out chlorofluorocarbons, nitrous oxides, methane, and the other contributors to global warming, we still end up with a CO_2 problem that must eventually be addressed. But by concentrating on the other pollutants now, we might buy a lot of time for reducing carbon dioxide, for getting information about the real consequences of global warming, and for allowing new technologies to be developed for carbon dioxide control.

Market Values

Finally, let me address market values. I have a fantasy that at one of these gatherings someone will describe some real failures in markets or institutions that give us reason to believe that real cost savings are possible, and not have elitist pronouncements about how most people cannot do arithmetic well enough to figure out lifecycle costs, or hear about how everyone has a MasterCard and so should be able to buy a new refrigerator. I think that many of these engineering calculations are flawed by what seems, in some cases, to be a willful refusal to recognize that consumers have legitimate interests in some aspects of a commodity other than its lifecycle energy costs.

Motor vehicles are a good case in point. Bob Leone of Boston University has done a good study comparing the costs of Corporate Average Fuel Economy (CAFE) standards to other measures for reducing motor vehicle emissions. He attempts to measure the value that consumers place on amenities in cars, including acceleration, safety, size, accessories, comfort – much more than just the cost of gasoline and the make and model. These kinds of amenities are changed by the changes required to reduce gasoline consumption at least in the short run, or can only be maintained by increasing the cost of the vehicle substantially in order to make very sophisticated and complex engine changes. Taking the changes into account, we may find that consumers lose a great deal when required to purchase more efficient vehicles, and that needs to be compared with the cost of the gasoline savings.

Finally, as was pointed out several times yesterday, economic behavior can almost always defeat the purpose of regulations. CAFE standards are a good case in point. We do know that vehicle miles traveled respond very strongly to the cost of driving. If by making vehicles more efficient we reduce the cost of driving, we can predict with confidence from our economic models the increase in driving, the resulting additional gasoline consumption. We could avoid the problem of increasing gasoline consumption by using economic incentives like carbon charges to bring about both changes in behavior in driving and changes in the design of vehicles as manufacturers respond to what would then be consumer sentiment, that is, a desire for more efficient vehicles.

Conclusion

I will end with a notion I heard a long time ago, a recommendation from a modeler: "Don't eat the menu." Don't believe that your model results represent the real world you want to regulate. It worries me that many proposals for efficiency standards and regulation of energy use are based on energy models that assume the typical consumer situation to be the same as that of the engineer doing the calculation. Too often such models are believed to be representative of the entire diversity of circumstances and preferences of consumers in this country.

Take inefficient air conditioners. For the consumer with the typical profile – time they spend in the house, family size, and so on – we can calculate the type of air conditioner they ought to have. Yet we find that lots of people do not buy the "right" one. Why? Are they stupid? Perhaps, but I don't think so.

For single people, one good reason for buying an inefficient air conditioner may be that they spend little time in their houses. They are quite content to turn the air conditioner on, have it cycle for an hour or two when they get home in the evening, turn it off when they go to sleep, and leave it off until they return home the next evening. That is a very different pattern of use than that of someone at home all day with three children.

I find it absurd to require that people buy the high-efficiency model when they do not need constant air conditioning. There is a good reason why the market produces the diversity of products that it does, and why consumers should be able to make a choice among these products.

The same holds for refrigerators. People may well decide that it's better to purchase an affordable refrigerator when they need one than to save for a more efficient model required by some efficiency standard.

I think that considerably more investigation of the attributes of commodities that matter to people, and the actual diversity of circumstances that exists in the world, is necessary before we claim that we can save people money by requiring them to do what they have already decided is not in their own economic interest.

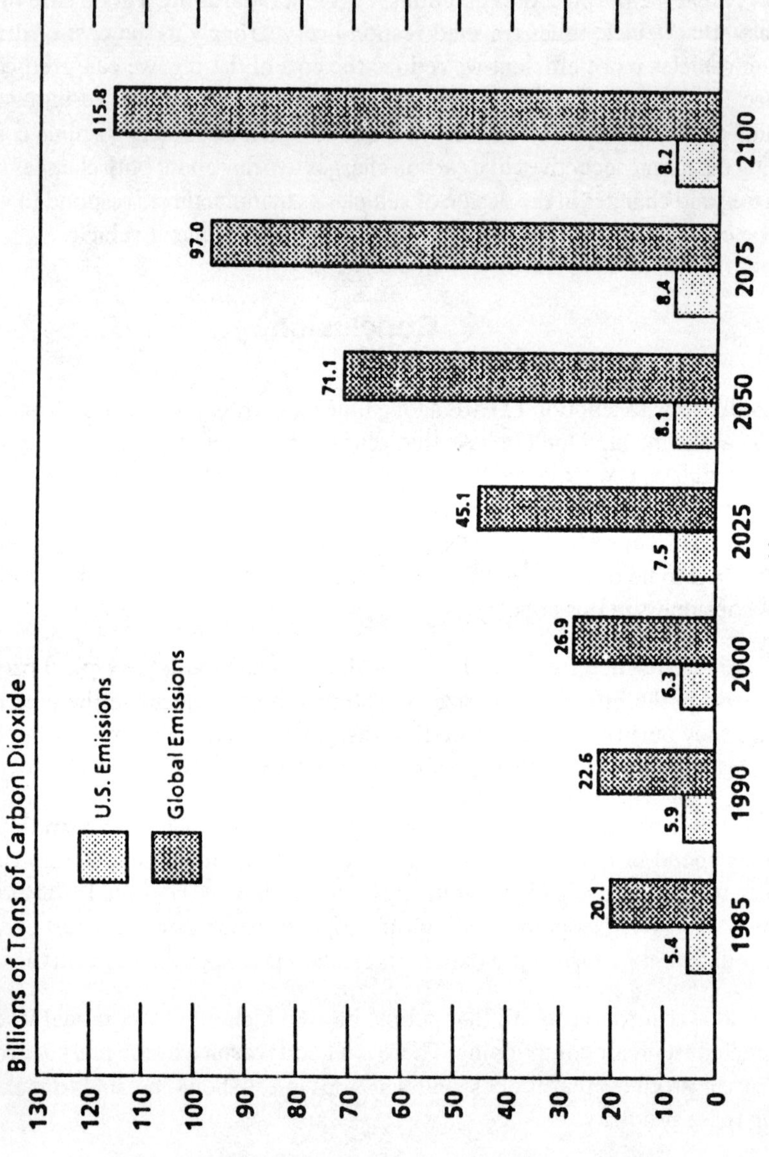

Figure 1

BASELINE CARBON DIOXIDE EMISSIONS, ASSUMING HIGH ECONOMIC GROWTH

SOURCE: Congressional Budget Office simulations using the Edmonds-Reilly model with the high-growth scenario developed by the Environmental Protection Agency.

Figure 2

EFFECTS OF CARBON CHARGES ON U.S. EMISSIONS OF CARBON DIOXIDE, ASSUMING HIGH ECONOMIC GROWTH

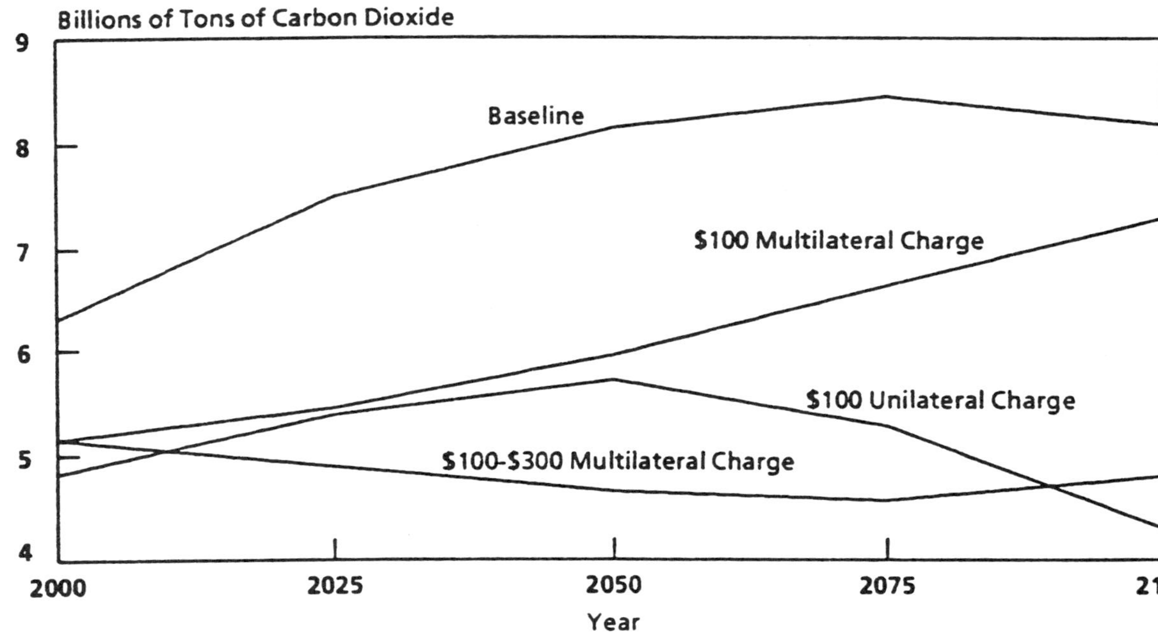

SOURCE: Congressional Budget Office simulations using the Edmonds-Reilly model with the high-growth scenario developed by the Environmental Protection Agency.

Figure 3

EFFECTS OF CARBON CHARGES ON GLOBAL CONCENTRATIONS OF CARBON DIOXIDE, ASSUMING HIGH ECONOMIC GROWTH

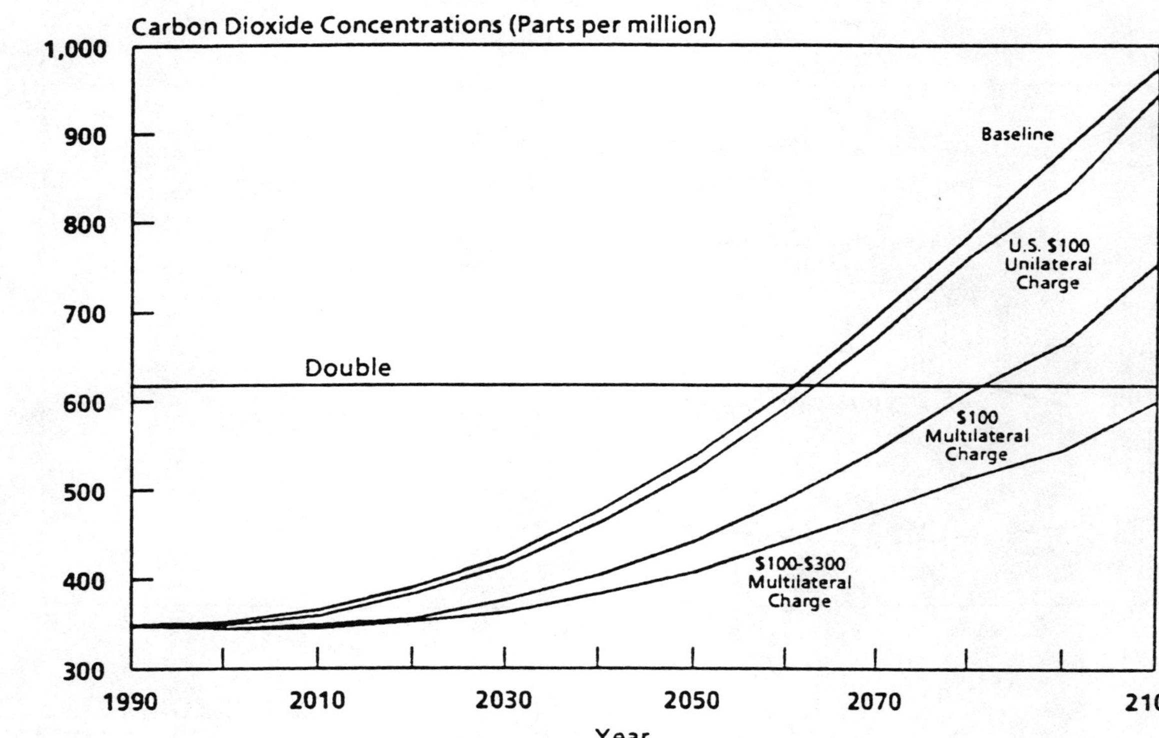

SOURCE: Congressional Budget Office simulations using the Edmonds-Reilly model with the high-growth scenario developed by the Environmental Protection Agency.

Notes

1. Congress of the United States, Congressional Budget Office. *Carbon Charges as a Response to Global Warming: The Effects of Taxing Fossil Fuels*. August 1990.

2. Cline, William R. "Methods of Analysis." Paper included in this volume.

3. Zimmerman, Mary Beth. "The Benefits of Mitigation." Paper included in this volume.

4. Congress of the United States, Congressional Budget Office. *Carbon Charges as a Response to Global Warming: The Effects of Taxing Fossil Fuels*. August 1990.

5. Richels, Richard G. "Global CO_2 Emission Reductions The Impacts of Rising Energy Costs." Paper included in this volume; coauthored by Alan S. Manne.

6. Manne, Alan S. and Richels, Richard G. "CO_2 Emission Limits: An Economic Cost Analysis for the USA." *The Energy Journal*, Vol. 11, No. 2., April 1990.

7. *Ibid.*

8. Nordhaus, William D. "To Curb or Not to Curb: The Economics of the Greenhouse Effect." Paper presented to the annual meeting of the American Association for the Advancement of Science, New Orleans, February 1990.

9. Manne, Alan S. and Richels, Richard G. "Buying Greenhouse Insurance." Preliminary draft of chapter in forthcoming monograph, *Global 2100: The Economic Costs of CO_2 Emission Limits*. November 1990.

10. Peck, Stephen C. and Teisberg, Thomas J. "A Framework for Exploring Cost Effective CO_2 Control Paths." Presented at a *Workshop on Economic/Energy/Environmental Modeling for Climate Policy Analysis*, Washington, D.C., October 22-23, 1990.

11. Clark, William C. "Adaptive Planetary Management." Paper included in this volume.

12. Cristofaro, Alex. "The Cost of Reducing CO_2 Emissions in the United States." Paper included in this volume.

ECONOMIC MODELING OF GLOBAL CLIMATE: COMPARING APPROACHES*

John Reilly

Resources and Technology Division
Economic Research Service
United States Department of Agriculture
1301 New York Avenue, NW
Washington, DC 20005-4788

Increasing consideration regarding the need for and cost of limits on trace gas emissions comes against a background of little if any consistent analytical efforts to address the economic costs of climate change or the benefits of slowing the rate of increase of trace gases. Almost universally, the trace gas limitations policies considered by analysts fall far short of the constraints needed to stabilize climate. Thus, these policies would only avoid a fraction of—or delay slightly—the expected damages of climate change. Studies that have addressed the costs of climate change have generally shown the costs to be quite small relative to estimates of the costs of limiting trace gas emissions. There continues to be a significant focus on climate change due to carbon dioxide emissions from fossil fuels to the exclusion of analyses of other radiatively active gases and to the exclusion of other effects of trace gases—for example, the positive effect of carbon dioxide on plant growth. It is possible to concisely represent the relative value of reducing different trace gases, but this requires more information on the economic damages due to climate change and the nonclimatic damages or benefits of trace gases and the relationship of damages to the atmospheric concentration.

A variety of models has been used to examine the economics of climate change. As important as the model structure and type is how the model will be used in the policy process. It is not possible to develop a single model that does everything well and to try to do so produces models that may be equally good at everything but not very good at anything. In reviewing policy modeling research on climate change, this paper begins by

---

* The views expressed in this paper are those of the author and do not reflect the views of the U.S. Department of Agriculture or the U.S. government.

Copyright 1991 by Elsevier Science Publishing Company, Inc.
Global Climate Change: The Economic Costs of Mitigation and Adaptation
James C. White, Editor

considering the role in the policy process the models were designed to play. Next, fundamental aspects of modeling natural resource/energy issues are identified to further understand differences among economic models of the climate change issue. Finally, several widely accepted propositions that are embodied in models, research, and policy discussions are questioned under the general rubric of "modeling fallacies."

Roles of Models in Policy Research

LeBlanc and Reilly provide a useful typology for evaluating models used for energy and natural resource issues, identifying three roles research modeling exercises play in the policy process:

- *Anticipatory research* uncovers potential policy issues and develops general expectations about the future level of relevant variables. Forecast models are the embodiment of research that fills this role.

- *Policy design research* aids in the specific design of policies. Models that illustrate the costs and benefits, that illustrate the consequences of climate change or policy actions, or that examine the effectiveness of specific policy measures are included here.

- *Evaluative research* evaluates the effectiveness of ongoing policies. Are the policies effective? Are the costs and benefits of the policies as expected? Are there unanticipated loopholes or effects of the policies?

Until very recently, economic research on climate change largely fell into the category of anticipatory research. Nordhaus and Yohe, Edmonds and Reilly, and Manne and Richels, for example, were primarily interested in describing likely future emissions and the uncertainty associated with emissions. There is some capability in these models to broadly examine the costs of control strategies but they provide only rough estimates of the effectiveness and cost of policy actions. These rough estimates help policymakers anticipate the magnitude of the cost of control and illustrate the broadest considerations regarding policy design. They are unable to provide evaluation of many specific proposals that are being and are likely to be developed. A principal advantage of these models in their anticipatory role, their ability to create scenarios that extend 50 to 100 years into the future, is a liability as the policy question turns toward policy design because the regional and sectoral detail needed to evaluate specific proposals and their effects over the next five to ten years is lacking. None of these models has been statistically estimated from historical data (they are models based on economic principals but are not econometric models), because such past data provide only the broadest guide to the long-term future.

The climate change issue is moving from the stage of creating broad expectations of the future to one of evaluating and designing specific proposals. The energy concerns of the past have generated a wide variety of models and modeling approaches that are being brought into service. The Jorgenson and Wilcoxen model and models at DRI and WEFA offer examples of traditional econometric models. Applied general equilibrium models (Shoven and Whalley, OECD effort underway, Australian model) are a somewhat different class. There are also a variety of non-economic models and some interesting hybrids that have been developed to deal with energy demand and supply analysis and could be used to assess climate change policy. Included among these models are sectoral econometric, time series models, both econometrically estimated and non-econometrically parameterized general equilibrium models and non-economic optimization and simulation models.

Even though climate change is a problem of global concern with a long time horizon, detailed policy is relatively short-term and will take place at the national level. Thus, it does not seem critical that models aimed at global climate change policy have a time horizon of more than five to ten years, whereas greater geographical disaggregation than available in the long-term models will be critical. At the same time, a significant shortcoming of existing short- and medium-run models is that they do not provide global coverage. The danger of this shortcoming was illustrated by Edmonds and Reilly (1983), when they demonstrated that emissions reductions policies implemented in one country would be fairly ineffective because energy consumption would move abroad. This occurs because world energy prices fall in response to reduced demand and because energy intensive activity may move abroad. National models, no matter how detailed and accurate, will overestimate the effectiveness of emissions limitations policies unless the policies are uniformly implemented across countries.

The work of Manne and Richels, and understanding the importance of expectations in economic decisions, mean that some care is needed in interpreting results from models with shortened time horizons. Clearly, such models do not provide insight into the benefit of taking action now versus waiting. Nor are they able to fully consider how the effectiveness of a policy will depend on whether private sector producers and consumers believe, for example, that emissions taxes will rise steadily in the future or after initial increase fall (or more accurately the energy price plus tax). If the private sector generally believes policy will increasingly press for emissions reductions, then long-term investment and research and development will begin to respond in anticipation of higher taxes. If the control policies are seen as a short-term policy "fad" then investment decisions will not be altered and conservation and fuel switching will be limited to adjustments in the use of capital. Consumers may drive a bit less but will not buy more fuel efficient vehicles. Companies may push to market energy-saving technology they have developed in the past but will not invest in a long-term research and development program to bring innovative products and approaches to market.

The danger in detailed policy design studies is that the broader insights developed by models that capture the big picture are lost and, as a result, the policies fail miserably. Thus, even as the problem of climate change moves to one of detailed policy design, the types of models developed in the anticipatory research stage will continue to be useful. Whatever the form of climate change policy (i.e. whether emissions taxes, tradable permits, or commitment to specific technology options or lifestyle goals), it is unlikely that the policy will have a lifetime of more than five or ten years. It will prove less effective or more costly than anticipated or we will find climate change to be more or less serious than we thought. Anticipation of the next round of adjustment will be necessary and the longer-term models will be useful for this purpose.

There is also a need to identify the energy problem as part of the broader trace gas problem, as part of a cost-benefit tradeoff, and as part of a broader set of environmental effects of trace gases and the human activities contributing to climate change. Nordhaus (1990), Reilly (1990), and Peck and Teisberg have models which attempt to assist in the comparison of benefits and costs and in how one might best trade-off between reductions in carbon dioxide from fossil fuels, and methane, CFC and nitrous oxide emissions from other sources. Since at least 1983 it has been recognized that other trace gases are of roughly equal importance to carbon dioxide (EPA). This has been confirmed by the recent IPCC Report. Deforestation/land use and cement manufacture are other sources of carbon dioxide. Thus, fossil energy is only 35 to 40 percent of the problem. Even if we eliminated all fossil energy carbon emissions we would still have climate change. Clark (1986) looks at multiple environmental effects of human activities. Reilly (1990) develops a model with both climatic and nonclimatic effects of trace gases.

Given that climate change policies have not been implemented, it may seem premature to discuss evaluative models. But these are frequently ignored. Policy is judged successful or not on broadly observable measures of success that may have little to do with the policy and much to do with private market forces or extraneous political events. How much of energy savings in the 1970s and 1980s was due to policy, and how much was due to market-induced conservation that led to the collapse of oil prices? Generally, econometric and statistically based models can be used to attempt to separate effects. Notably, the energy policies of the 1970s and 1980s provide a rich set of "experimental" data. Many of the same policies are being resurrected as solutions to climate change but their previous effectiveness has not been carefully evaluated. As we move forward with climate change policy, it will be important to evaluate the effectiveness and cost per unit of reduction of control policies and the avoidance of climate change damage.

What makes a good model?

LeBlanc and Reilly suggest three key considerations in modeling energy and natural resource issues:

- Physical characteristics and constraints – the amount of resource available, physical capital constraints, and new technology options.

- Institutional concerns – market structure considerations (OPEC), political concerns (stability of governments, acceptability of approaches), and regulatory considerations (e.g. existing environmental constraints, public utility regulation).

- Economic behavior—how consumers and producers will respond to changing prices, regulations, or constraints.

Few models capture all of these aspects and there are some distinctly different approaches. The long-term models (Nordhaus and Yohe, Edmonds and Reilly, and Manne and Richels) represent physical characteristics and constraints and economic behavior in different degrees of detail but generally ignore institutional concerns.

Among the medium-term models, there is a significant disjuncture between those which provide detail on physical/technological constraints and those who focus on economic behavior. Technological simulation models frequently incorporate the assumption that the optimal solution is the least-cost solution as costs are described by the technology descriptions in the model and as limited by specific constraints. Clearly, economic agents do not choose solely on the basis of minimum cost. With adequate income and choice, consumers do not buy the cheapest foods that meet daily nutritional requirements or the minimum-cost housing nor do they choose transportation that provides the cheapest per mile travel. Economic models assume that consumers maximize utility rather than minimize cost, and take their revealed choices in the market place as evidence of how they respond to changes. In these aggregate models, institutional considerations can be handled through scenario construction or constraints, or they are implicit in the revealed behavior of individuals.

In general, the modeling and analysis of institutional and regulatory considerations are detailed and limited in scope and of a very different nature than aggregate policy models. There is a tendency to demand that aggregate models represent these features. This is probably neither possible nor necessary. Improvements in design of utility regulation, information for consumers to make better decisions regarding energy use, zoning and transportation infrastructure development need study and may be effective, but aggregate econometric evidence implicitly includes the willingness to undertake such change through energy demand responsiveness to price. Carbon taxes or other broad policy instruments will increase the incentive to undertake such disaggregate responses whether the response is through purely private sector channels or through public agencies or regulatory bodies.

Ongoing Debates—Modeling Fallacies

There is a view that greater disaggregation improves models and understanding of aggregate behavior. A strong case can be made that greater disaggregation leads to poorer aggregate forecasts. As an example, there has been an ongoing debate regarding whether consumers increase energy consumption as their income increases. Aggregate econometric evidence suggests that there is a significant positive relationship, recognizing that other factors such as rising energy prices may offset the income effect. Yet, evidence on very disaggregate household data has been used to suggest that residential energy consumption of higher income consumers was no higher than lower income consumers.

In a cross-section econometric study, Reilly and Shankle found very low, and sometimes insignificant or negative income elasticities of energy demand in residential uses. However, they separately controlled for such variables as number of appliances, whether the home was single or multifamily, and square footage. Rising income probably does not lead consumers to heat their homes ever hotter or leave their refrigerator doors intentionally open, but rather consumers purchase larger, single family homes, with more appliances. In a similar vein, detailed studies of the industrial sector have sometimes underestimated the aggregate energy price response by failing to recognize that energy savings may occur not only through greater energy efficiency in, for example, iron and steel production, but also through substitution of materials in, for example, automobile production that are less energy intensive for iron and steel. While aggregate econometric evidence may give scant evidence on the specific mechanisms through which energy use changes, the reduced form nature of the relationships makes it less likely that important relationships are omitted.

The need or desire for disaggregation is one of wanting or needing disaggregated effects. If the distributional consequences are results of importance to policymakers and/or one adopts a policy approach with specific sectoral or technology goals, the disaggregation is necessary to identify how such goals will be achieved. For example, if climate change policy involves targets for auto fuel efficiency, converting fossil electric energy generation to nonfossil, and minimum efficiency standards for homes, then models with such detail are needed to assess what standards are needed to meet target emissions. If instead policy takes the form of fossil fuel taxes, allowing vehicle fuel efficiency and use, electric power generation and home heating to respond to these market signals, then the disaggregated results are unnecessary.

Interest group pressure for distributional consequences of policies typically creates a strong demand for disaggregated results. Unfortunately, many studies of sectoral consequences of policies are wrong or misleading. Independent sector studies may fail to consider that relative effects among sectors may be more important than the absolute effect on the sector. The plastics industry may suffer higher energy costs but if plastics are substituted for steel in manufactured goods, the sector may be a net beneficiary of higher energy prices even though the costs of plastics production rises. Even getting sectoral effects correct using a general equilibrium approach may have little to do with distributional consequences. Corporations and capital ownership are diversified sectorally and internationally and individual firms in industries may be better or worse positioned to take advantage of policy change. The Standard Industrial Codes (SIC) coding of data by sector makes it difficult to construct models that accurately treat distribution.

Thus, the aggregation versus disaggregation debate can be seen as having little to do with the quality of forecast result but rather is a reflection of policy philosophy. The danger in disaggregation is that it tends to create simple representations that are mathematically tractable. In the effort to estimate and parameterize hundreds or thousands of equations, the necessary scrutiny of the estimates and consideration of important interrelationships is secondary to consistency of approach and completing the effort.

Small sectoral changes in one country add up to something big without having effects on energy markets. Detailed studies and small country models appropriately make the assumption that changing a few light bulbs or reducing energy consumption in the country will not affect prices of energy. But studies that go on to add up global savings if these actions occur in many countries and over many specific technological options will overestimate savings because they will fail to incorporate the effects of lower energy prices as a result of reduced demand for energy. Because there is probably a significant rent in fossil fuel prices representing the mineral scarcity and exhaustibility, reductions in demand may have a much more significant effect on price than on quantity. In the end, the current price of fossil fuels is not the important comparison because fossil resources will be used unless conservation and fuel alternatives are cheaper than exploration/extraction costs. Similarly, Kane, Reilly, and Tobey have shown that international trade in agricultural commodities significantly redistributes the welfare changes due to yield losses or gains that may occur because of climate change.

Climate control strategies are neutral or beneficial with respect to all other environmental and human health and safety concerns. Aggregate energy models generally do not do a good job of considering joint benefits and costs of emissions reductions. However, there is a folklore that renewable energy and conservation are environmentally attractive whereas fossil energy is environmentally degrading. Fossil energy has serious environmental consequences but many of the conservation and alternative fuel strategies have negative environmental consequences as well. Large-scale biomass production would make extensive use of land and water resources, put added pressure to bring more land under cultivation with increases in erosion and pesticide and fertilizer use. Small, fuel efficient cars are less safe. Air-tight homes increase the potential for indoor air quality problems. And, compact fluorescent bulbs and photovoltaic technology could present significant toxic waste disposal problems.

Global warming has negative consequences. Much of the demand for control of emissions of trace gases is based on general circulation models and other climatological evidence that warming will be unprecedented. It remains undemonstrated whether warming and climate change will be unprecedentedly bad or unprecedentedly good (or not unprecedented at all) with regard to effects on society. There has been little work on the economic and social effects of the climate and resource consequences of atmospheric trace gas accumulation, and most of the work that has been done has focused on losses and has not identified gains. The U.S. has experienced a several decade long trend of population migration from the cold Northeast and Middle West toward the South, Southeast, and Southwest including the desert Southwest providing evidence that heat and low precipitation may be desirable attributes or at least climatic conditions that can be overcome by modern technology.

Technological change occurs exogenously and would not respond to fuel prices/taxes. Most models treat new technology development as exogenous. Rising taxes and other policy instruments that restrict fossil fuel use will cause increased incentives for research and development to find new technology. Failure to consider this effect probably leads to an upward bias in many estimates of cost. Similarly, the effects of climate change frequently fail to consider full use of existing technology or the response of agents to changes in the implicit price of resources through changes in their scarcity.

GNP does not measure welfare or environmental damage. While economic theory and applied general equilibrium analysis identify economic welfare as a separate concept from GNP, many policy discussions treat positive and negative changes in GNP as the equivalent of welfare. GNP is simply a measure of economic activity that is channeled through markets. Environmental degradation can lead to increased GNP with greater employment and output, but where the added employment is offsetting environmental degradation without an increase in real standards of living.

Developing countries demand less environmental quality. While not a fallacy directly, care must be taken in this generalization. The usual view is that the environment is a luxury that the poor cannot afford. But, climate change and other environmental issues as well can affect human health (labor productivity), the ability to produce food and fiber, tourism industries, and other economic sectors. In these cases, low effective demand for environmental quality may not be due to the luxury nature of avoiding environmental damage but to ineffective and poorly developed governmental institutions for managing environmental externalities.

A significant disagreement exists in how to trade-off between future and present generations. Both economists and noneconomists have focused on the discount rate as an important consideration regarding climate change with arguments that discounting is inappropriate for very long-term issues or that a high (or positive) discount rate shortchanges future generations. Elementary calculations demonstrate that changing the discount rate can have an important effect on any problem where costs and benefits are dissimilarly distributed over time. While there are theoretical arguments why the time preference rate of society could be negative, a riskless market interest rate represents the market interaction of investment opportunity and society's time preference as shaped by public policy. Voter preferences in the U.S. during the 1980s appear to have decidedly pushed for a high real discount rate through repeatedly encouraging public deficit financing. And, while the benefits and costs of investment in the traditional economy are frequently cast as a two-period problem of consumption today versus consumption tomorrow, the full solution is an infinite time problem. If we invest less today, output in the next period is affected, leaving less for both consumption and investment for the succeeding period. In other words, a different level of investment today will change the output path for all time. Thus, it is inaccurate to suggest that climate change is unique because of its long time horizon.

It is useful to view climate policy as an alternative investment to that which produces traditional market goods. Both sets of investments have implications for distant-future consumers even though in both cases the physical capital may have a life of only five, ten, or twenty years. If we are concerned that the discount rate as evidenced by market interest rates is too high, then social policy ought to assure that both private and public investment planning decisions use a lower rate, i.e. the market interest rate should be lowered.

The debate about discount rates and undervaluing future generations more likely stems from different views about the economic implications of climate change (before they are discounted) as well as different perceptions about how future consumers will value traditional economic goods and how they will value the environment and resources changed because of climate change. Catastrophic consequences, even discounted, will lead to a cost-benefit calculation that suggests action.

The other missing element in the future versus present generation debate is that the problem is cast as either we do something now or we accept a seriously degraded climate in the year 2075. The other alternative is to do little now and begin taking serious action in 25 years at which point the 2075 damages will be discounted by 25 fewer years and will therefore be larger and may justify significant investment. The critical parameter in making this decision is how costly it is to change rapidly. Large fossil power plants have lives of about 40 years, therefore there is considerable flexibility in fossil carbon dioxide emissions if the costly consequences are more than 40 years in the future. However, system considerations, such as urban design that may last centuries and therefore shape transportation needs, must be considered even though individual automobiles only last ten years.

The pressure and stress due to change and resource limits is always costly. Fundamental to a dynamic economic paradigm is that rapid change is costly but such models do not include induced technological change. Landes (p. 42), in analyzing the cause of the industrial revolution and increased standard of living in Britain, identified technological change as requiring "an opportunity for improvement or a need for improvement created by autonomous increase in factor cost." In this view, exhaustion of wood supplies and consequent increase in cost contributed to the switch to coal which made possible the development of more powerful engines and a wide range of machines that could take advantage of concentrated power. The point is that resource constraints contributed to significant advance by calling forth innovative solutions. It is relatively easy to predict the shorter dislocation costs that might occur from higher fuel prices or from climate change, but it is quite difficult to predict completely new innovations that might come from dealing with these added constraints.

Conclusions

There is much we do not know about climate change or the costs of controlling it. A positive move in the direction of the economic research on climate change is broadening the focus beyond carbon dioxide and beyond the consideration of only the climatic effects of trace gases. Research on the costs of climate change or on the benefits of avoiding climate change has not proceeded far. Further quantification of effects that have reasonably direct effects on economic well-being is needed. While further research in these areas will help provide general guidance, the implications of climate change and the costs of controlling climate will remain highly uncertain. The problem, however, is frequently cast as one of taking significant costly actions immediately or accepting irreversible and damaging climate change. Fortunately, this characterization overstates the case. Even as we begin addressing climate change there will be the opportunity to assess the costs and effectiveness of policies we put in place as well as to better understand the consequences of climate change. Climate change is not an issue of brinkmanship. The world will not be thrown to catastrophe based on what we do or do not do over the next year but neither will it be saved by a single agreement.

References

Edmonds, J. and J. Reilly. 1983. "Global Energy and CO_2 to the Year 2050," *The Energy Journal*, 4(3): 21-47.

Jorgenson, D. W. and P. J. Wilcoxen. 1990. "The Cost of Controlling U.S. Carbon Dioxide Emissions," presented at a *Workshop on Economic/Energy/Environmental Modeling for Climate Policy Analysis*, Washington, D.C., October 22-23.

Kane, S., J. Reilly, and J. Tobey. 1990. *Climate Change: Economic Implications for World Agriculture*, draft, Economic Research Service, U.S. Department of Agriculture, Washington, D.C.

Landes, David S. 1969. *The Unbound Prometheus*, Cambridge University Press: London.

LeBlanc, Michael and John Reilly. 1988. "Energy Policy Analysis: Alternative Modeling Approaches," in George Johnston, David Freshwater, and Philip Favero (eds.), *Natural Resource and Environmental Policy Analysis: Cases in Applied Economics*, Westview Press: Boulder, CO.

Manne, A.S. and R.G. Richels. 1990. "CO_2 Emissions Limits: An Economic Cost Analysis for the USA," *The Energy Journal*, 11(2):51-75.

Manne, A.S. and R.G. Richels. 1990. *The Costs of Reducing U.S. CO_2 Emissions—Further Sensitivity Analysis*, Electric Power Research Institute, Palo Alto, CA.

Morris, S.C., B.D. Solomon, D. Hill, J. Lee, and G. Goldstein. Forthcoming. "A Least Cost Energy Analysis of U.S. CO_2 Reduction Options," *Energy and Environment in the Twenty-first Century*, MIT Press, Cambridge, MA.

Nordhaus, W. D. 1989. *A Survey of Estimates of the Costs of Reduction of Greenhouse Gases*, Department of Economics, Yale University, New Haven.

Nordhaus, W. D. 1990. "A Sketch of the Economics of Global Climate Change," Paper presented at the *Allied Social Science Association Meetings*, (Dec.), Washington, D.C.

Nordhaus W. D. and G. W. Yohe. 1983. "Future Carbon Dioxide Emissions from Fossil Fuels," in *Changing Climate*, pp. 87-153, Washington, D.C.: National Academy Press.

Peck, S.C. and T.J. Teisberg. 1990. *A Framework for Exploring Cost Effective CO_2 Control Paths*, Progress Report, Electric Power Research Institute and Teisberg Associates, Palo Alto, CA, November.

Reilly, J. 1990. *Climate Change Damage and the Trace Gas Index Issue*, (November 1), draft, Economic Research Service, U.S. Department of Agriculture, Washington, D.C.

Reilly, J.M., J. A. Edmonds, R. H. Gardner, and A.L. Brenkert. 1985. "Uncertainty Analysis of the IEA/ORAU CO_2 Emissions Model," *The Energy Journal*, 8(3).

Reilly, J.M. and S.A. Shankle, "Auxiliary Heating in the Residential Sector," *Energy Economics*, 10(1):29-41.

OPENING REMARKS: INTERNATIONAL PANEL

James P. Bruce

Canadian Climate Program Board
1875 Juno Avenue
Ottawa, Canada K1H 6S6

Scientific concerns have been growing for several decades about the rapidly rising concentrations of greenhouse gases – CO_2, methane and the CFCs – in the global atmosphere due to human activities, and the predicted global warming. Studies of the technical feasibility and economic costs of reducing emissions have only become a growth industry in the past few years. You have heard about the U.S. studies in the presentations of the past day and a half. This afternoon we will focus on similar evaluations in other countries.

Two major stimuli for studies around the world of the economics of greenhouse gas emission reductions occurred in 1988. The first was the International Conference on the Changing Atmosphere, held in Toronto, in June 1988 in the midst of a severe drought and hot period in central North America. That conference first proposed that, as in initial goal, industrialized countries should reduce CO_2 emissions by 20% of 1988 levels by the year 2005.

Later in 1988, the World Meteorological Organization and U.N. Environment Programme formed the Intergovernmental Panel on Climate Change. This panel was charged with reviewing the science of global warming, the impacts, and the response strategies that might be followed. More than a thousand specialists from seventy countries participated in producing the panel's report, which was completed in August, 1990. This comprehensive three-volume report, with a fourth overview volume, was a key document for discussion at the Second World Climate Conference.

The Response Strategies Working Group, chaired by Dr. Fred Bernthal of the U.S.A., stimulated a number of national studies of the feasibility of reducing greenhouse gas emissions. To try to compare and consolidate these in some way, the Working Group noted that CO_2 emission changes could be considered to be a product of three factors:

1. Changes in carbon intensity of energy production;

2. Changes in energy intensity (energy use per unit of GNP);

3. Changes in GNP or GDP.

No country was asked to limit growth in GNP, but many were willing to consider the rates at which reductions in energy intensity and in carbon intensity could be achieved. There is at present a large discrepancy in energy intensity among the industrialized countries, with economic superstars Japan and Germany being already highly efficient in energy use and promising more, and with Canada and the U.S.A. on the other end of the scale.

None of the industrialized countries found the Toronto Conference target easy to achieve, but a number found it technologically feasible. In Canada, for instance, a number of studies have been done by and for government agencies, both environment and energy, and by university groups. They all conclude that the 20% reduction target could be achieved but, depending on the assumptions of the studies, the costs of doing so varied from -0.3% of GNP to +0.5%.

The studies which showed huge benefits from the improvements in energy intensity and fuel switching were those that analyzed the economics of individual energy efficiency measures. They find most of these measures to be attractive from an economic and societal perspective. They assume that the measures are implemented as part of the normal replacement cycle. This yields an overall savings as a result of implementing the CO_2 emissions reduction measures.

The other approach used to estimate the cost of CO_2 emissions abatement in Canada was a macro-economic model. A carbon tax is assumed in order to achieve a target reduction in emissions. The tax revenues are recycled so that there is no net loss to the economy. Studies yield positive costs for CO_2 emissions reduction. My personal view is that these studies do not prove that it will be costly to reduce CO_2 emissions, but that carbon taxes may be an inappropriate way to do it.

The Second World Climate Conference, October 29-November 7, 1990, provided a much broader international review of the science and possible responses since 138 countries participated. The conference was in two parts, a scientific and technical portion which produced a Conference Statement, and a ministerial part, led off by Prime Ministers Thatcher and Rocard, King Hussein and several other heads of state. A Ministerial Declaration was agreed to at the end of this part of the conference.

The scientific Conference Statement indicated that the scientific consensus articulated by the Intergovernmental Panel on Climate Change is broad enough for nations to take action *now* to reduce atmospheric concentrations of greenhouse gas. It noted that stabilization of CO_2 emissions could be achieved by a one to two percent per year reduction in emissions. In reviewing the studies of feasibility of such reductions in various nations, it concluded that "technologically feasible and cost-effective opportunities exist to reduce CO_2 emissions in all countries. Such opportunities...are sufficient to allow many industrialized countries...to reduce these emissions by at least 20% by 2005."

For developing countries, the conference considered that "if they are to avoid the potentially disastrous course followed by industrialized countries in the past, they need to 'leapfrog' over the heavily polluting energy technologies and move quickly to the most clean and efficient technologies available." For this they will need major assistance.

The Ministerial Declaration confirmed the commitments of 22 countries, nearly all the OECD members, to limit CO_2 emissions by the year 2000 or 2005. The most common commitment was to limit emissions to 1990 levels by the year 2000, but some went much further and committed themselves to substantial reductions, some of which you will hear about from Mr. Metz of the Netherlands.

The ministers also agreed it was timely to begin negotiations of a Global Convention on Climate Change, embodying some real commitments. Negotiations will open in early 1991 here in Washington, at the invitation of President Bush. It is hoped that an agreement can be reached by the time of the 1992 World Conference on the Environment and Development in Brazil.

This would be appropriate because, if the agreement does contain commitments on emission limitations, reforestation, and assistance to developing countries, it could be one of the most important global agreements ever developed. It will be of great importance to energy policies and national economies, to land use policies, to relations between North and South, and above all it will be of enormous importance to the future environment of our small planet.

GLOBAL FUNDING FOR THE ENVIRONMENT: PERSPECTIVES OF DEVELOPING COUNTRIES AND RELEVANCE OF ECONOMIC COSTING

Erik Helland-Hansen

United Nations Development Programme
Bureau for Programme Policy and Evaluation
Technical Advisory Division
One United Nations Plaza
New York, New York 10017

Abstract

New global funding facilities are being established to provide grants or concessional loans to developing countries to help them implement programs that protect the environment. Combatting climate change is a major thrust of this new funding mechanism being organized by the World Bank, United Nations Environmental Programme (UNEP) and United Nations Development Programme (UNDP) as collaborating agencies with the former two agencies in special administrative functions. The presentation will discuss the advantages offered by such a fund from a developing-country perspective. The developmental dimension of the issue will be stressed by underscoring the need to blend such new and additional funding with national policies and development planning in recipient countries.

Introduction

It is commonly accepted that the threat of global climate change is primarily a result of human activities in the industrial world. Frequently this is the starting point of developing-country officials when the theme of development and environment is being discussed. It is a fact, however, that developing countries today produce a considerable portion of greenhouse gases through such activities as land clearing, paddy rice growing, energy generation, transport, etc. The World Resources Institute estimates that about 45 percent of the present net additions to the greenhouse heating effect stem from developing regions of the world. But the importance to the global community of the greenhouse gas emission issue in developing countries is associated with their quest for development measured in conventional economic terms; thus it is their future emissions that really matter as economic development continues and populations increase.

Copyright 1991 by Elsevier Science Publishing Company, Inc.
Global Climate Change: The Economic Costs of Mitigation and Adaptation
James C. White, Editor

The need for additional financial resources to be made available to developing countries on concessional terms to address global environmental issues at the country and regional levels has long been recognized by the industrial countries. The establishment in late November 1990 of the Global Environment Facility (GEF) and, in June 1990, of its related forerunner under the Montreal Protocol, the interim Multilateral Fund (or Ozone Fund), are steps toward fulfilling such needs. Activities supported by these two funds, which will both be executed under collaborative agreements between the World Bank, UNEP and UNDP, will target aspects of global warming, biodiversity, international waters and ozone layer protection in developing countries. The World Bank will administer the GEF with an anticipated disbursement of US$1.0-1.5 billion during 1990-1992, while the UNEP-initiated $160-million Multilateral Fund will be administered by a new secretariat to be set up in Montreal under the auspices of the Executive Committee established by the parties to the Montreal Protocol. UNEP is designated as the Treasurer of this ozone fund. As the GEF also incorporates ozone protection, and the political intentions are to eventually merge the two funds following the three-year pilot period, this paper focuses on GEF alone.

As is well known, a number of developing countries struggle with short-term economic survival problems. However much they may recognize the global threats from climate change, and they all generally do, they are unable to invest in projects for the sake of producing benefits of a common global nature. Their cost-benefit analysis of, for example, coal versus geothermally powered electricity production will probably never reflect the global warming factor as a cost element of coal-fired plants, or the absence of global warming potential as a benefit accruing to geothermal plants. Neither is it realistic to hope that afforestation projects or transport efficiency initiatives will be financed locally on the basis of their inherent climatic benefits. The GEF is being established precisely to address such decision making in an attempt to give developing countries short-term economic rationales for making decisions that are environmentally benign in a long-term global context.

This paper attempts to clarify UNDP's role as the agency responsible for technical assistance activities under the GEF and it evaluates the utility of considering project alternatives in economic costing terms from the perspective of the developing countries.

UNDP's Mandate and Role

To set the scene correctly, a quick overview of UNDP may be valuable. The United Nations Development Programme is the central funding, planning and coordinating organization for "technical assistance" in the United Nations system. UNDP currently supports more than 5,000 projects valued at about $8 billion in some 150 developing countries and territories; the annual project expenditure is more than $1 billion. The primary objective of UNDP is to support the efforts of the developing countries to accelerate their economic and social development by providing systematic and sustained technical assistance meaningfully related to their national development plans and priorities, and for the benefit of their entire populations. In more specific terms, the assistance provided by UNDP promotes increased self-reliance in the developing countries. Help is given to train managers to guide governments and industries. Knowledge and technology are transferred where countries say they need them. Countries receive help to build the institutions they need to stand on their own two feet.

UNDP is unique in its nonpolitical and sectorally neutral role, offering assistance through grant funding. With offices in 112 developing countries, UNDP has the greatest on-the-scene representation of any development organization.

Overall, UNDP is small compared to the World Bank, and in many developing countries UNDP's program is dwarfed by large single-donor country programs like USAID, the Canadian International Development Agency (CIDA), Nordic aid agencies, etc. But UNDP is "everywhere" in the developing world and it administers grant funds on behalf of each country.

Economic Costing in Technical Assistance

It is important at this stage to distinguish between UNDP's technical assistance activities and capital investment activities as carried out by the international investment banks. UNDP principally provides grant resources towards human and institutional development that supposedly brings about economic and social development. In practical terms this is achieved through projects and programs where people's activities and their associated needs for equipment are paid for by UNDP when so requested by the recipient country. UNDP carries out preinvestment and feasibility studies and as a general rule UNDP avoids funding physical infrastructure, leaving the financing of construction and procurement of capital assets to other funding mechanisms such as its Capital Development Fund and World Bank loans. The Bank's requirements for economic analysis of alternative investment options will of course regularly be based on economic costing principles. This is contrasted by UNDP's form of technical assistance in which projects are compared and selected for implementation solely on the basis of their developmental features; this concept of the developmental dimension is further discussed later in this paper.

The Global Environment Facility

The GEF was established at a meeting in Paris on 27-28 November 1990, attended by 25 developed and developing countries. The US$1-1.5-billion facility will provide concessional funding for investment projects and related activities in developing countries with four broad objectives:

1. To support energy conservation, the use of energy sources which will not contribute to global warming, forestry management, and reforestation to absorb carbon dioxide, in order to limit the increase in greenhouse gas emissions;

2. To preserve areas of rich ecological diversity;

3. To protect international waters, where transboundary pollution has had damaging effects on water purity and the marine environment;

4. To arrest the destruction of the ozone layer, by helping countries make the transition from the production and use of CFCs, halons and other gases to less damaging substitutes.

Rather than establishing new institutional structures, earlier preparatory meetings of the participants in the GEF had indicated that the Facility should be administered by the World Bank and executed collaboratively between UNEP, UNDP and the World Bank, drawing upon the respective areas of expertise and comparative operational advantages of each agency. Briefly stated, the agencies' roles are as follows:

1. UNEP will help ensure that the policies of the Facility are consistent with existing global conventions and related legal frameworks. UNEP will provide scientific and technical guidance, facilitate research and disseminate information on relevant technological developments;

2. UNDP, which links with governments at developing-country level, will undertake pre-investment studies, technical assistance and training for global environmental policies, programs and projects, and should help ensure compatibility with national development and environment policies; it will also coordinate with other donors;

3. The World Bank will administer the Facility and manage the project cycle for global environmental investments, associated technical assistance and training supported by the resources of the GEF.

In order to achieve complementarity, coherence and cost-effectiveness for the fulfillment of the roles of the agencies, planned programs of activities will be prepared for technical areas and countries and submitted to the participants for policy guidance by the Scientific and Technical Advisory Panel. Project approvals will be made at working levels collaboratorily between the three agencies. NGOs will be brought into the GEF management process through participation in such activities as:

- Association to the Advisory Panel;
- Identification of projects;
- Implementation of projects when feasible and as approved by governments.

Criteria for Support from GEF

The GEF is supposed to support programs and activities for which benefits would accrue to the world at large while the country undertaking the measures would bear the cost, and which would not otherwise be supported by existing development assistance or environment programs. Country eligibility will in principle be according to the UNDP definition of developing countries.

The GEF has identified the existence of three broad types of investment projects for consideration:

1. Type 1, projects are economically viable on the basis of domestic benefits and costs to the country itself.

2. Type 2, projects are those which would not be justified in a country context if the full costs are borne by the implementing country. However, if part of the costs can be offset by concessional or grant assistance from the GEF, then the project's overall rate of return would be raised enough to be attractive to the implementing country and substantial global benefits would be realized.

3. Type 3, projects are justified in a country context, but the country would need to incur additional costs to bring about additional global benefits. The additional costs of accommodating global concerns would be eligible for GEF funding.

Types 2 and 3 are considered candidates for funding under the GEF provided they are within predefined development sensitive cost-effectiveness guidelines. Such guidelines still remain to be defined.

The cost-effectiveness of the global component(s) will be determined initially on the basis of physical rather than monetary measures of global benefits. Guidelines will need to be established during the envisaged pilot period to help ensure that an excessive price is not paid to achieve given global physical benefits. Since there is currently no analytical basis to compare one type of global benefit with another, priority among the four global environmental objectives will require judgment.

Project selection under the GEF must also satisfy some criteria of a more cross-cutting and developmental nature, tentatively listed in the GEF document as:

1. Consistency with global environment conventions;

2. Consistency with country-specific environment strategies or policies;

3. Utilization of appropriate technologies;

4. Being both cost-effective and of high priority from the global perspective.

The Development Dimension and Perspectives of Developing Countries

The philosophy of the GEF brings an entirely new dimension to some development projects. Eligible countries may now access additional financial resources on concessional terms and as grants, as they have repeatedly been requesting, provided their projects fulfill the type of criteria outlined above. But what is really to be gained by these countries in addition to the moral high ground of, for example, not pushing global warming infinitesimally further on a global scale, which is already beyond their control? The answer must be found in the developmental dimension of the undertaking. If assessed economic costs and benefits, as they eventually shall be defined in cost-effectiveness guidelines, adequately reflect the country's own judgment of development factors, the financing modality offered by the GEF may appear beneficial also in narrow nationalistic views.

The kind of domestic noneconomic criteria that may be of importance in this connection are, *inter alia*:

1. Compliance with national development plans and policies, such as local/district focus, self-sufficiency, national security.

2. Consideration of recurrent expenses and human and institutional resources to properly maintain and operate project installations, for instance switch from centralized coal-fired electricity generation to dispersed optimal generation of wind-, hydro- and diesel-powered electricity with complex transmission networks.

3. User-friendliness, ability and willingness to pay, etc. of development intervention alternatives.

4. Community involvement, incentives and management responsibilities.

5. Proper valuation of cultural aspects, religious rituals, political influences, etc. that may be changed as a result of project alternatives.

6. Considerations of interministerial cooperation, benefits and problems.

7. Gender concerns and employment opportunities.

8. Local or downstream environmental effects and opportunities.

Such elements need to be addressed from the outset, whether through economic costing, cost-efficiency guidelines or awareness analysis, to convince domestic decision-makers that GEF financing also brings real short-term benefits to their country.

Conclusions

As new global funds to address climatic change problems at the developing country level are being established within the UN family, their aim must be to combine global environmental benefits and national development. The capital investment decisions must be preceded by comprehensive analyses of national/local socioeconomic and environmental factors in addition to project economics and global environmental concerns. This phase of the project development cycle will in principle be the domain of UNDP-managed assistance. UNDP's comparative advantage is its representation at developing-country level with its close contact with governments and proven track record of delivering effective and sensitive development aid from the perspective of the recipient countries. On this basis it should follow as a conclusion that economics is not the tool for selection between competing technical assistance projects of relevance to the GEF. But economics will guide the capital investment decisions as they are being analyzed in the technical assistance phase. The formulation of usable and developmentally sensitive economic costing principles to frame the cost-efficiency guidelines so that they become palatable to developing-country decision-makers will be a challenge as we initiate activities under the Global Environment Facility.

THE PROSPECTS AND ECONOMIC COSTS OF THE REDUCTION OF CO_2 EMISSIONS IN THE PEOPLE'S REPUBLIC OF CHINA (PRC)

Yingzhong Lu

Institute for Techno-Economics and
Energy System Analysis
Beijing, China

Introduction

The energy demand growth rate in developing countries exceeds that in developed countries and, as a result, the share of CO_2 emissions from the Third World will exceed that from the developed countries as forecasted by Manne and Richels (1990).[1] (See Figure 1.) More and more concern has been given to the potential of reducing CO_2 emissions in developing countries by energy conservation and the transition to noncarbon energy sources. However, the cost of these energy sources is often much higher. As a result, the constraints of financial sources will limit the exploitation of these alternatives and hence the increasing of CO_2 emissions will continue in the developing countries. Manne and Richels give the relative costs of various energy sources as shown in Table 1. The estimated annual losses in aggregated gross domestic product (GDP) due to a carbon emission limit has been found intolerable in the case of China, in which 70% of the primary energy comes from coal, as shown in Figure 2. Such a penalty is much higher than the tolerable limit of developing economies. An analysis is therefore carried out to re-evaluate the potential of various energy conservation and noncarbon resources in China, and the possible strategy of deployment of these sources, as well as the trends of the costs involved.

Long-Term Energy Demand of the PRC

A forecast of the long-term energy demand was made in 1988 as shown in Table 2.[2] The per capita GDP growth in the forecast was assumed to attain US$11,000-12,000 by the year 2050, or a 40-50 fold increase as compared with the base year 1980. Although the growth rate may not be so fast as forecast, the trend remains the same if one extends the time span a little bit further, e.g., the GDP will attain some $10,000 per capita by 2070 instead of 2050.

Copyright 1991 by Elsevier Science Publishing Company, Inc.
Global Climate Change: The Economic Costs of Mitigation and Adaptation
James C. White, Editor

The energy conservation potential is already considered in this forecast, since technical progress is taken into account in each sector. The elasticity remains well below 0.6 for the next four to five decades which implies that continuous efforts must be put into reducing the energy intensity every year by a factor of 1.6 or more.

The CO_2 contribution of Chinese energy systems to global emissions is disproportionate to its energy share due to the predominate role of coal. In Table 3 it is noted that the share of Chinese CO_2 will even exceed its population share which represents a real threat to the world environment.

Energy Conservation

In the forecast made above, much emphasis has been put on energy conservation. Since the economy will grow some forty- to fifty-fold in the following six to eight decades, it is incredible to assume a similar growth rate in energy supply. Both economic structural change and technology progress have been taken into account to reduce the energy intensity of the Chinese economy. It is noted in this forecast that the energy intensity has been reduced by a factor of 5.3 based on past experience and future prospects as described below.

Energy Consumption Potential

An important approach to reduce CO_2 emissions is energy conservation. It is generally believed that in developing economies like China the conservation potential is great due to the backwardness of its technology. In fact this argument is convincing, since the obsolete technology as well as the management imply enough room for the increase of efficiency of energy utilization. However, the very low level of current per capita consumption and the rapid rate of economic growth still require a tremendous increase of energy supply even if all the potential of conservation is realized. We will discuss these two aspects separately.

The Chinese economy is considered an energy intensive one as compared with either developed countries or other developing countries as shown in Table 4. It is seen that the per GDP energy consumption of the PRC in 1980 is about twice that of Canada and India, about three times as much as the United States and South Korea, or four times that of Japan, most European countries, and Argentina. In order to reduce the energy intensity and achieve the target of "quadrupling the industrial and agricultural products" with only "doubling the energy consumption," great effort has to be made for energy conservation.

However, some misunderstandings could be caused in such a comparison based on per dollar value energy intensity. The volatile foreign exchange ratios and the irrational domestic pricing system not only cause the wasting of energy, but also lead to some paradoxes in the comparison of energy intensities between different countries. Since prices of the same commodity in different countries may vary substantially, the comparison of energy intensity per unit value created often loses its common ground. For instance, the per GDP energy intensity in China was about 230% that of India in 1985. However, when comparing the sectoral energy intensities based on physical output, it has been found that the energy intensities of Chinese products are in general lower than those of India. The low prices of many commodities tend to exaggerate the difference of energy intensity or even reverse the facts. This situation is best illustrated by Table 5.[5] It is noted that, although the per dollar value of energy intensity of Chinese industry is about 50% higher, the specific energy consumptions per physical unit output in China are in general lower than those in India except for cement production.

Special attention should be paid to the exchange rates used in these comparisons, since many countries have to adjust their exchange rates periodically to control trade deficits. China adjusted its exchange rate to U.S. dollars from 1.5:1 in 1980 to 4.71:1 in 1990. If the current exchange rate is used, the energy intensity of the Chinese economy in 1987 and 1980 would be 3.36 and 2.45 kilograms of coal equivalents/dollar (kgce/dollar), respectively, or a 37% increase! This is entirely contrary to the fact. The energy elasticity has been decreased substantially to 0.51 from 1980 to 1987. The comparison between different countries is therefore better carried out in physical units instead of monetary value of products. Thus the high value of per capita energy intensity in China is partially due to the statistic paradox caused by the irrational low prices and varying exchange rates.

Policy and Achievements of Energy Conservation

However, in view of the abnormally high energy intensity as compared with most countries, much emphasis has been put on energy conservation since 1980. The official *General Policy of Energy Conservation* was stated as "to put equal emphasis on energy conservation and energy provision." Following this general policy, four specific policies on energy conservation were formulated:

(1) *Adjustment of economic structure* toward a less energy-intensive economy, including the adjustment of the relative shares of various sectors and changing the product mix within each sector. Service sectors should be developed faster than industrial sectors, and light industries faster than heavy industries. Within each sector, deep processing or refining of products should be encouraged to create more value from the same amount of energy or raw material consumed.

(2) *Innovations in technologies*, including adopting energy-saving technologies, processes, and equipment; phasing out obsolete energy-wasting technologies, processes and equipment.

(3) *Improvements in management*, including the establishment of the hierarchy of management, assignments of responsibilities and authorities as shown in Figure 3, setting up energy audits and accounting systems, promulgating energy conservation norms and regulations, implementing energy conservation programs, and the use of rewards and punishments.

(4) *Incentives with economic measures*, including the partial deregulation of energy prices and the provision of energy conservation investments, loans and subsidies.

The energy savings from 1981-1987 based on the energy intensity of 1980 are shown in Table 6. It is noted that the savings in these seven years totaled to some 1127 million tonnes of coal equivalents (Mtce).

The energy intensity changes in various sectors are shown in Table 6, and the annual savings in these sectors based on energy intensities of the previous year are shown in Tables 7 and 8. It is interesting to note that the energy savings in heavy industry were predominant in the Sixth Five Year Plan. Since the savings in these tables were calculated on the basis of annual change of sectoral energy intensity, the continuous efforts made in each sector could be envisaged. However, due to the limitations of data availability, the effect of changes in product mix could not be singled out. Thus the contributions entitled "structural change" in these tables are underevaluated since they only take into account the changes of sectoral shares of production.

The Economics of Energy Conservation

A survey of the effect of energy conservation investment during the Chinese Sixth Five Year Plan is given in Table 9.[6] It is interesting to note that the specific cost to create 1 tonnes of coal equivalents/year (tce/year) conservation capacity averaged just 347 Yuan, substantially lower than the specific investment of any other type of primary energy supply in China. However, such a "cheap" approach to energy conservation has its limit. The estimated potential of energy conservation by the year 2000 and up to the year 2050 will be discussed in the next two sections, respectively. It will be noted that, even if the total potential for savings will attain about 46% of the total consumption by 2050, i.e., over 4 billion tce, the remaining 54%, or about 5 billion tce will still have to be supplied by exploiting fossil and nonfossil primary energy resources.

The Medium-Term Energy Conservation Potential in China

According to the technological development target set by the Chinese government, the level of energy intensities for producing major industrial products should attain the 1980s level of those of the advanced developed countries. At the same time, the waste of energy shall be prohibited. These general targets imply the following specific estimations in energy conservation potentials:

(1) The energy savings in the most energy-intensive products, steel, cement, synthetic ammonia, and thermal electricity, are listed in Table 10 together with the respective outputs and energy intensities. The energy conservation potential of these products totals 90 Mtce.

(2) The improved designs of seventeen categories of mechanical and electrical equipment will result in a total savings of about 200 Mtce.

(3) Raising the efficiency of industrial boilers from 60% to 70% will contribute some 37 Mtce to energy conservation.

(4) Recovery of waste heat in steel mills, chemical plants, the petrochemical industry, and light industries will account for some 30 Mtce.

(5) Improvements in design and insulation of buildings will save about 8 Mtce. However, due to the enormous additional investments needed, further energy conservation potential in buildings will not be fully realized.

The Forecast of the Long-Term Potential of Energy Conservation

For the estimation of long-term energy conservation potential, the comparison of historical records and developing trends of other countries has been carried out, and reasonable assumptions have been made. The results are summarized in Table 11.

It is noted in this table that, while technical advance and product-mix changes contribute the most in the industrial sector (the energy intensity drops from 3.01 kgce/US$ to 0.522 kgce/US$ in 2050), the structural change also contributes substantially to the long-term energy conservation potential. The share of the energy-intensive industrial sector decreases from 54% in 1980 to 36% in 2050, while the share of the much less energy-intensive service sector increases from 6.6% to 52.9%. The overall energy intensity decreases to 18% by 2050 from its initial value of 1.955 kgce/US$ in 1980. Even so, the total primary energy supply increases to 864% of its 1980 value of 0.6 billion tce. This will be a common feature of the future energy system in most of the developing countries.

The Potential of Noncarbon Energy Sources in China

It is seen in Table 11 that, although energy conservation in every economic sector will be actively pursued, the energy demand by 2050 will still attain a mammoth figure of 5.2 billion tons. Over 80% of this demand will have to be supplied with coal, if no other alternatives are available by the earlier part of next century. In order to explore the potential of the transition to noncarbon energy, the possible noncarbon resources in the PRC will be first evaluated.

Hydropower

Hydropower resources are abundant in China as shown in Table 12. The total exploitable capacity amounts to 360 gigawatts of electricity (Gwe) with an annual output of 1900 tetrowatt hours (Twh). Presently only 6% has been exploited. The technology of hydropower is mature and well commercialized. The capital investment is high but the power cost is competitive. However, there exist three major drawbacks in the development of hydropower in China.

(1) The uneven geographic distribution of hydropower resources as shown in Figure 4. More than 70% of the total reserve concentrates in mountainous south-

west China which is far from the industrialized coastal provinces. In particular, about 26% of these resources are farther in distant Tibet and the Yunnan Plateau. As a consequence, additional investment will be needed for constructing long-distance ultra-high voltage transmission lines. Some of the resources cannot be utilized at all.

(2) The higher specific capital (in general more than 50% higher than that of fossil-power plants) and longer lead time make hydropower plants less attractive to planners in spite of their lower operating costs, since both capital and power are in short supply for the rapidly growing Chinese economy.

(3) The more direct environmental impacts, such as the inundation of large areas of arable land, the relocation of millions of people, and the obvious disruption of the ecological environment, often introduce unsolvable controversies which have shelved some major hydropower projects for decades. The most striking example is the Shanxia Dam across the Yangtze River, which has been debated for the past four decades.

Of course, all the above obstacles will be overcome eventually if the time span is sufficiently long. However, even if most of the economically justifiable potential hydropower is developed, it is still small compared with the tremendous energy demand of the PRC in the 21st century. The annual total energy consumption will attain four to five billion tce as shown in Table 2, and then hydropower will only account for some 6.2% assuming two-thirds of all the potential resources are exploited. Presently the hydropower share in primary energy supply is about 4.5% at an exploitation rate of 6%. The role of hydropower is thus much less important than it seems at first sight.

Nuclear Energy

The controversy over nuclear energy in China is different from that in the developed countries. Nuclear energy is considered cleaner than coal, even taking account of the radioactive emissions. The amount of radioactivity thrown into the atmosphere from coal-fired power plants is proved to be more than that from the nuclear power plants due to the uranium and thorium contents in most coal varieties. Besides, the adverse effects of the nonradioactive emissions from coal-fired plants are a thousand times higher than its radioactive emissions. As a result, many environmentalists in China advocate using nuclear energy to replace coal, and the public is not hostile to nuclear installations. Consequently, ambitious nuclear programs have been drafted for providing electricity as well as supplying district heating for Chinese cities. However, there are several serious obstacles to be overcome before nuclear energy becomes a real substitute for fossil energy in China.

The "Absolute" Safety Guarantee and the Requirements of Inherently Safe Reactors[9]

Since the substitution in the Chinese energy system is measured by billion tce, or in 10,000 Twh range, the consideration of the safety requirements of a nuclear reactor should be different from small-scale deployment. An "absolute" safety guarantee is required if China plans to rely on nuclear energy for replacing a billion tons of coal. Hundreds of nuclear power and heating stations need to be built all over the country, and the fate of such a choice depends not only on domestic safety records but also on exogenous events of the global nuclear community. In case there will appear another Chernobyl or Three Mile Island anywhere in the world, the nuclear energy based on present technology and safety philosophy may not be able to survive any more. Then China will have to look for another alternative. In view of this, at the beginning of such a large-scale nuclear program, it is logical to look first for an invulnerable design which can exclude the catastrophic accident of core-melting. The inherently safe reactor, either using water or gas coolant, would be the ideal choice. This is the very focus of the R&D nuclear energy program at present.

The Less Expensive "Economic" Reactor

Another obstacle bears some similarities all over the world. Nuclear energy is well known to be capital-intensive as compared with fossil-power generation. However, it is particularly critical for a rapidly developing economy like that in China. The specific cost of installed capacity of a nuclear power plant is estimated to be 50-100% more expensive than a conventional fossil plant built in China. As a result, it cannot be used to replace coal without a serious financial penalty. Such a high cost may eventually make the nuclear option unfeasible. A careful analysis indicates that the share of the reactor investment is already less than the turbo-generator and the auxiliary systems. When inherent safety is realized in reactor systems, then the adoption of conventional quality equipment for the turbo-generator and auxiliary system may be acceptable. This will reduce the capital investment substantially. The mass production of reactor equipment may reduce both the direct cost of field installation and the indirect cost caused by a long lead time. Taking all these into consideration, it seems not impossible to work out an "inherently safe and economic reactor" design for the future Chinese nuclear program.

The Controversy Over Waste Treatment and Disposal, and the Issue of Nonproliferation

This is an unsolved issue in current technology which China must face if it is going to engage in large-scale nuclear development. Some preliminary work has been done in reprocessing of the spent fuel and the vitrification of the high-level waste (HLW). By taking out the long-life actinides from the HLW, it is believed that the present storage technology is able to guarantee the safe storage of rad-wastes for several hundreds of years and reduce the radioactivity of shorter life isotopes to unharmful levels. Further R&D should be carried out, however, before the technologies involved in this process can eventually be put into commercial operation.

As to the issue of nonproliferation, China holds its own stand. China has already been a nuclear power and is obliged not to use a nuclear weapon "first." Under the world's current political condition, the probability of nuclear war between the superpowers seems to be reduced. However, local conflicts may trigger the temptation of acquiring and even using nuclear weapons on a smaller scale, but it has nothing to do with nuclear energy development. Currently advanced technologies of isotope separation (e.g., with centrifuges or laser lights) could produce fissile materials at a comparatively small scale, totally without the need of the construction and operation of commercial nuclear reactors. Also, civil application of nuclear energy provides the best way to consume the dangerous fissionable material. It is believed that the issue of nonproliferation could only be solved politically, instead of limiting the utilization of nuclear energy.

The current nuclear energy program in the PRC includes both nuclear power and heating reactors as shown in Table 13.[6] The first prototype pressurized-water reactor (PWR) is scheduled to be completed at the end of 1990. The French designed 900 million watts of electricity (Mwe) commercial reactor will be on line by 1992. However, the domestic commercial reactor is still in the planning stage, and the unit power level is 600 Mwe, too small as compared with current practice. A limitation exists in the manufacturing capability of heavy components needed for PWR technology. On the other hand, the development of nuclear heat is actively pursued in the PRC. The prototype 5 MWt district heating reactor started operation in November, 1989. The commercial 200 Mwt heating reactor in Jilin Chemical Company has just been approved. The prototype high-temperature reactor (HTGR) is under planning. The application of simpler pool-type heating reactors is also under preparation. A map showing the location of these projects is included in Figure 5.

Solar Energy

Solar radiation is as plentiful in China as it is in most parts of the world. In Figure 6 the solar energy distribution in China is given.[7] For these regions where the annual average radiation is higher than 120 kcal/m^2, solar energy collector or photovoltaic cells may hopefully be utilized with certain economic benefit. Currently solar water heaters are produced on a commercial scale in China. Solar stoves are disseminated in desert regions. Over 100,000 units have been installed in Gansu province alone.

Solar energy on a large scale cannot compete with carbon-based fuels at current technology and cost level in most cases. However, since the plans under consideration extend well into the next century, the great potential of solar devices must be taken into account. The large-scale demonstration of a solar thermal power station is under way in several sites in Europe and in the United States. It is hoped that it will attain the target of 11 cents per Kwh with current technology at a scale of 100 Mwe. The photovoltaic cells are developing very fast. The highest efficiency with multi-junctions and with concentrators has reached 30% in experiments. The cost at the present level is around 12 cents per Kwh, and in the next decade it is expected to be reduced to 6 cents per Kwh. Extrapolating to four or five decades further, there is reason to expect the solar electricity becoming competitive in some parts of China and provide sizable power for the national grid around 2050 at reasonable cost.

Another approach for utilization of solar energy in conjunction with ecological environmental improvement is the planting of fast growing trees or other energy crops. China has also an ambitious reforestation program to increase its forest coverage from the current value of 12% to 18% by 2000, and more in next century. Such a program will help in solving the energy problem in vast rural regions characterized by a widely scattered huge population.

Other Renewable Resources

Other renewables distribute very unevenly and hence only local importance can be assigned. Wind energy is exploited in the prairies of Inner Mongolia and on some islands, and geothermal energy in Tibet and some other regions. In the long run, if the conduction type of geothermal resources (also called the dry hot rock) can be economically exploited, the potential will be much greater as shown in Table 14-A. The ordinary convection type of geothermal resources are also listed in Table 14-B for reference. Tide energy is limited to particular sites along the sea coast. All these above resources are too small to have significant impact on fossil fuel reduction in future Chinese energy system.

Concluding Comments

The development of the Chinese economy requires tremendous increases in energy supply even if its full potential of energy conservation is exploited. As a result, the development of noncarbon energy for reducing CO_2 emissions from future Chinese energy systems will be indispensable. It is noticed that, in order to reduce the CO_2 emissions to a significant extent, great effort must be made to develop more advanced technologies not readily available at present. The plentiful resources of hydropower should be exploited first, but it is far from enough to create any notable impact on CO_2 emissions. Nuclear energy is hopeful only when inherently safe and economic reactors have been fully developed and commercialized. Solar energy, with its excellent environmental contribution, is still farther from deployment at gigawatts of electricity (Gwe) level. The economic penalty of such a large-scale energy transition will be high, or even intolerable, if no further breakthrough of cheaper advanced technology is realized. The CO_2 emissions from Chinese energy systems will therefore steadily increase until these new technologies reach their mature stage. International cooperation in the development of inexpensive new energy supply technology will therefore be indispensable for world-wide CO_2 reduction.

Table 1. The Cost of Various Energy Options [1]

Code	Description	Heat Rate	Capital Cost	Carbon Emission Coefficient	Fuel Cost	Cost Components, mills/Kwh				Comments
		Btu/Kwh	$/Kw		$/MBTU	Fuel	O&M	Capital	Total	
HYDRO	Hydro - Existing						2.1		2.1	
GAS-X	Gas Boiler - Existing	10800					2.6		2.6	Excludes fuel cost
GAS-N	Adv. Combine Cycle	7500	580				2.7	10.4	13.1	Excludes fuel cost
OIL-X	Oil Boiler Existing	10600					3.4		3.4	Excludes fuel cost
COAL-X	Pulverized Coal Existing	12000		0.301	1.8	15.6	4.5		26.1	
COAL-N	Pulverized Coal New	9800	1155	0.246	1.8	14.7	8.2	21.3	47.1	AFBC&IGCC with CO_2 control
LWR	Nuclear Existing	10805			0.8	8.6	9.6		18.2	
ADV-HC	Advanced Solar (30% CF max)		2600				10	100.9	110.9	
ADV-LC	Advanced Nuclear (600 MW)	10200	1400			8.2	9.2	25.1	42.5	

Table 2. The Energy Forecast of the PRC [2]

Item	Unit	2000	2020	2050		
Total	Mtce	1500	2400	5200		
Oil	Mt	200	250	116		
	Mtce	286	358	166		
	%	19.1	14.9	3.2		
Nat. Gas	Gm^3	40	100	200		
	Mtce	53	133	266		
	%	3.5	5.5	5.1		
Hydro	GWe	83	174	263		
	TWh	291	609	921		
	Mtce	100	210	320		
	%	6.7	8.8	6.2		
New Energy	Mtce	0	10	250		
	%	0	0.4	4.8		
				Case A	Case B	Case C
Coal	Mt	1470	2280	5100	4370	3640
	Mtce	1051	1626	3640	3120	2600
	%	70.1	67.8	70	60	50
Nuclear	GWe	5	30	291	563	835
	TWh	30	180	1750	3378	5008
	Mtce	10	63	558	1078	1598
	%	0.7	2.6	10.7	20.7	30.7

Table 3. The CO_2 Emission from China [3]

Item	Unit	2000	2025	2050*			
				Case 0	Case A	Case B	Case C
Global:							
Carbon	10^9ton	7.2	10.3	14.5	14.5	14.5	14.5
China:							
Population	%	19.17	20.91	19.92	19.92	19.92	19.92
Carbon	10^9ton	1.131	1.717	3.952	3.455	2.991	2.627
	%	15.71	16.67	27.26	23.83	20.63	17.43

Note: Case 0: No nuclear energy. Case A: Low nuclear energy. Case B: Medium nuclear energy. Case C. High nuclear energy.

Table 4. The Comparison of Energy Intensities [4]

Item	Primary commercial energy consumption per unit GDP	Total primary energy consumption per unit GDP *	Resid./commercial energy consump. per unit *
Unit	(kgce/US$)	(kgce/US$)	(kgce/US$)
Developing Countries			
China	2.13	2.90	1.14
Argentina	0.44	0.49	0.1
Brazil	0.61	0.88	0.19
Mexico	0.80	0.84	0.11
India	1.05	1.77	0.83
South Korea	1.06	1.12	0.48
Developed Countries			
Canada	1.39	1.39	0.45
France	0.45	0.45	0.14
FRG	0.49	0.49	0.18
Italy	0.53	0.53	0.16
Japan	0.51	0.51	0.13
UK	0.57	0.57	0.22
US	1.05	1.05	0.35

* Including the non-commercial energy such as biomass.

Table 5. The Comparison with India Based on Physical Units [5]

Item	Unit	China	India
Industrial energy consumption	MTOE	28.8	60.6
Industrial energy consumption intensity	TOE/1000 USD	1.83	1.28
Crude steel	TOE/ton	1.0	1.32
Coke consumption	kg coke/ton	558	723-1056
Cement	kcal/t clinker	400	1330
Wet process	Kcal/t clinker	1480	1500
Dry process	Kcal/t clinker	1340	1100
Paper & paperboard	kgOE/ton	0.88	0.95
Ammonia			
Feed stock: Natural gas	Gcal/ton	9.5	12.3
Naphtha	Gcal/ton	10.2	13.4
Fuel oil	Gcal/ton	11.7	17.2
Electricity generation	Kcal/KWh	2786	3201

Table 6. Chinese Energy Saving: 1981-1987

Item	GDP	Energy consumption	Energy Intensity	Annual Energy Saving
Unit	10^9/USD	Mtce	Kgce/USD	Mtce
1980	245.9	602.8	2.451	----
1981	257.9	594.5	2.305	37.6
1982	286.0	626.5	2.191	74.5
1983	306.7	660.4	2.153	91.3
1984	348.1	709.0	2.037	144.2
1985	393.9	770.2	1.955	195.2
1986	424.7	880.8	1.904	232.1
1987	464.4	845.0	1.820	293.2
Total	2481.7	5014.4	2.021	1068.1

Source: Derived from data in [7].

Table 7. The Change of Sectoral Energy Intensity

(Unit: kgce/capita)

Year	Total	Agriculture	Industry	Construction	Transportation	Commerce
1980	547.5	224.0	742.0	125.2	1174.1	118.2
1981	504.7	223.1	675.9	109.9	1157.5	114.3
1982	482.6	214.7	657.9	100.0	1087.4	114.7
1983	462.9	205.4	627.6	101.0	1044.9	115.3
1984	436.0	198.7	585.9	94.7	1005.8	116.2
1985	410.0	197.6	533.0	102.0	968.7	111.1

Source: Derived from data in [7].

Table 8. Annual Energy Savings of Heavy and Light Industries (Mtce)

Sector	1981	1982	1983	1984	1985	Total
Industry Total	34.23	10.04	18.68	29.31	43.88	136.14
Intensity Reduction	9.39	15.92	24.58	31.14	46.6	127.41
Light Industry	7.27	0.08	1.01	5.89	6.01	20.18
Heavy Industry	8.15	14.48	21.73	23.76	37.71	105.83
Structural Change	18.81	-4.06	-4.06	-0.34	0.16	10.05

Source: Derived from data in [7].

Table 9. The Effect of Energy Conservation Investments [6]

Project Category	Investment (10^6 Yuan)	Annual capacity (10^3 tce)	Specific Investment (Yuan/tce)
A. Capital Construction:			
1. Cogeneration	1814	1380	1314
2. District Heating	312	604	516
3. Retrofitting Cement Kiln	66	197	330
4. Town Gas Plant	535	982	545
5. Continuous Casting	289	106	273
6. Coal Washing	143	756	189
7. Scrap Recovery	190	1130	168
8. Retrof.Small Fert. Plant	542	1140	475
Sum	3891	6305	617
B. Tech. Innovation			
9. Furnaces Improvement	579	1640	353
10. Waste Heat Recovery	781	3080	254
11. Boiler Improvement	750	3030	248
12. Coal Blending	46.7	280	167
13. Gangue Utilization	104	740	141
14. Urban Coal Utilization	91.7	940	98
15. New Designs	254.7	750	340
16. Biogas	17.4	110	158
17. Coordination	24.2	50	484
18. Others	982.9	NA	NA
Sum	3680	>10620	<347

Table 10. Energy Conservation Potentials of Selected Products [6]

Product	Output by 2000	Energy Intensity	Energy Saving
Steel	90 Mt	1400 kgce/t	21 Mtce/y
Cement	110 Mt	122 kgce/t	8 Mtce/y
Syn. Ammonia	31 Mt	1150 kgce/t	12 Mtce/y
Thermal Elect.	950 Twh	0.38 kgce/Kwh	49.4 Mtce/y

Table 11. Forecast of Long-term Energy Conservation Potential

Item	Unit	1980	2000	2020	2030	2050
Population	10^9	0.987	1.254	1.393	1.441	1.416
GDP	10^9USD	246	1003	2780	4320	11300
GDP/capita	USD/a	249	800	2000	3000	8000
1. Industry						
Value Added	10^9USD	123.6	502	1250	1730	4080
Share	%	54	50	45	40	36
Consumption	Mtce	391.43	1017	1218	1335	2129
Intensity	kgce/USD	3.01	2.03	0.974	0.772	0.522
2. Agriculture						
Value Added	10^9USD	88.4	252.6	435	530	787
Share	%	36	25.2	15.6	12.3	6.96
Consumption	Mtce	46.92	59.8	66.8	72.3	97.5
Intensity	kgce/USD	0.53	0.237	0.153	0.140	0.124
3. Transportation						
Value Added	10^9USD	8.3	38	111	173	452
Share	%	3.4	3.8	4.0	4.0	4.0
Consumption	Mtce	29.02	118.0	259	390	990
Intensity	kgce/USD	3.45	3.11	2.33	2.25	2.19
4. Services						
Value Added	10^9USD	16.5	210.4	984	1807	5980
Share	%	6.6	21	35.4	43.7	52.9
Consumption	Mtce	5.18	60.0	212.7	308.7	725.2
Intensity	kgce/USD	0.315	0.285	0.216	0.171	0.121
5. Total (production)						
Consumption	Mtce	480.55	1254	1765	2108	3942
Intensity	kgce/USD	1.955	1.660	0.631	0.487	0.348
6. Total (Incl. Livelihood)						
Consumption	Mtce	602.75	1500	2400	3000	5200
Intensity	kgce/USD	2.450	1.495	0.863	0.694	0.460
Elasticity		1.622	0.555	0.451	0.443	0.535

Source: Derived from data in [2].

Table 12. The Exploitable Hydropower Resources in China [8]

River System	Potential Capacity (Gwe)	Annual Output (Twh)	Percentage of Total (%)
Chang Jiang (Yangtze R.)	197.25	1,027	53.4
Huang He (Yellow River)	26.00	117	6.1
Zue Jian (Pearl River)	24.85	112	5.8
Hai and Lan He	2.13	5	0.3
Huai He	0.66	2	0.1
Northeastern Rivers	13.71	44	2.3
Southeastern Rivers	13.90	55	2.9
Southwestern Rivers	37.68	210	10.9
Yalu Zangbo Jiang	50.38	297	15.4
Northwestern Rivers	9.97	54	2.8
National Total	378.53	1,923	100.0

Table 13. The Nuclear Program in the PRC [10]

Location	Type	Capacity	Vendor	Status
1. Nuclear Power				
Qingshan 1	PWR	300	Domestic	Under Const. Oper. Sched. end 1990
Daya Bay	PWR	2X900	Framatom (R) GEC (T)	Under Const. Oper. Sched. 1992
Qingshan 2	PWR	2X600	Domestic (Coop. FRG)	Planned
Liaoning	PWR	2X1000	USSR	Planned, Negotiating
Local	PWR	300	Domestic	Negotiating
2. Nuclear Heat				
Beijing	LTR	5	Domestic	In Operation since Nov., 1989
Jilin	LTR	200	Domestic	Planned
Beijing/Tianjing	LTR	100	Domestic	Negotiating
Beijing	HTGR	10	Domestic (Coop. FRG)	Planned

Table 14-A. The Geothermal Resources (Conduction) in China [12]

(Unit: 10^8 tce)

No.	Region	Exploitable Resource(TCE)	No.	Region	Exploitable Resource (TCE)
I	Songliao Basin	30.714	16	Linying	1.918
1	Thickness > 3000 m	15.316	17	Danhao	1.306
2	Thickness < 3000 m	15.398	18	Xixitai	1.693
II	North China Plain (North)	48.068	IV	North Jiangsu Plain	8.844
3	Liaohe Delta	0.610	19	Yianfu	1.285
4	Central Hebei	13.408	20	Jianhu	0.766
5	Huanghua	4.619	21	Dongtai	6.793
6	Jiyang	9.954	V	Jianghan Plain	9.484
7	Lingqing	10.145	22	Depth>3000m	1.349
8	Changxian	8.151	23	Depth<3000m	8.135
9	Chengning	1.181	VI	Sechuan Basin	25.550
III	North China Plain (South)	20.722	VII	Shaangan'ning Basin	52.355
10	Kaifeng	2.159	24	Depth>3000m	2.565
11	Dongming	1.542	25	Depth 2000-3000m	8.872
12	Zhoukou	1.021	26	Depth 1000-2000m	40.918
13	Shengqiu	1.357	VIII	Fengwei Basin	2.920
14	Hezai	5.403			
15	Taikang	4.323		Total	198.657

Table 14-B. The Geothermal Resources (Convection) in China [12]

Province	Number	Quantity	Number by Temperature Range			
		kgce/sec	>80	80-60	60-40	40-20
Zhejiang	6	0.12		1	3	2
Shandong	19	1.68	3	7	8	1
Hubei	32	3.77	1	4	7	20
Shanxi	10	14.26		1	2	7
Jiangsu	10	0.92		1	4	5
Neimongu	44	0.23		2	1	41
Qinghai	30	1.41	3	5	8	14
Hebei	253	15.48	19	33	30	171
Tibet	293	98.17	41	73	90	89
Shaanxi	24	8.55		2	6	16
Henan	16	1.33		7	4	5
Anhui	12	0.47		3	4	5
Helonjing	2				1	1
Guangxi	24	0.45	1	4	9	10
Ningxia	4	0.09				4
Gansu	23	2.24			4	19
Jiling	4	0.93		2		2
Jiangxi	96	1.98	1	15	36	44
Hunan	112	6.03	2		29	81
Yunnn	280	48.16	21	25	69	165
Beijing	20	1.11		1	14	5
Tianjing	192	11.74	2		40	150
Sechuan	264	6.73	9	18	96	141
Guizhou	64	1.8			21	43
Fujian	147	3.29	7	41	80	19
Liaoning	42	1.83	2	11	15	14
Guangdong	195	4.96	13	51	72	59
Xinjiang	7	0.28			3	4
Total	2225	237.68	125	307	656	1137

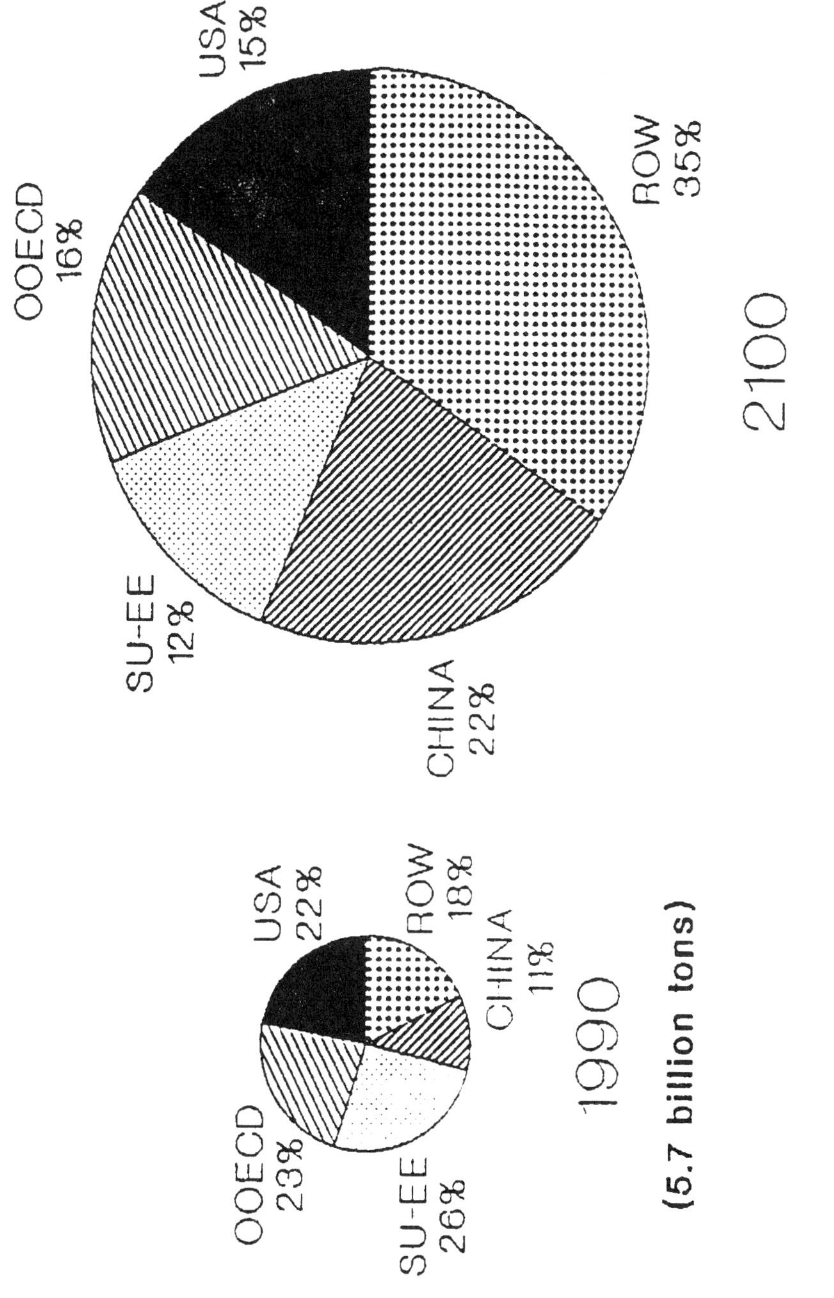

Figure 1. The Increasing Share of CO_2 Emissions from the Developing Countries [1]

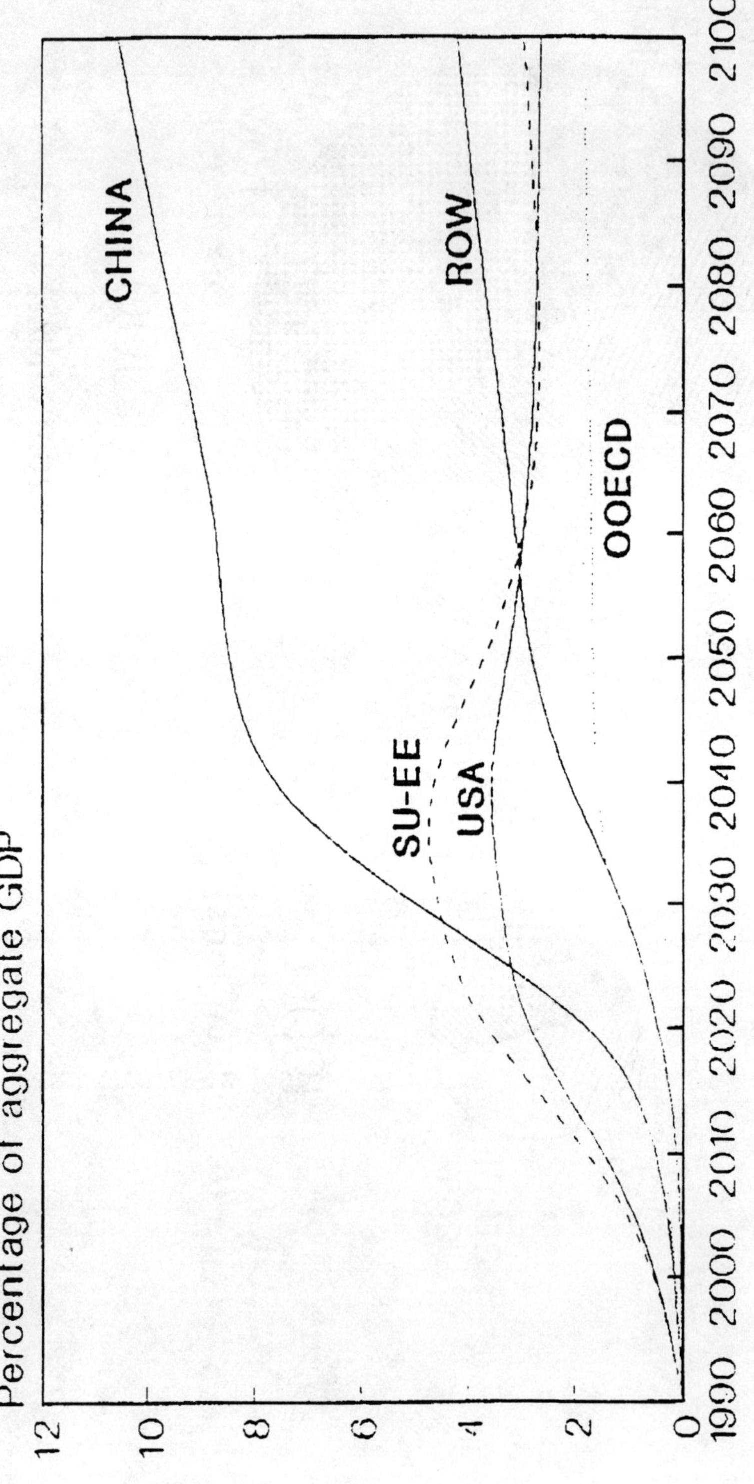

Figure 2. Economic Loss Due to Limit of CO_2 Emissions [1]

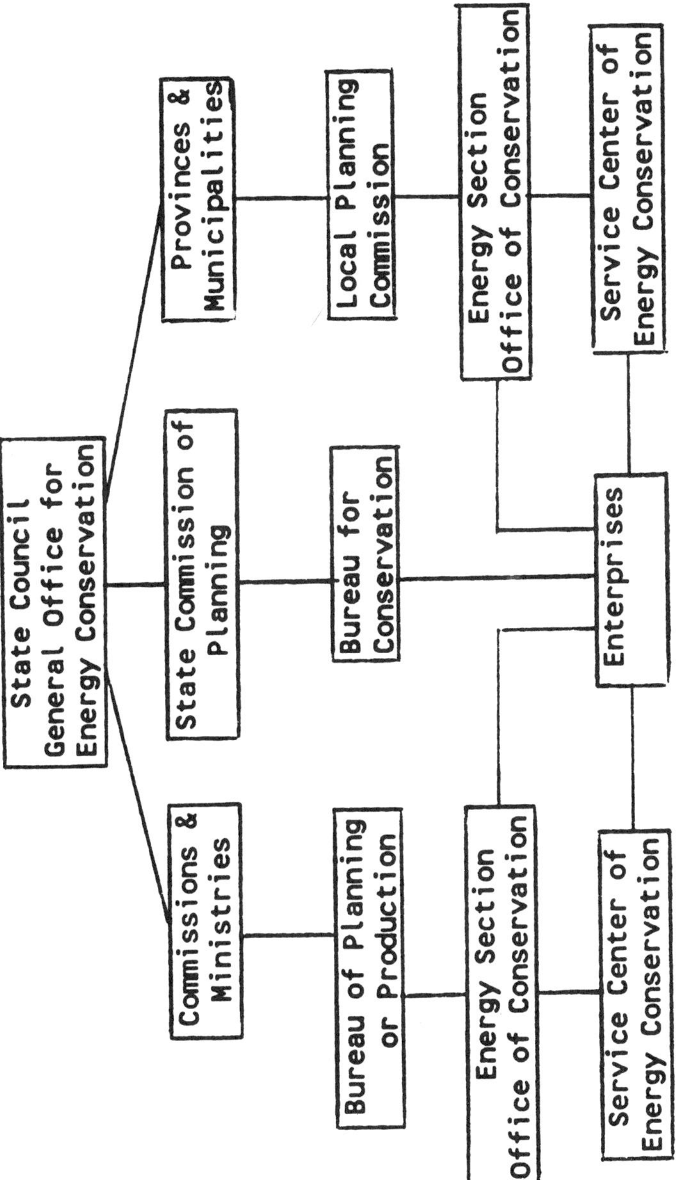

Figure 3. Hierarchy of the Energy Conservation Administration [6]

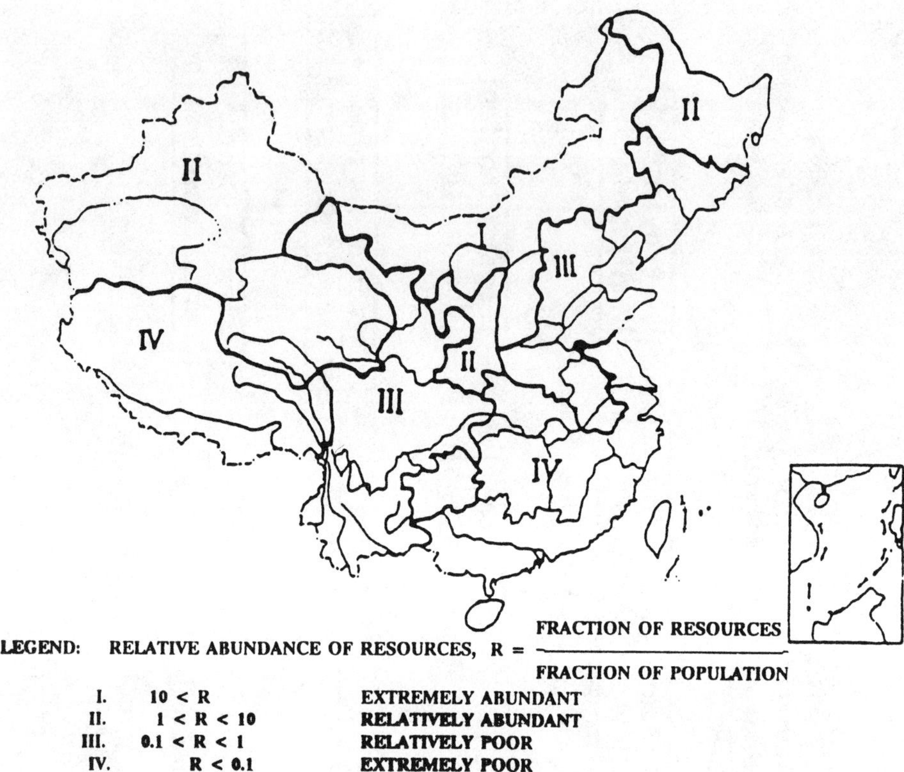

Figure 4. The Distribution of Hydropower Resources in China [8]

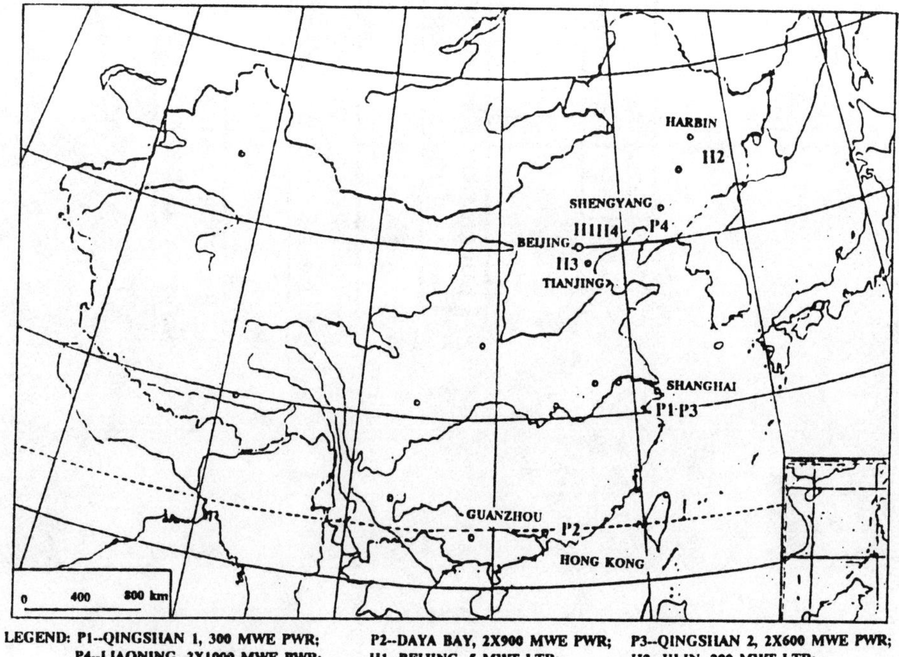

LEGEND: P1--QINGSHAN 1, 300 MWE PWR; P2--DAYA BAY, 2X900 MWE PWR; P3--QINGSHAN 2, 2X600 MWE PWR; P4--LIAONING, 2X1000 MWE PWR; H1--BEIJING, 5 MWT LTR; H2--JILIN, 200 MWT LTR; H3--BEIJING/TIANJING, 100 MWT LTR; H4--BEIJING, 10 MWT HTGR.

Figure 5. The Current Chinese Nuclear Projects

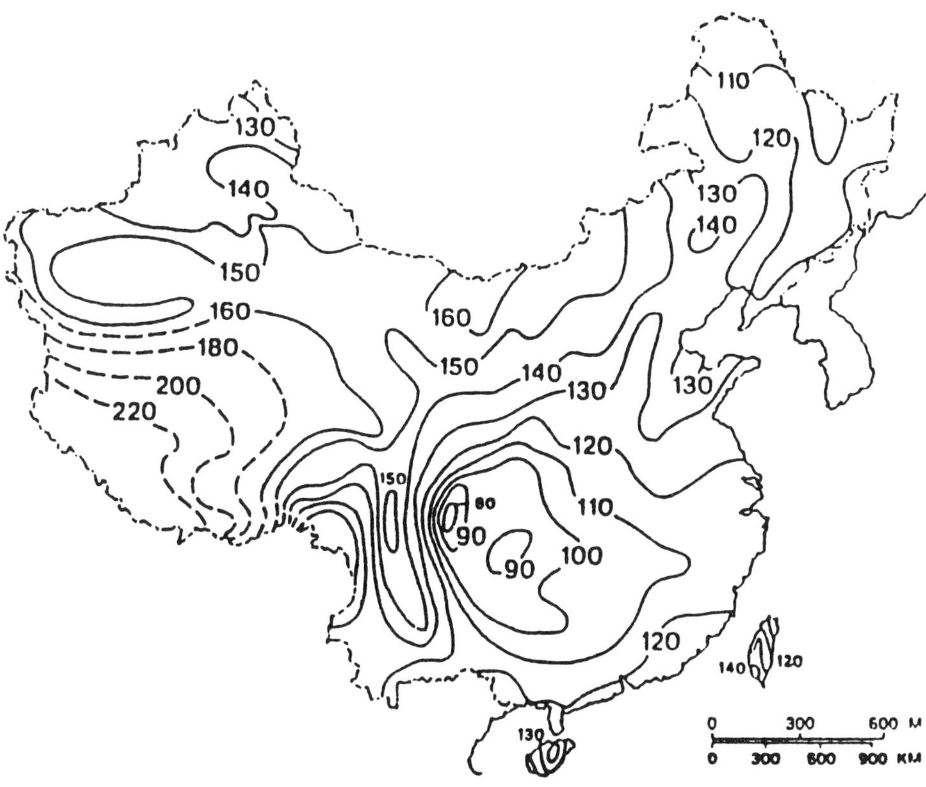

Figure 6. The Distribution of Solar Radiation in China [11]

REFERENCES

1. Manne, Alan, and Richels, Richard, (1990) "Global CO_2 Emission Reductions – The Impacts of Rising Costs," *The Energy Journal*, in press.

2. Wu, Zongxin, et al., (1988) *The Energy Demand of the PRC up to 2050*, Internal Report, INET, Beijing.

3. *Water Power Generation*, (1982) n8, special insert.

4. World Bank, (1985) *China: the Energy Sector.*

5. Lu, Yingzhong, (1989) *Comparison of Energy Consumption, Supply and Policy of the People's Republic of China and Some Developing Countries*, The Washington Institute for Values in Public Policies, Washington DC.

6. Zhou, Dadi, et al., (1989) "The Experiences and Prospects of Energy Conservation in China," Presented at Workshop at Lawrence Berkeley Laboratories, Berkeley.

7. *China Statistical Yearbooks, 1980-1988*, China Statistical Information and Consultancy Service Center, Beijing, China.

8. Lu, Yingzhong, (1985) "The Important Role of Nuclear Energy in the Future Energy System of China," *Energy*, v9, n9/10, pp 761- 771.

9. Lu, Yingzhong, (1987) "The Ordeals of Chernobyl and the Rejustification of Inherently Safe Reactors," *Proceedings, The International Forum on Fueling the 21st Century*, September 29-October 4, Moscow.

10. Lu, Yingzhong, (1990) "The Challenge and the Hope – The Status and Prospects of Nuclear Energy in Asia," presented at the ANS 1990 Annual Meeting, Nashville, June 10-14.

11. Wang, Bingzhong, et al., (1980) "Solar Energy Resources in China," *Acta Energia Solaris Sinica*, v1, n1, pp 1-9.

12. Cai, Yihan, (1982) *The Geothermal Resources in China*, private communication.

SUSTAINABLE ENERGY-ENVIRONMENTAL POLICY OPTIONS FOR THE DEVELOPING WORLD

Mohan Munasinghe

World Bank
1818 H Street, NW
Washington, DC 20433

Introduction

Energy became a major international issue in the 1970s, following the oil crisis. More recently, global climate change induced by excessive greenhouse gas accumulation in the atmosphere has emerged as a potential problem that further complicates energy issues. In this paper, we will focus on energy-related environmental issues in the developing countries. The development-related energy needs of the Third World and their financial implications, the potential for better energy management, barriers to reducing greenhouse gas emissions, and financial mechanisms and policies that can improve the performance of developing countries in this respect will be explored below.

The experience of the industrialized countries emphasizes that a reliable supply of energy is a vital prerequisite for economic growth and development. For example, the observed trends relating to electricity demand in developing countries (which indicate annual growth rates in the region of six to twelve percent) are consistent with the development objectives that these countries all share. Up to the present time, many developing countries have been struggling with the formidable difficulties of meeting these demands for energy services at acceptable costs. If such needs cannot be met, economic growth is likely to slow down and the quality of life will fall.

Given these already existing handicaps, the growing additional concerns about the environmental consequences of energy use considerably complicate the policy dilemma facing the developing countries. In the past, industrial countries that faced a trade-off between economic growth and environmental preservation invariably gave higher priority to the former. These richer countries have awakened only recently to the environmental consequences of their economic progress, and only after a broad spectrum of economic objectives has been reached. This model of economic and social development has been adopted by many Third World regions. Therefore, until both developed and developing countries find a less material-intensive sustainable development path, environmental protection efforts will be hampered.

Published 1991 by Elsevier Science Publishing Company, Inc.
Global Climate Change: The Economic Costs of Mitigation and Adaptation
James C. White, Editor

The less developed countries (LDCs) share the deep worldwide concerns about environmental degradation, and some have taken steps already to improve their own natural resource management as an essential prerequisite for sustained economic development. However, they also face other urgent issues like poverty, hunger and disease, as well as rapid population growth and high expectations. The paucity of resources available to address all these problems constrains the ability of LDCs to undertake costly measures to protect the global commons.

The crucial dilemma for LDCs is how to reconcile development goals and the elimination of poverty – which will require increased use of energy and raw materials – with responsible stewardship of the environment, and without overburdening already weak economies. The per capita GNP of low income economies (with half the world population) averaged US$290 in 1987, or under one-sixtieth of the US value ($18,530). In the two largest developing countries, India and China, per capita GNP was $300 and $290 respectively. Correspondingly, the U.S. per capita energy consumption of 7265 kilograms of oil equivalent (kgoe) in 1987 was 35 and 15 times greater than the same statistic in India and China respectively.

The disparity in both per capita income and energy use among different countries also raises additional issues in the context of current global environmental concerns, and the heavy burden placed on mankind's natural resource base by past economic growth – fossil fuel-related CO_2 accumulation in the atmosphere is a good example. The developed countries accounted for over 80% of such cumulative worldwide emissions during 1950-86: North America contributed over 40 billion tons of carbon; western and eastern Europe emitted 25 and 32 billion tons, respectively; and the developing countries share was about 24 billion tons. On a per capita basis the contrasts are even more stark, with North America emitting over twenty times more and the developed countries as a whole being responsible for over eleven times as much total cumulative CO_2 emissions as the LDCs. The LDC share would be even smaller if emissions prior to 1950 were included. Clearly, any reasonable growth scenario for developing nations that followed the same material-intensive path as the industrialized world would result in unacceptably high levels of future greenhouse gas accumulation as well as more general depletion of natural resources.

Up until now, scientific analysis has provided only broad and rather uncertain predictions about the degree and timing of potential global warming. However, it would be prudent for mankind to buy an "insurance policy" in the form of mitigatory actions to reduce greenhouse gas emissions. Ironically, both local and global environmental degradation might affect developing countries more severely, since they are more dependent on natural resources while lacking the economic strength to prevent or respond quickly to increases in the frequency, severity and persistence of flooding, drought, storms, and so on. Thus, from the LDC viewpoint, an attractive low-cost insurance premium would be a set of inexpensive measures that could address a range of national and global environmental issues, without hampering development efforts.

The recent report of the Bruntland Commission (WCED, 1987), which has been widely circulated and accepted, has presented arguments along the theme of sustainable development, which consists of the interaction of two components: needs, especially those of the poor segments of the world's population, and limitations, which are imposed by the ability of the environment to meet those needs. The development of the presently industrialized countries took place in a setting which emphasized needs and de-emphasized limitations. The development of these societies has effectively exhausted a disproportionately large share of global resources – broadly defined to include both the resources that are consumed in productive activity (such as oil, gas and minerals) as well as environmental assets that absorb the waste products of economic activity and those that provide irreplaceable life support functions (like the high altitude ozone layer). Indeed, some analysts argue that this development path has significantly indebted the developed countries to the larger global community.

The division of responsibility in this global effort is clear from the above arguments. The unbalanced use of common resources in the past should be one important basis on which the developed and developing countries can work together to share and preserve what remains. The developed countries have already attained most reasonable goals of development and can afford to substitute environmental protection for further growth of material output. On the other hand, the developing countries can be expected to participate in the global effort only to the extent that this participation is fully consistent with and complementary to their immediate economic and social development objectives.

In the context of the foregoing, this paper identifies critical energy-environmental issues, using examples from the highly capital-intensive and pivotal power sector to illustrate specific points. It also explores some policy implications, constraints and opportunities at the national and global level for both the developing countries as well as the wider international community, and examines the role of emerging mechanisms, such as the global environmental fund and ozone defense fund, in the allocation of resources for addressing transnational environmental problems.

Electric Power Needs of the Developing Countries

Despite some anomalies, the link between energy demand and gross domestic product (GDP) is well established. Electric power, in particular, has a vital role to play in the development process, with future prospects for economic growth being closely linked to the provision of adequate and reliable electricity supplies. Figure 1 indicates the relationship between electricity use and income for both developed and developing countries. A more systematic analysis of World Bank and U.N. data over the past two or three decades indicates that the ratio of percentage growth rates (or elasticity) of power system capacity to GDP is about 1.4 in the developing countries.

Assuming no drastic changes in past trends with respect to demand management and conservation, the World Bank's most recent projections indicate that the demand for electricity in LDCs will grow at an average annual rate of 6.6% during the period 1989-99 (World Bank, 1990). This compares with actual growth rates of 10% and 7% in the seventies and eighties, respectively. Such rates of growth indicate the need for total capacity additions of 384 GW during 1990-99 (see Figure 2), and annual energy consumption of 3844 TWh by 1999. As indicated in Figure 3, the Asia region's requirements dominate with almost two-thirds of the total, and coal and hydro are the main primary sources – both of which have specific environmental problems associated with their use.

The investment needs corresponding to these indicative projections are also very large. Table 1 shows the projected breakdown of LDC power sector capital expenditures in the 1990s. Of a total of $745 billion (constant 1989 US$), Asia (which includes both India and China) again dominates, accounting for $455 billion or over $45 billion annually. In comparison with the total projected annual requirement for LDCs of $75 billion, the present annual rate of investment in developing countries is only around $50 billion. Even this present rate is proving difficult to maintain. Developing country debt, which averaged 23% of GNP in 1981, increased dramatically to 42% in 1987 and has not declined significantly since then. In low income Asian countries, outstanding debt doubled, from 8% in 1981 to 16% in 1987. Capital-intensive power sector investments have played a significant role in this observed increase.

If the developing countries follow this projected expansion path, the environmental consequences are also likely to increase in a corresponding fashion. There is already a growing concern at the national level about the environmental consequences of energy use in developing countries. At a recent workshop on acid rain in Asia, participants reported on a wide range of environmental effects of the growing use of fossil fuels, especially coal, in the region (Foell, 1989). For example, total 1985 sulfur dioxide emissions in Asia were estimated at around 22 million tons, and these levels, coupled with high local densities, have led to acid deposition in many parts of Asia.

The developing countries feel that any attempts to mitigate these environmental effects, however, cannot jeopardize the critical role played by electric power (and more generally, energy) in economic development. Similarly, the allocation of resources to environmental programs in developing countries cannot diminish the resources needed to fund projected expansion of supply. Energy and environmental policymakers in both developing countries and the global community are, indeed, confronted with a formidable dilemma.

The Economics of Energy-Environmental Issues

The foregoing discussion has helped to establish a rational and equitable basis for addressing the problems of energy-environmental impact mitigation. In this section we present an economic efficiency framework which ties together the issue of environmental protection with the existing energy sector goals of energy efficiency and economic growth.

It is convenient to recall here that traditionally, the specific prerequisites for economic efficiency have included both (Munasinghe, 1990b):

(1) Efficient consumption of energy, by providing efficient price signals that ensure optimal energy use and resource allocation; and

(2) Efficient production of energy, by ensuring the least-cost supply mix through the optimization of investment planning and energy system operation.

A new issue which has emerged in recent decades as an area of particular concern is the efficient and optimal use of our global natural resource base, including air, land and water. Since there has been much discussion also about the key role that energy efficiency and energy conservation might play in mitigating environmental costs, it is useful first to examine how these topics relate to economic efficiency. Specific issues dealing with the formulation and implementation of economically efficient energy policies are presented in the next section.

Major environmental issues vary widely, particularly in terms of scale or magnitude of impact, but most are linked to energy use. First, there are the truly global problems such as the potential worldwide warming due to increasing accumulation of greenhouse gases like carbon dioxide and methane in the atmosphere, high altitude ozone depletion because of release of chlorofluorocarbons, pollution of the oceanic and marine environment by oil spills and other wastes, and excessive use of certain animal and mineral resources. Second in scale are the transnational issues like acid rain or radioactive fallout in one European country due to fossil-fuel or nuclear emissions in a neighboring nation, and excessive downstream siltation of river water in Bangladesh due to deforestation of watersheds and soil erosion in nearby Nepal. Third, one might identify national and regional effects, for example those involving the Amazon basin in Brazil, or the Mahaweli basin in Sri Lanka. Finally, there are more localized and project-specific problems like the complex environmental and social impacts of a specific hydroelectric or multipurpose dam.

While environmental and natural resource problems of any kind are a matter for serious concern, those that fall within the national boundaries of a given country are inherently easier to deal with from the viewpoint of policy implementation. Such issues that fall within the energy sector must be addressed within the national policymaking framework. Meanwhile, driven by strong pressures arising from far-reaching potential consequences of global issues like atmospheric greenhouse gas accumulation, significant efforts are being made in the areas of not only scientific analysis, but also international cooperation mechanisms to implement mitigatory measures.

Given this background, we discuss next some of the principal points concerning energy use and economic efficiency (Munasinghe, 1990a). In many countries, especially those in the developing world, inappropriate policies have encouraged wasteful and unproductive uses of some forms of energy. In such cases, better energy management could lead to improvements in economic efficiency (higher value of net output produced), energy efficiency (higher value of net output per unit of energy used), energy conservation (reduced absolute amount of energy used), and environmental protection (reduced energy-related environmental costs). While such a result fortuitously satisfies all four goals, the latter are not always mutually consistent. For example, in some developing countries where the existing levels of per capita energy consumption are very low and certain types of energy use are uneconomically constrained, it may become necessary to promote more energy consumption in order to raise net output (thereby increasing economic efficiency). There are also instances where it may be possible to increase energy efficiency while decreasing energy conservation.

Despite the above complications, our basic conclusion remains valid—that the economic efficiency criterion which helps us maximize the value of net output from all available scarce resources in the economy (especially energy and the ecosystem in the present context), should effectively subsume purely energy oriented objectives such as energy efficiency and energy conservation. Furthermore, the costs arising from energy-related adverse environmental impacts may be included (to the extent possible) in the energy economics analytical framework, to determine how much energy use and net output that society should be willing to forego in order to abate or mitigate environmental damage. The existence of the many other national policy objectives—including social goals that are particularly relevant in the case of low income populations—will complicate the decision making process even further.

The foregoing discussion may be reinforced by the use of a simplified static analysis of the trade-off between resource use and net output of an economic activity (see Figure 4).

Energy Efficiency

Y represents the usual measure of net output of productive economic activity in a country, as a function of some resource input (say energy)—considering only the conventional internalized costs, i.e. not accounting for environmental impacts. Due to policy distortions (for example, subsidized prices), the point of operation in many developing countries appears to be at A, where the resource is being used wastefully. Therefore, without invoking any environmental considerations, but merely by increasing economic and resource use efficiency (i.e. energy efficiency), output as usually measured could be maximized by moving from A to B. A typical example might be improving energy end-use efficiency or reducing energy supply system losses (see next section for practical examples).

Quantifiable National Environmental Costs

Now consider the curve EC_{NQ} which represents economically quantifiable national environmental costs associated with energy use. The latter might include air pollution-related health costs of a coal power plant or the costs of environmental protection equipment (like scrubbers and electrostatic precipitators to reduce noxious gas and particulate emissions) installed at such a plant, or the costs of resettlement at a hydropower dam site. The corresponding corrected net output curve is: $YE_{NQ} = Y - EC_{NQ}$, which has a maximum at C that lies to the left of B, implying lower use of (more costly) energy.

Nonquantifiable National Environmental Costs

Next, consider the "real" national output YE_{NT}, which is net of total environmental costs, whether quantifiable or not. The additional costs to be considered include the unquantified yet very real human health and other unmonetized environmental costs. These total (quantifiable and nonquantifiable) costs are depicted as EC_{NT}; and once again $YE_{NT} = Y - EC_{NT}$. As shown, the real maximum of net output lies at D, to the left of C.

Transnational Environmental Costs

Finally, EC_G represents the globally adjusted costs, where the transnational environmental costs (to other countries) of energy use within the given country have been added to EC_{NT}. In this case, $YE_G = Y - EC_G$ is the correspondingly corrected net output which implies an even lower level of optimal energy use.

For example, consider the costs imposed on other countries (such as transborder impacts of a major dam or global climate impacts of carbon dioxide emissions). If it is decided to reduce resource use within this country further in order to achieve the internationally adjusted optimum at E, then a purely national analysis will show this up as a drop in net output, i.e. from D to E. Since other countries benefit, this drop in net output may justify compensation in the form of a transfer of resources from the beneficiary countries. Note that the (reverse) transnational costs imposed by other countries on the nation in question will be a function of regional or global resource use rather than the national resource use shown on the horizontal axis.

The additional curve YT shows net output for a technologically advanced future society that has achieved a much lower resource intensity of production.

Policy Measures for the LDCs

The foregoing analysis illustrates the crucial dilemma for developing countries. In Figure 4, all nations (including the poorest) would readily adopt measures that will lead to shift (1) which simultaneously and unambiguously provides both economic efficiency and environmental gains. Most developing countries are indicating increasing willingness to undertake shift (2). However, implementing shift (3) will definitely involve crossing a "pain threshold" for many Third World nations, as other pressing socioeconomic needs compete against the costs of mitigating nonquantifiable adverse environmental impacts.

We note that real economic output increases with each of the shifts (1),(2) and (3) – shown by the movement upward along the curve YE_{NT} from G to D. However, these shifts are mistakenly perceived often as being upward only from A to B (energy efficiency improvements), followed by downward movements from B through C to D. It is, therefore, important to correct any misconceptions that environmental protection results in reduced net output – this can be achieved through institutional development, applied research, strengthening of planning capabilities etc. However, it is clear that shift (4) would hardly appeal to resource constrained developing countries unless concessionary external financing was made available, since this movement would imply optimization of a global value function and costs that most often exceed in-country benefits. In the foregoing, we have neglected considerations involving reciprocal benefits to the given country due to energy use reductions in other countries.

Therefore, we may conclude briefly that, while the energy required for economic development will continue to grow in the developing countries, in the short to medium run there is generally considerable scope for most of them to practice better energy management, thereby increasing net output, using their energy resources more efficiently, and contributing to the effort to reduce global warming. In the medium to long run, it will become possible for the developing countries to adopt newer and more advanced (energy efficient) technologies that are now emerging in the industrialized world, thus enabling them to transform their economies and produce even more output using less energy.

In other words, the developing countries could be expected to cooperate in global environmental programs only to the extent that such cooperation is consistent with their national growth objectives. The role of the developed countries, on the other hand, is to incur the risks inherent in developing innovative technological measures which are the prerequisites for the next level in environmental protection and the mitigation of adverse consequences. These risks include the possibility that the more extreme measures may turn out to be unnecessary or inapplicable after all, given the prevailing uncertainty about the future impact of current environmental developments.

Framework for Energy-Environmental Policy Analysis and Options at the National Level

We have introduced a way of considering energy-environmental issues in terms of four shifts. In this section a rational framework for energy-environmental policy analysis within a country is presented in terms of these shifts. This coverage will be expanded to the global level in the next section, focusing particularly on the interaction between developed and developing countries.

Developed countries generally differ from the developing countries in the extent to which the shifts have already been made. The LDCs are more likely to be characterized as being at point A in Figure 4, while the developed countries are somewhere between C and D. This means that for developing countries there is still considerable scope for environmental improvement by undertaking programs that are consistent with the national objective of increasing overall output. The challenge for national decision makers and the international community is to find as many areas as possible where such consistency and complementarity exist between growth and environmental protection goals.

Advantages of an Integrated Approach

Successful policy analysis requires an integrated approach, unified in an economic sense so that all feasible options can be balanced and traded off if necessary in the search for an overall optimal strategy. It is within such a comprehensive framework that barriers to and opportunities for making various choices will become apparent. For successful energy policy analysis, a better understanding of economy-wide linkages is useful, whatever the prevailing political system. It will help decision makers in formulating policies and providing market signals and information to economic agents that encourage more efficient energy production and use, as well as better protection of the environment. We summarize in Figure 5, a hierarchical framework for integrated national energy planning (INEP), policy analysis and supply-demand management to achieve these goals (Munasinghe 1990a).

Although the INEP framework is primarily country focused, we begin by recognizing that many energy-environmental issues have global linkages. Thus individual countries are embedded in an international matrix, while economic and environmental conditions at this global level will impose a set of exogenous inputs or constraints on decision makers within countries. The next hierarchical level in Figure 5 treats the energy sector as a part of the whole economy. Therefore, energy planning requires analysis of the links between the energy sector and the rest of the economy. Such links include the energy needs of user sectors, input requirements of the energy sector, and impact on the economy of policies concerning energy prices and availability.

The next level of the integrated approach treats the energy sector as a separate entity composed of subsectors such as electricity, petroleum products and so on. This permits detailed analysis, with special emphasis on interactions among the different energy subsectors, substitution possibilities, and the resolution of any resulting policy conflicts. The final and most disaggregate level pertains to analysis within each of the energy subsectors. It is at this lowest hierarchical level that most of the detailed energy resource evaluation, planning and implementation of projects is carried out by line institutions (both public and private). In practice, however, the three levels of INEP merge and overlap considerably. Thus the interactions of electric power problems and linkages at all three levels need to be carefully examined.

In practice, the three levels of INEP merge and overlap considerably. Energy-environmental interactions (represented by the vertical bar) tend to cut across all levels and need to be incorporated into the analysis. Finally, spatial disaggregation may be required also, especially in larger countries. Such an integrated framework facilitates policymaking and does not imply rigid centralized planning. Thus, such a process should result in the development of a flexible and constantly updated energy strategy designed to meet the national goals mentioned earlier. This national energy strategy (of which the investment program and pricing policy are important elements), may be implemented through a set of energy supply and demand management policies and programs that make effective use of decentralized market forces and incentives.

Policy Tools and Constraints

To achieve the desired national goals, the policy instruments available to Third World governments for optimal and energy management include: (a) physical controls; (b) technical methods; (c) direct investments or investment-inducing policies; (d) education and promotion; and (e) pricing, taxes, subsidies and other financial incentives. Since these tools are interrelated, their use should be closely coordinated for maximum effect.

The chief constraints that limit effective policy formulation and implementation are: (a) poor institutional framework and inadequate incentives for efficient management; (b) insufficient manpower and other resources; (c) weak analytical tools; (d) inadequate policy instruments; and (e) other constraints such as low incomes and market distortions.

Technological Options

The INEP framework is particularly appropriate for considering shift (1) in Figure 4, which implies improvement in overall economic efficiency but without any explicit consideration of the external (environmental) costs. Such efficiency improvements require better energy supply and demand management. More specifically, the former category includes more accurate demand forecasting, improved least-cost investment planning, and the optimal operation of energy systems – implying that plant performance, operating and maintenance procedures, loss levels, etc. are optimized. The latter comprises efficient electricity end use, load management and pricing (described in the next section). All the above options constitute an attractive policy package for most power utilities in both the developing and developed countries.

There is a spectrum of technological options which the developing countries could potentially utilize in order to improve energy efficiency and thereby reduce environmental effects arising from energy sector activity. These range from simple infrastructural retrofits to the use of advanced energy supply technologies. Among the short-term technological options for the developing country power sector, reducing transmission and distribution losses, and improving generation plant efficiencies appear to be the most attractive. Recent studies show that, up to a certain point, these supply efficiency enhancing measures yield net economic savings or benefits that are several times the corresponding costs incurred (Munasinghe, 1990b). While estimates of such power system losses vary, they all point to levels which are far in excess of accepted norms. Table 2 presents estimates for some Asian countries in comparison with industrialized countries. While acceptable loss levels may be about 6-8% in transmission and distribution as a percentage of gross generation, these losses in Third World power systems are estimated to average in the 16-18% range (of which about one-third could be theft).

The consequences of reducing these losses can be quite important. On the basis of our previous estimates of capacity requirements, a one percentage point reduction in losses per year would reduce required capacity by about 5 GW annually in the developing countries. The estimated saving in capital investment would be around 10 billion dollars per year. Meanwhile, the Agency for International Development (USAID, 1988) has estimated that the average heat rate of LDC power plants is around 13,000 Btu/kWh, compared to 9,000-11,000 Btu/kWh if these plants were operated efficiently. The energy savings (and positive environmental consequences) implied in these figures are quite significant also.

Similar gains are possible by conservation on the demand side. Johansson et al. (1989) provide an insightful review of the developments that have been taking place in end-use technologies which can have a major impact on energy efficiency. These technologies (which developed in the industrialized countries as a response to the oil price escalation in the seventies) can be easily applied towards more efficient lighting, heating, refrigeration and air conditioning around the developing world, as described below.

Substitution of primary energy sources in power generation is another potential means of achieving dual benefits. In the developing world, natural gas is the most likely candidate for coal or oil substitution. The economic benefit of natural gas substitution comes from either import substitution for petroleum products or releasing these products for export. On the environmental front, natural gas firing typically achieves reduction in carbon emissions of 30-50%. Many Asian countries are endowed with significant resources of natural gas, including Malaysia, Indonesia and Thailand.

In the longer term, the developing countries will need to rely on more advanced technological options which are currently being developed in the industrialized countries. As we have discussed above, power generation capacity in developing countries is expected to nearly double by the turn of the century, and will increase further thereafter. This provides opportunities to add state-of-the-art technologies which have been designed with regard to both economic and environmental criteria. Clean coal technologies, cogeneration, gas turbine combined cycles, steam-injected gas turbines, etc. are all part of this menu of technologies which have important potential in developing countries. Similar applications will become available for emission control technologies. However, as we have argued previously, the developing countries will look to the industrialized nations to provide the leadership in refining and proving these technologies before they are implemented in the developing world.

As one indicative example of the state-of-the-art in supply planning and demand management possibilities, we summarize the results of a recent Swedish study (Johannson et al., 1989). The power sector in Sweden, currently supplied half by hydro and half by nuclear generation, faces the following severe restrictions: (a) hydro expansion limited by environmental constraints; (b) mandatory phasing out of all nuclear units by 2010; and (c) no increase permitted above present CO_2 emission levels. The demand for electricity derived services is projected to increase by 50% from 1987 to 2010. If end use efficiency remained unchanged, then, under this "frozen efficiency" scenario, the electrical load also would increase by 50%, from 129 TWh in 1987 to 195 TWh in 2010, at an average annual growth rate of 1.8%.

The *same* output of electrical services could be provided, but with steadily declining electricity input needs and load levels – based on the increasingly energy efficient scenarios A, B, C and D, shown in Figure 6. The corresponding loads in 2010 would be

140, 111, 96 and 88 TWh, respectively. Only options C and D permit the load to be met after all the nuclear plants are retired in 2010. Figure 7 indicates that the total costs of energy supply also fall progressively under the scenarios A through C (some of the costs are undefined for scenario D). In addition, there are three supply scenarios based on different selection rules for generation; the supply costs rise steadily as we move through the economic dispatch, natural gas/biomass, and environmental dispatch options. These costs exclude taxes and subsidies and are based on a 6% discount rate, 1987 world oil and coal prices, and coal equivalent gas prices for steam power generation.

Economic Incentives and Related Options

Providing the correct economic signals, or more specifically price rationalization, offers the most attractive demand-side option for improving energy sector efficiency that also corresponds to the shift (1) in Figure 4. While the economic principles of energy pricing are now well understood, pricing policy in developing countries is guided by a trade-off between economic efficiency on the one hand and a series of financial and socioeconomic considerations on the other. It is widely accepted that energy is fundamental to productivity and economic growth. However, the strong perception among decision makers that access to energy is a basic need that improves living standards of the people has driven a policy in which affordability competes with economic efficiency as the criterion for energy pricing. Furthermore, in practice it has been difficult to separate social and economic criteria within the same pricing structure, leading to poorly designed policies that may be both economically inefficient and socially regressive (or perverse).

A recent study of over 350 electric power utilities in developing countries (Munasinghe et al., 1988) indicates that electricity tariffs have not kept up with cost growth. The operating ratio (defined as the ratio of operating costs before debt service, depreciation and other financing charges, to operating revenue) for the extensive sample studied deteriorated from 0.68 in the 1966-73 period to 0.80 between 1980 and 1985. At the same time, the financial rate of return on fixed assets has decreased steadily from over 10 percent in the mid-1960s to around 5 percent in the mid-1980s. In some countries, these declines are significantly greater.

This study and other available evidence indicates that a significant shift towards economic criteria in electricity pricing would be possible without creating undue hardship to the poorer segments of the population. While extensive information (in the form of price elasticities) is not available for most developing countries, several recent studies provide a reasonable basis for projections. Assuming the price elasticity for electric power to be -0.3, a 20% real increase in electricity prices (which would restore the above operating ratio to its 1960/70s level) would result in a 6% reduction in electricity demand.

Apart from price rationalization, there is also scope for applying a coordinated package of other measures aimed at improving the efficiency of energy use. These would include taxes and subsidies based on fuel type, technology, R&D, retrofits, conservation programs, etc. In most instances these programs are likely to achieve desirable effects on both energy use and environmental impacts. Fiscal instruments such as emission fees and carbon-based user fees can be used to control environmental impacts more directly. In many LDC applications, however, problems of implementing and monitoring such mechanisms are significant.

To conclude, generally the first priority is to improve the efficiency of energy supply and use through technical means and pricing energy at marginal economic cost. Improvements in energy intensive industries could yield significant efficiency gains. In transport, considerable opportunities exist to reduce fuel consumption by improving traffic management methods, using less energy intensive travel modes, and phasing out inefficient vehicles. Substitution of fossil fuels with less polluting fuels (like natural gas, where available) and increasing the efficiency of fuelwood use could also be environmentally and economically sound. Agriculture presently contributes about 14% of all CO_2 emissions, mainly due to forest clearing and burning of wastes. Strengthening property rights, protecting forest lands, and improving land use planning and management (e.g., through agro-ecological zoning) are some key actions in this sector.

Examples of more expensive options are nonconventional energy sources, large-scale reforestation, advanced energy technologies, and substitution for chlorofluorocarbons (CFCs). LDCs will need significant financial assistance to implement such measures, once the less costly options have been exhausted.

National Level Organizational/Institutional Options

The energy sector in developing countries is typically owned and controlled by the government, and is characterized by large monolithic organizations. While there is some rationale for this centralization, it could be a critical barrier in the path of greater efficiency and improved flexibility. The desperate circumstances of many developing country energy-supplying enterprises have generated pressures for new approaches to organizing the sector. In particular, there appears to be considerable interest in the scope for more decentralization and greater private participation. Developing countries power sector officials have been very active in studying this option, and some countries have already prepared the necessary legislative and institutional groundwork for this transition. India plans to install as much as 5000 MW of private power capacity over the 1990-95 period, and similar plans are underway in Indonesia, Malaysia, Thailand, Philippines and Pakistan. In Sri Lanka, a private company has been distributing power since the early eighties and significant efficiency and service improvements have been observed during this period.

Despite these trends, enhanced private participation in the energy sector is likely to be more successful when it is one element in a broader economic package involving policy reforms in other parts of the economy. Market forces confined to the power sector in a highly distorted economy may not necessarily improve the power sector situation since private participants will try to maximize financial rather than economic costs. Thus, private sector participants would make full use of cheaper generation inputs such as coal even when this is potentially detrimental (both economically and environmentally). Even in a reasonably market-oriented economy, the introduction of private participation in the energy sector is unlikely to lead to environmental benefits, unless the costs of pollutants can be fully captured (i.e. internalized) in the financial cost to the participant. Thus, while private participation is likely to bring significant gains by the infusion of new capital and innovative management methods, it is likely to remain one of several methods aimed at restructuring the sector.

Environmental Costs

With regard to environmental issues, the national environment in which the utility functions is likely to play an equally important role. Actions of the utility need to be backed up by a set of consistent national policies and legislative support. The development of environmental standards and regulations is likely to (and should) take place outside the utility, and the public needs to understand the importance of a commitment to a program of environmental mitigation.

Our integrated framework also provides an appropriate starting point for consideration of shift (2) in Figure 4 that seeks to incorporate the quantifiable environmental costs. A number of techniques exist for valuing environmental impacts of energy projects (Munasinghe and Lutz, 1990). Such evaluation approaches may be used to incorporate environmental costs into methodologies mentioned above for least-cost planning and estimating marginal costs of energy production. However one should be aware of the uncertainties in such estimates of value and perform sensitivity tests where appropriate.

Going beyond the quantifiable environmental costs is of course problematical but, to the extent that these costs are significant, an attempt must be made rather than implicitly assuming that these costs are negligible. Nonquantifiable environmental costs can be incorporated in various ways, such as adding new constraints on the optimization that reflect social concerns or absolute environmental standards, or even by using an entirely different methodology than least-cost planning, e.g. a type of multi-attribute assessment. This is still consistent with INEP, though the various trade-offs would be made explicitly on social-environmental criteria rather than implicitly in economic terms.

Global Environmental Issues in the LDC Context

The developed and the developing countries are at different points in resolving domestic energy-environmental interactions, and this is an important difference which must be taken into account when devising forms of cooperation for solving transnational and global energy-environment problems. Developing countries still have considerable scope for environment-improving activities that are economically attractive for them, e.g., energy conservation and ameliorating the domestic environmental consequences of energy use. These actions will, of course, have positive global environmental benefits also. While no country can be said to have exhausted the potential for shifts (1), (2) and (3), the developed countries are generally closer to D in Figure 4, at which point they need to explicitly consider trade-offs in domestic policy options to improve the global commons.

A second aspect is that developing countries are less able to afford actions to protect the global environment. This is also an equity issue, since much of the responsibility for cumulative damage to the global environment lies with the developed world.

A third aspect is the extent to which developing countries contribute to and are affected by global environmental problems. This is not a clear-cut issue since some countries may feel that they are highly vulnerable although contributing little to global climate changes (e.g., Bangladesh or Maldives who will suffer from sea level rise), while others will contribute far more and experience varying degrees of impact (e.g. China and India).

The foregoing sets the context within which the developing countries are capable of participating in environmental mitigation efforts at the global level. It is quite obvious that LDCs do not have the ability to contribute financially for global environmental cleanup efforts where the measurable benefits to the national economy are too low to trigger investment. Indeed, this paper has argued that many LDC projects which do have positive measurable benefits at the national level are being by-passed on account of capital constraints.

The principle of assistance to developing countries for environmental mitigation efforts, in terms of technology transfer, financial support and other means, is already well established. The Montreal Protocol, which was adopted in 1987 as a framework within which reduction in the consumption and production of certain types of chlorofluorocarbons (CFCs) is to be achieved, recognized the need for global cooperation and assistance to the developing countries. Subsequent Ministerial Conferences on various aspects of global environmental issues have reinforced the idea of protecting the global commons.

Currently, discussions are underway among world bodies and governments to define effective criteria and mechanisms for both generating and disbursing funds from a global environmental fund. While a broad workable agreement will not be easy to reach, global financing issues might be analyzed and resolved through a trade-off involving several criteria: *affordability/additionality*, *fairness/equity* and *economic efficiency*.

First, since LDCs cannot afford to finance even their present energy supply development, to address global environmental concerns they will need financial assistance on concessionary terms that is additional to existing conventional aid. Second, as noted in the recent Brundtland Commission report, past growth in the industrialized countries has exhausted a disproportionately high share of global resources, suggesting that the developed countries owe an "environmental debt" to the larger global community. This approach could help to determine how the remaining finite global resources may be shared more fairly and used sustainably. Finally, the economic efficiency criterion indicates that the "polluter pays" principle may be applied to generate revenues, to the extent which global environmental costs of human activity can be quantified. If total emission limits are established (e.g. for CO_2), then trading in emission permits among nations and other market mechanisms could be harnessed to increase efficiency.

One specific proposal that is being reviewed calls for a core multilateral fund of about SDR one billion – the Global Environmental Fund (GEF) – to be set up as a pilot over the next three years. This fund would finance investment, technical assistance and institutional development activities in four areas: global climate change, ozone depletion, protection of biodiversity, and water resource degradation. A more narrowly focused Ozone Defense Fund (ODF) of about US$160 to $240 million is also under discussion, to help implement measures to reduce CFC emissions under the Montreal Protocol. Both funds are likely to be managed under a collaborative arrangement between the UNDP, UNEP and the World Bank. In particular, they would fund those investment activities that would provide cost-effective benefits to the global environment, but would, however, not be undertaken by individual countries without concessions. Thus, these funds are being specifically designed to fill the void which is created by the lack of individual national incentives for those activities which would, nonetheless, benefit us all.

Conclusions

International pressures to implement environmentally mitigatory measures place a severe burden on developing countries. The crucial dilemma this poses to LDCs is how to reconcile development goals and the elimination of poverty – which will require increased use of energy and raw materials – with responsible stewardship of the environment, and without overburdening economies that are already weak. This paper has argued that, in view of the severe financial constraints that developing countries already face, the response of these countries in relation to environmental preservation cannot extend beyond the realm of measures that are consistent with near-term economic development goals. More specifically, the environmental policy response of LDCs in the coming decade will be limited to conventional technologies in efficiency improvement, conservation and resource development.

The developed countries are ready to substitute environmental preservation for further economic expansion and should, therefore, be ready to cross the threshold, providing the financial resources that the LDCs need today and developing the technological innovations and knowledge base to be used in the 21st century by all nations. The Global Environmental Fund and Ozone Fund, presently being established, will facilitate the participation of LDCs in addressing issues at the global level.

Table 1
Regional Breakdown of LDC Power Capital Expenditures in the 1990s

	Asia	EMENA*	LAC**	Africa	Total
Generation	277	82	83	6	448
Transmission	39	8	32	2	81
Distribution	100	23	27	2	152
General	39	11	13	1	64
TOTAL	455	124	155	11	745
Percent	61.1	16.6	20.8	1.5	100

* Europe, Middle East and North Africa (Meditarranean region)
** Latin America and the Caribbean

Source: The World Bank

Table 2
Electrical Transmission and Distribution Losses
(% of gross generation)*

Pakistan	28%
India	22%
Bangladesh	31%
Sri Lanka	18%
Thailand	18%
Philippines	18%
South Korea	12%
Japan	7%
US	8%

* These loss estimates include non-technical losses (i.e. due to deficient metering and theft).

Sources: The World Bank and USAID

Figure 1

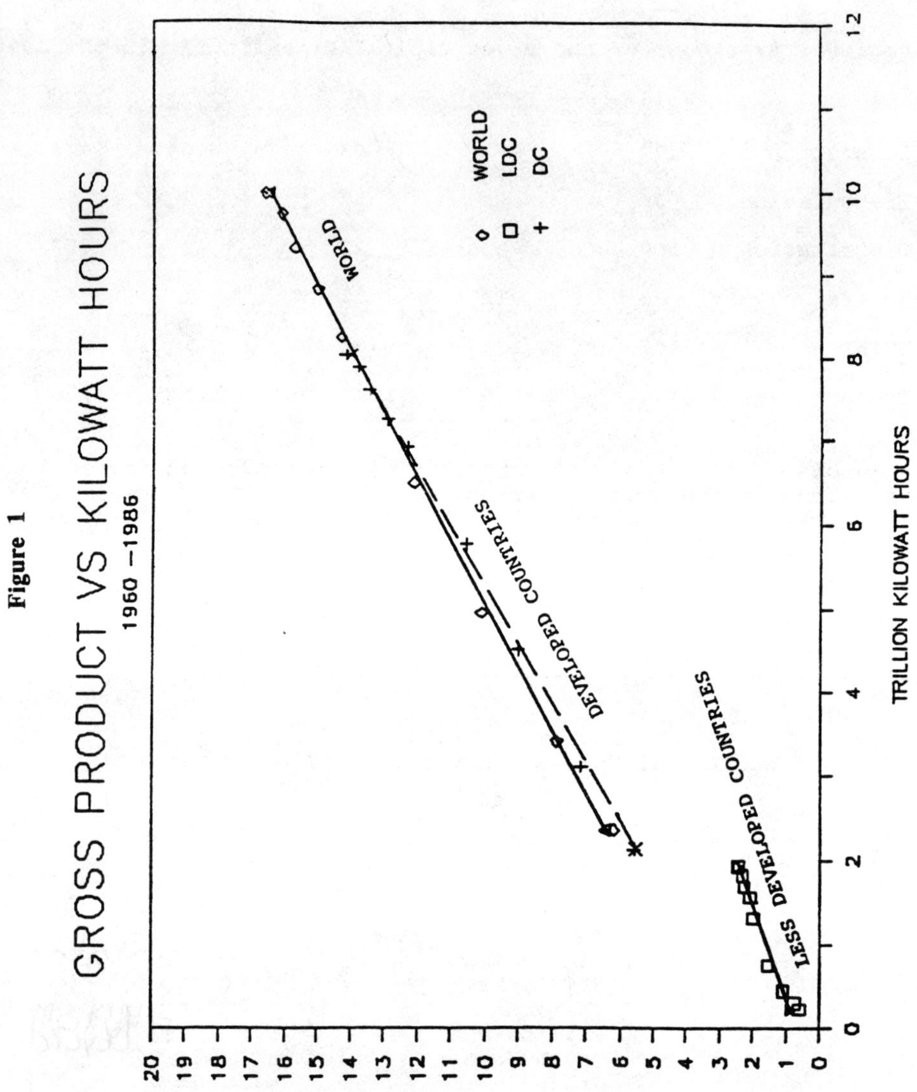
GROSS PRODUCT VS KILOWATT HOURS
1960 -1986
WORLD
DEVELOPED COUNTRIES
LESS DEVELOPED COUNTRIES
WORLD
LDC
DC
TRILLION 1987 US DOLLARS
TRILLION KILOWATT HOURS

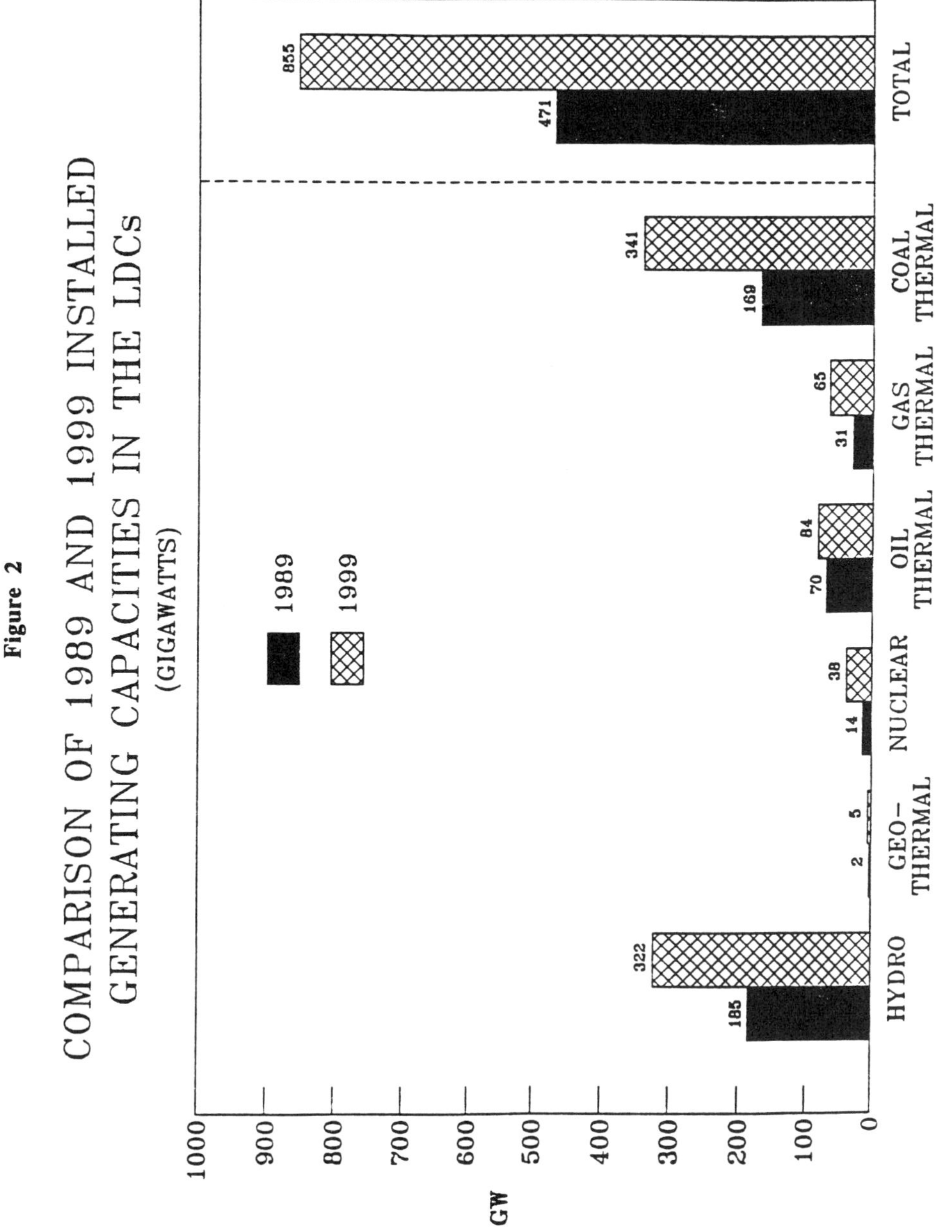
Figure 2
COMPARISON OF 1989 AND 1999 INSTALLED
GENERATING CAPACITIES IN THE LDCs
(GIGAWATTS)
1989
1999
GW
1000
900
800
700
600
500
400
300
200
100
0
185
322
2
5
14
38
70
84
31
65
169
341
471
855
HYDRO
GEO-THERMAL
NUCLEAR
OIL THERMAL
GAS THERMAL
COAL THERMAL
TOTAL

Figure 3

BREAKDOWN BY PLANT TYPE AND REGION OF CAPACITY EXPECTED TO BE ADDED IN THE LDCs IN THE 1990s (384 GW TOTAL)

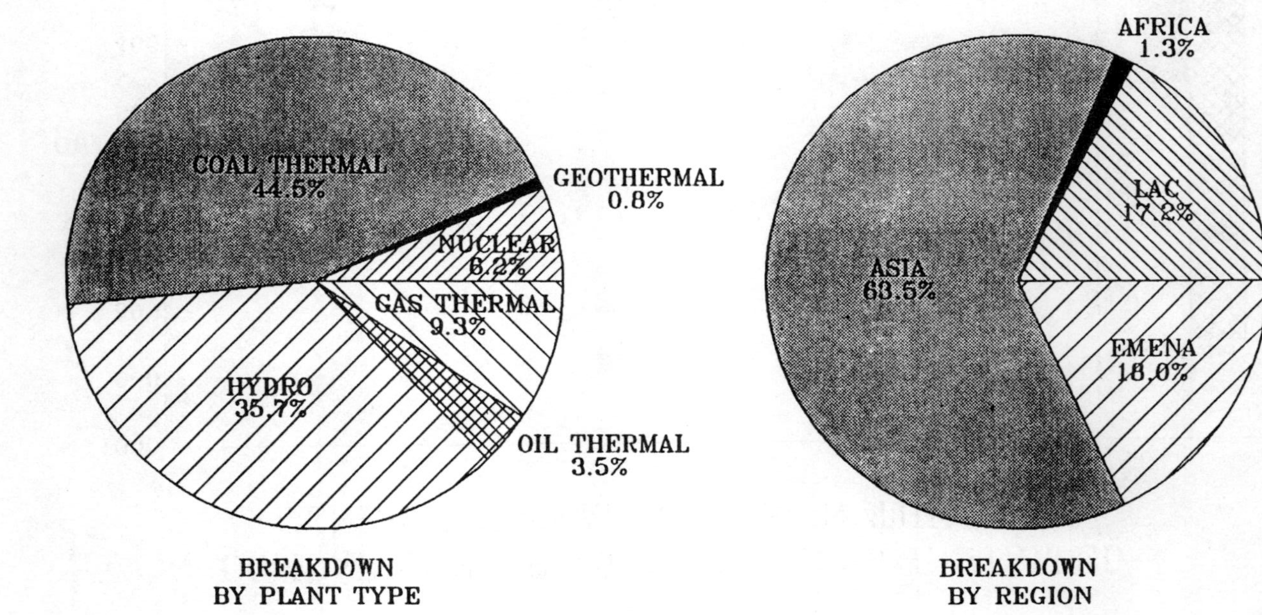

Figure 4

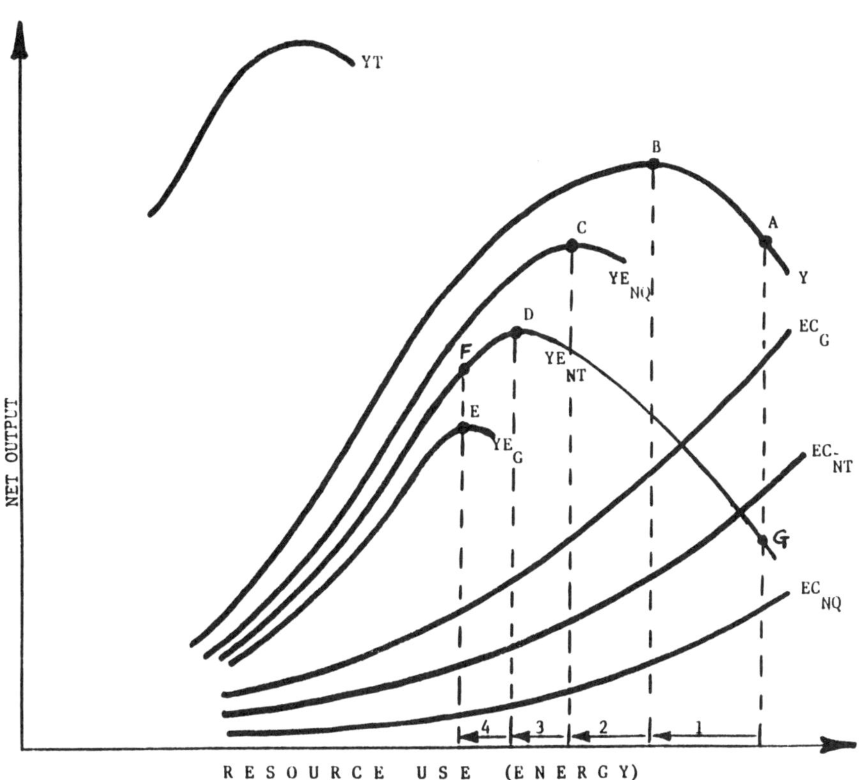

Net output, resource use and environmental cost.
$YE_{NQ} = Y - EC_{NQ}$; $YE_{NT} = Y - EC_{NT}$; $YE_G = Y - EC_G$.

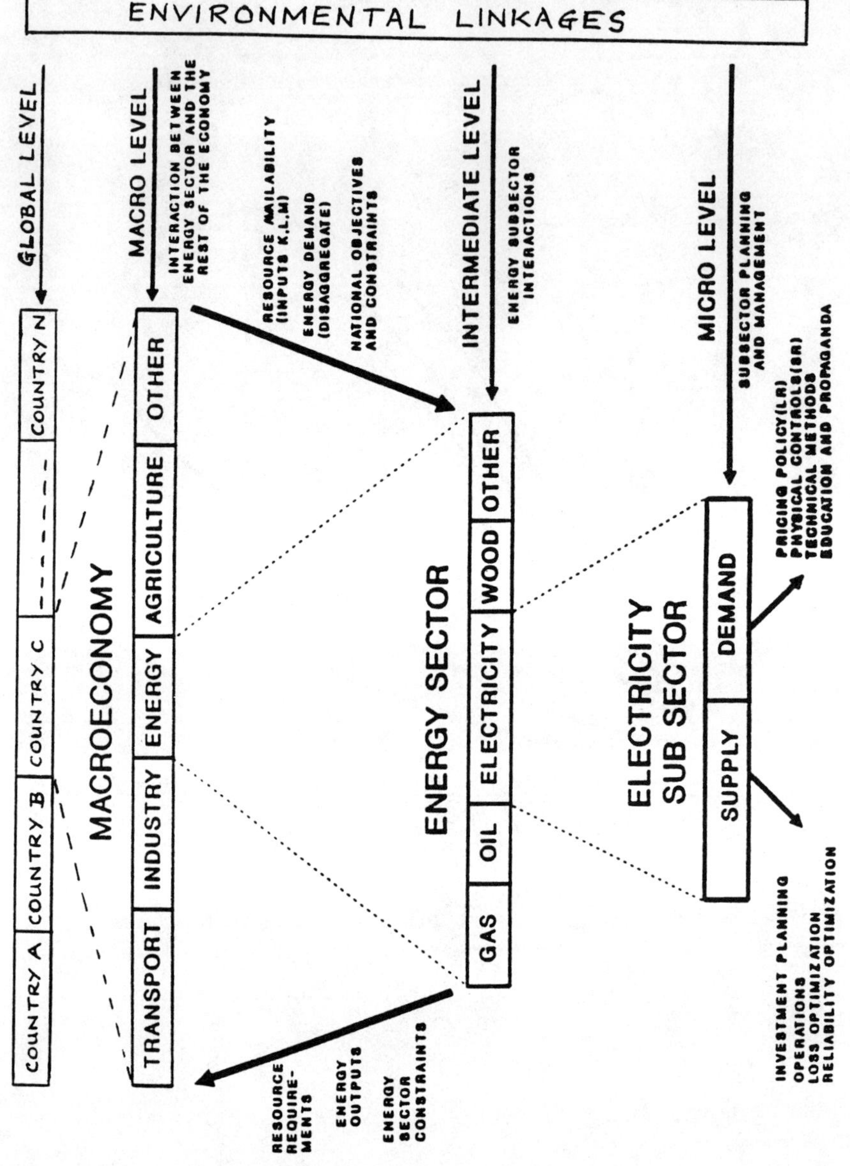

Figure 5: Integrated Conceptual Framework for Energy and Environmental Policy Analysis

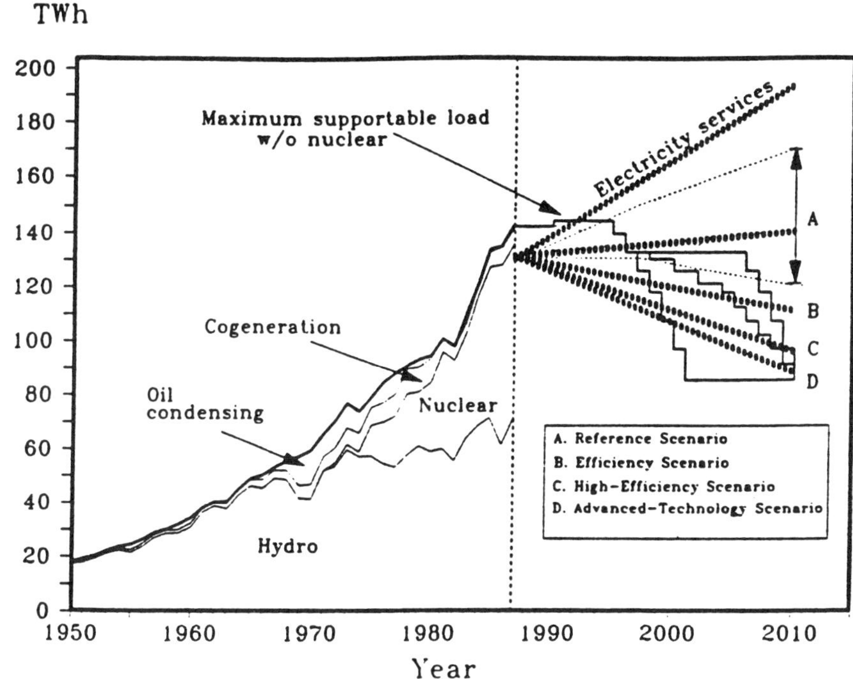

Figure 6. Generation Capability and Load Under Various Scenarios.

Notes:

1. Electricity services = end use efficiency frozen at 1987 level
2. Scenario A = normal penetration of energy efficient end use technologies (vertical arrows show range of uncertainty)
3. Scenario B = high penetration of energy efficient end use technologies that are cost effective and commercialised
4. Scenario C = same as scenario B, but includes uncommercialised newly developed technologies
5. Scenario D = same as scenario C, but includes advanced technologies still in the R&D stage
6. Downward stepped curves indicate generation capability with different options for phasing out 12 nuclear plants

Source: Bodlund et al (1989)

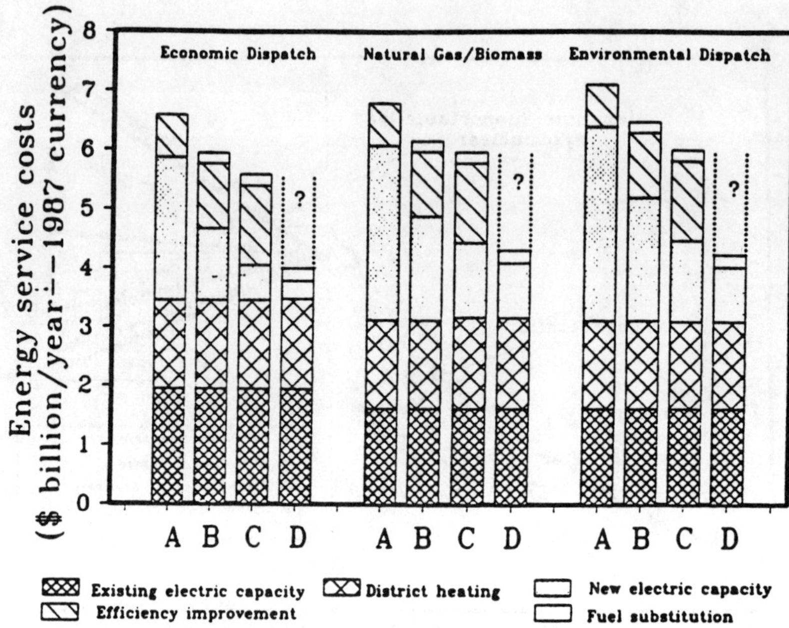

Figure 7. Total Energy Supply Costs Under Various Scenarios.

Notes:

1. Scenarios A to D are the same as in Figure 1.6, but efficiency improvement costs are unknown for scenario D
2. Economic despatch = traditional least cost generation expansion and operation
3. Natural gas/biomass = intensive use of gas and biomass, with coal use banned (to limit CO_2 emissions)
4. Environmental despatch = generation expansion and operation in order of increasing CO_2 emissions per kWh produced

Source: Bodlund et al (1989)

Acknowledgment

This paper was prepared for the AEI Bellagio Conference on Energy and the Environment in Developing Countries, November 26-30, 1990. The author is Chief of the Environmental Policy Division, The World Bank, Washington DC Formerly, he was Division Chief for Energy and Infrastructure Operations. Until recently, he served as Senior Energy Advisor to the President of Sri Lanka. The opinions expressed in this paper are those of the author, and do not necessarily represent the views of any institution or government. Assistance provided by Chitru Fernando and Ken King is gratefully acknowledged.

References

Asian Development Bank, *Private Sector Participation in Power Development*, Manila, Philippines, November 1988.

Economic and Social Commission for Asia and the Pacific (ESCAP), *Structural Change and Energy Policy*, United Nations, May 1987.

Flavin, C., *Slowing Global Warming: A Worldwide Strategy*, World Watch Institute Paper No. 91, Washington, DC, October 1989.

Foell, W., *Report on the Workshop on Acid Rain in Asia*, Asian Institute of Technology, Bangkok, Thailand, November 1989.

Johansson, T. et al., *Electricity*, Lund Univ. Press, Sweden, 1989.

Krause, F., Bach, W. and Koomey, J., *Energy Policy in the Greenhouse*, Vol. 1, International Project for Sustainable Energy Paths (IPSEP), El Cerrito, CA 94530.

Moore, E., *Capital Expenditures for Electric Power in the Developing Countries in the 1990s*, Industry and Energy Department Working Paper No. 21, The World Bank, Washington DC, Feb. 1990.

Munasinghe, M., Gilling, J. and Mason, M., *A Review of World Bank Lending for Electric Power*, Industry and Energy Department, Energy Series Paper No. 2, The World Bank, Washington DC, March 1988.

Munasinghe M., *Energy Analysis and Policy*, Butterworths, London, 1990a.

Munasinghe, M., *Electric Power Economics*, Butterworths, London, 1990b.

Munasinghe, M., and E. Lutz, *Economic Evaluation of the Environmental Impact of Investment Projects and Policies*, ENV WP40, The World Bank, Washington DC, Dec. 1990.

Starr, C. and Searl, M., "Global Projections of Energy and Electricity," Paper presented at *American Power Conference Annual Meeting*, Chicago IL, April 1989.

USAID, *Power Shortages in Developing Countries*, Washington DC, 1988.

Wilbanks. T. and Butcher, D., "Implementing Environmentally Sound Power Sector Strategies in Developing Countries," Paper prepared for the *Annual Review of Energy*, January 1990.

World Commission on Environment and Development (WCED), *Our Common Future*, Oxford Univ. Press, London, 1987.

World Bank, *Capital Expenditures for Electric Power in the Developing Countries in the 1990s*, Industry and Energy Department Working Paper No. 21, Washington DC, February 1990.

World Bank, *World Development Report 1989*, Oxford Univ. Press, 1989.

CLIMATE POLICY IN THE EUROPEAN COMMUNITY AND ITS ECONOMIC ASPECTS

Bert Metz

Counselor for Health and Environment
Royal Netherlands Embassy
4200 Linnean Ave. N.W.
Washington DC 20008

Abstract

A comparison is made between the climate policy of the United States and the European Community. The biggest difference lies in addressing CO_2 emissions, which the EC has decided to stabilize in the year 2000 at present levels. The background for this policy is explained. As an illustration of the practical implementation of such policies the climate action program of the Netherlands, leading to a CO_2 stabilization in 1995 and a 3-5% reduction in the year 2000, is described. The analysis of the economic impact of this heavily energy conservation oriented program shows that economic impact is relatively small. Some economic indicators show slight positive, others slight negative effects over a 20-year period of time.

The European Community and the USA

The European Community (EC) consists of twelve countries: Belgium, Denmark, the Federal Republic of Germany (including the former East Germany), France, Greece, Ireland, Italy, Luxembourg, the Netherlands, Portugal, Spain and the United Kingdom. The population of the EC is now approximately 350 million, compared to about 240 million for the U.S. The population density in the EC is about five times as high as in the U.S. (See Figure 1.) In terms of the size of the economy, the EC's GNP is about three-quarters that of the U.S.

A comparison of the contributions to the greenhouse effect shows that the U.S. emits more greenhouse gases than the EC. Figure 2 represents the "weighted" emissions of the various gases that have a "warming potential," mainly CO_2, CFCs, methane and nitrous oxides. Expressed as per capita emissions, the EC emits about half the amount per capita as the U.S. (See Figure 3.) For CO_2 the picture is about the same. (See Figure 4).

Copyright 1991 by Elsevier Science Publishing Company, Inc.
Global Climate Change: The Economic Costs of Mitigation and Adaptation
James C. White, Editor

The question is often raised whether this low CO_2 emission is caused by a very high share of nuclear energy in Europe. Nuclear energy currently represents 19% of the total energy consumption in the EC, with gas at 24%, oil at 49% and coal at 27% (1). Although the distribution of energy sources in the EC is somewhat less carbon intensive than that of the U.S., it does not explain the difference. (See Figure 5.) It is simply the amount of energy used that is responsible for the difference; the U.S. uses almost two times as much total energy as the EC and more than two times as much per person. Although the economy of the U.S. is more advanced and more service oriented, this is not reflected in a higher energy efficiency. On the contrary, the amount of energy per unit of GNP generated is higher in the U.S. than in the EC (2)(3).

One reason for this higher energy efficiency in Europe is of course that energy has been more expensive in Europe. A striking example is the price of gasoline. Figure 6 shows a comparison of gasoline prices in a number of EC countries and the U.S.; the reason the price in the EC is about three times as high is simply the high tax on gasoline.

Climate Policy in the EC

There is little difference between the U.S. and the EC regarding CFCs, which are responsible for the destruction of the stratospheric ozone layer and are also important greenhouse contributors. Both have subscribed to the Montreal Protocol of 1987, which was adopted in 1990 in London, and which requires a phase-out of CFCs and halons by the year 2000. The EC has decided to finish the CFC phase-out by the end of 1997 (4). On the other hand, the U.S., through the amendments of the Clean Air Act, has addressed the issue of the partially halogenated CFCs, the replacements of choice for many applications of the banned fully halogenated CFCs. These chemicals will be frozen by 2015 and banned by 2030, a decision the EC has not yet made.

The most important greenhouse gas is CO_2. It is responsible for about half of the warming potential of today's emissions. In October 1990 the EC Council of Ministers adopted a resolution to stabilize the emissions of CO_2 of the EC as a whole in the year 2000 at the level of 1990. This stabilization is considered as a first step toward further reductions of CO_2 emissions in the period after 2000. The U.S. Federal Government has so far refused to commit to a similar target and only agreed to the general need for stabilizing greenhouse gas emissions at some time in the future at a level that would prevent dangerous interference with the natural climate (5).

What remains to be done in the EC is to decide upon a formula for distributing the contributions of each Member State towards the stabilization goal. Given the differences in economic development between, in particular, the southern and northern Member States of the EC, there will be differentiated requirements under which some Member States will be able to increase their CO_2 emissions and others will have to realize a net reduction. Given the fact that decisions in the EC are taken by the Council of Ministers – one per Member State, with a voting system requiring a large majority to pass a proposal – it is important to look at the commitments already made by various EC Member States to address the CO_2 problem.

Table 1 gives an overview of the commitments made by various EC Member States individually. It shows that in particular Germany and Denmark and, to a more modest degree the Netherlands, have decided to make a net reduction in CO_2 emissions by the year 2000, thereby making some room for countries like Ireland, Spain, Portugal and Greece to increase their energy use and their CO_2 emissions.

Other Organization for Economic Cooperation and Development (OECD) countries, such as Japan, Sweden, Norway, Canada and others have also made commitments to stabilize or reduce CO_2 emissions. In fact, apart from the U.S., only Turkey has not taken such a decision.

Why is there such a support in Europe for making these commitments on CO_2, while energy conservation in Europe is more difficult and more expensive than in the U.S., given the already much lower energy consumption per capita? First of all, there are other benefits from energy conservation, reduction of fossil fuel use and shifts towards less carbon intensive fuels. Acid rain and urban smog are serious problems in Europe. Actions towards reduction of CO_2 emissions will contribute to the abatement of these problems. The second reason is that there is a widespread awareness in Europe that major changes in fossil fuel usage will be required in the long run to avoid a catastrophical change of the climate. It means drastic changes in technology as well as adaptations in life-style. These processes of social and technical change are slow. Disseminating new technologies requires a long period of time. So the sooner you start with creating incentives for energy conservation, development of renewable energy sources and life-style adjustments, the better it is.

The third reason for Europe to move forward on CO_2 policies now is the conviction that by the time we have scientific evidence of a changing climate it will be too late to act. Each year we continue our "business as usual" policies we reduce our own flexibility to make the required policy response and we will further limit the freedom of developing countries to make use of their righteous share of the world's resources. We must therefore buy an "insurance policy" against a changing climate. Last but not least, there is a strong feeling in Europe that it is extremely important for the industrialized countries of the world to show that they are willing to set the first step in addressing the greenhouse problem. The increased concentration of CO_2 in the atmosphere is predominantly caused by the industrialized countries. In order to get the necessary cooperation of the developing countries – who will contribute a major share to future emissions – a signal from the developed world is a must.

The Netherlands

To illustrate the way CO_2 stabilization and reduction is implemented in Europe, the policy approach of one EC Member State, the Netherlands, will be discussed in more detail.

The Climate Policy

The Dutch policy calls for a stabilization of CO_2 emissions in 1995 at the average of 1989 and 1990 emissions, followed by a net reduction of 3-5% in the year 2000, compared to the same 1989/1990 levels (6)(7). The program that is being implemented to achieve the required reductions is focusing primarily on energy related measures, such as energy conservation, fuel switching to less carbon intensive fossil fuels and promoting renewable energy generation (8). The transportation and the waste management sector will contribute about one-quarter to the reductions through the increase in public transportation, reduction in private vehicle use, reducing the need for transportation by means of land-use planning and strict requirements for access to public transportation systems and – in the waste management sector – increase of waste-to-energy systems, extracting methane from landfills and a strong emphasis on recycling. (See Figure 7.) In absolute terms the required reductions of CO_2 emissions amount to about 50 million tons per year in the year 2000, approximately 27% of our current emissions. As far as the contribution from the energy sector is concerned, about 60% will come from energy conservation and 40% from shifting to natural gas and away from coal. (See Figure 8.) In terms of energy use this will mean a reduction in energy use compared to present trends of about 20% in the year 2000, an average efficiency improvement of more than 2% per year over a ten-year period.

Tools for Implementation

The tools that will be used to achieve these reductions are a mix of regulatory instruments and economic and other incentives. The following regulations will be implemented:

- Stricter standards for thermal insulation of new buildings;
- Energy efficiency standards for household appliances, hot water boilers and central heating furnaces;
- Public transport access requirements for development of office buildings. (No more office buildings in areas without easy access to public transport) (9);
- (Within the framework of the European Community) Standards for the fuel efficiency of automobiles;
- Mandatory fuel switching for the utility sector.

Where possible these standards will be developed via a process of negotiation with industry. A study will be undertaken to investigate the potential of energy efficiency requirements as part of environmental permits for industrial establishments. The economic instruments to be used are:

- Subsidies for thermal insulation of existing buildings;
- Subsidies for investment in cogeneration, wind turbines, solar water heaters, waste-to-energy facilities, recycling projects, partially on the basis of competitive bidding;
- Accelerated tax write-off of investments in energy conservation;
- Subsidies on energy efficiency consulting services;
- Matching funds for research and development projects;
- Incentive programs run by energy distribution companies;
- Increased costs of using the automobile via increased fuel taxes, increased parking costs, introduction of road pricing (tolls).

Further study will be done on the feasibility of a substantial increase of the so-called CO_2 tax that is already in place for fossil fuels.

On top of this an important role will be played by public education and consumer information, both through mass media campaigns sponsored by the national government as well as through the energy distribution companies. Other important incentives will be created by a major government investment program in the mass transportation system and improvement of the information, ticketing systems and connection with biking and taxi facilities.

The contribution of the various sectors of society towards the overall energy efficiency improvement target does vary. Table 2 gives an overview of the sectoral breakdown.

The energy distribution companies, for electricity, gas and district heating, have responded to the government program in the form of a detailed Environmental Action Program (10). Careful analyses were made of the potential for and the costs of specific energy conservation measures. (See Figure 9.) A similar analysis was done for measures contributing to the reduction of emissions of acid rain precursors. Figure 10 shows an interesting analogy with the picture for CO_2 reduction. Both go to a large extent hand in hand, as was mentioned before.

The Economic Impacts

The impact on the economy of both the energy conservation program and the overall long-term environmental policy plan (11)(12)was studied carefully 13)(14)(15). A number of economic indicators was evaluated, such as volume of GNP, labor costs, employment, volume of exports, etc. Two cases were considered: one in which the rest of the world does not adopt similar policies, and the other in which it does.

The general conclusion of these studies is that the economic impact of the climate policy is small. Even for the comprehensive long-term environmental policy plan, which goes far beyond the climate program, estimates are that GNP over a 20-year period will be three percentage points less than under a business as usual scenario, that is instead of a 98% growth during that period only a 95% growth will be realized, and that only if the rest of the world does not adopt similar policies. In case other countries do follow the Dutch policies, there would be a positive effect on GNP of +2 percentage points.

For the energy conservation program the effects show a similar picture. For some economic indicators, such as employment and exports, slight positive effects can be expected. This is the case for both scenarios for the policy of other countries. For national income, investments and labor costs, the effect varies from small positive to small negative effects. (See Figures 11, 12, 13, 14, 15).

In summary, the carbon dioxide stabilization and reduction program can be effectuated without jeopardizing the Dutch economy, even when other environmental programs are taken into account and the assumption is made that other countries will not follow a similar track. Many of the investments for energy conservation will pay for themselves through reduced energy bills.

Conclusions

The case of the Netherlands is one example of the European attitude towards an active CO_2 reduction and stabilization policy. Although detailed information on the German and Danish policies are not available to me at this moment, there is basically the same attitude in these EC Member States; economic effects are not a constraint; in certain cases a competitive advantage of the industry will be the result of these policies.

Table 1. Carbon Dioxide Policies of Some EC Member States

Country	Policy
Denmark	20% reduction of current levels by 2005.
Germany	25% reduction of 1987 levels by 2005.
Italy	Stabilize at 1990 levels by 2000 and 20% reduction by 2005.
Netherlands	Stabilize by 1995 at 1989/1990 levels, and 3 to 5% reduction of 1989/1990 levels by 2000.
United Kingdom	Stabilize at 1990 levels by 2005, if other countries take similar action; agreed with EC target of stabilization in 2000 at 1990 levels for EC as a whole.

Table 2. Expected Efficiency Improvements According to Sector, in the Period 1990-2000

(The figures assume the annual 2% energy efficiency improvement objective for the Dutch economy)

Sector	Share of Total Consumption, 1989 (%)	Efficiency Improvement, 2000/1990 (%)
Manufacturing industry (incl. feedstock)*	47	15
(Ditto excl. feedstock)*		(20)
Agriculture	6	30
Nonresidential buildings	9	30
Households	20	25
Transport	14	20
Other industrial enterprises	4	20
Total Netherlands	100	20

*Consumption of primary fuels on raw materials for production is referred to as "feedstocks."

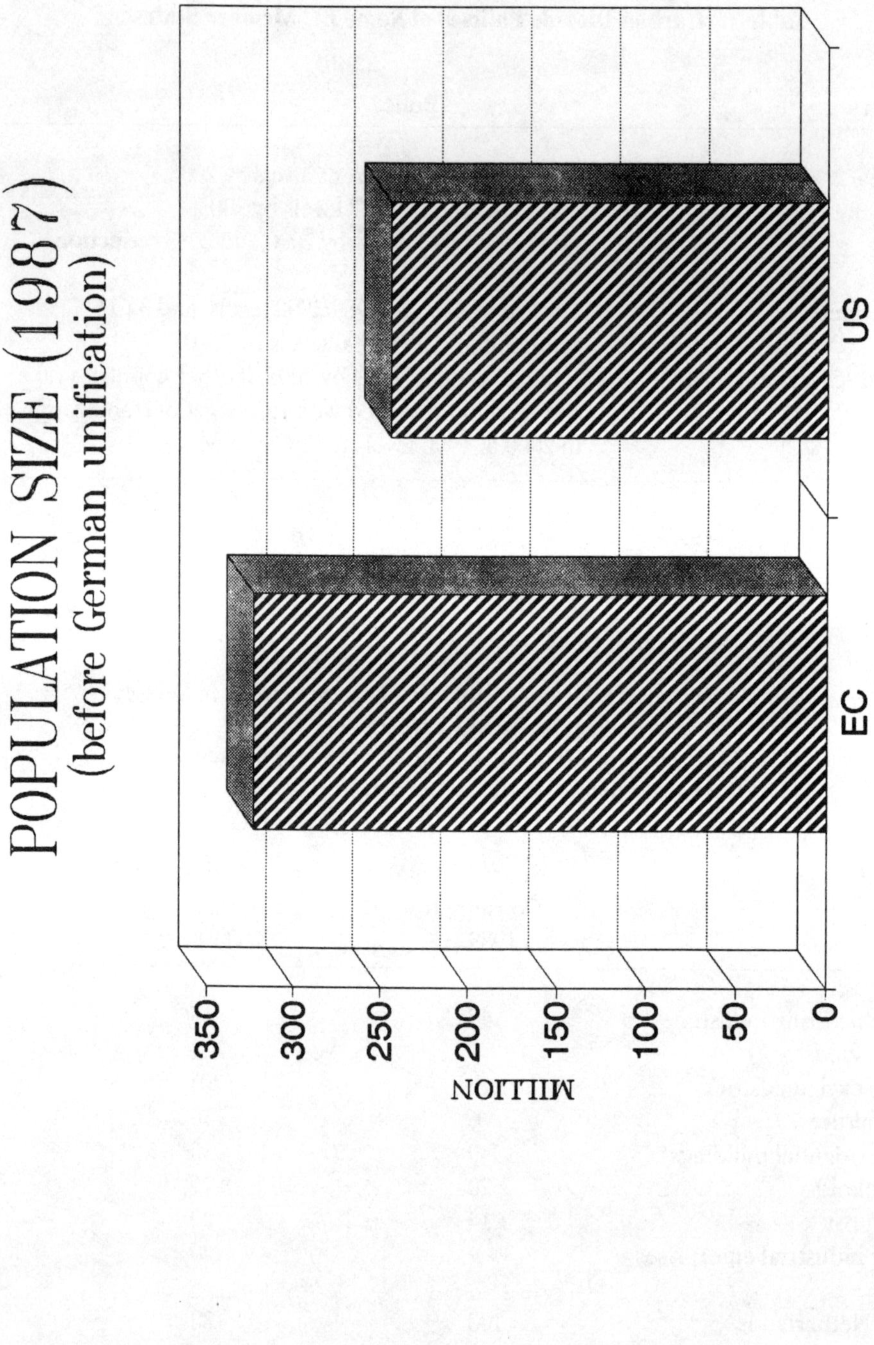

Figure 1 Population size

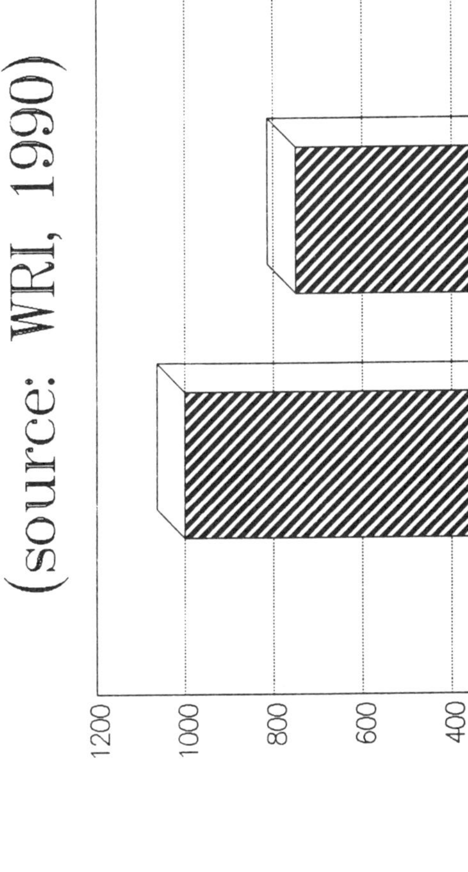

Figure 2 Net greenhouse gas emissions in 1987 (Source: *World Resources 1990-1991*, World Resources Institute, Washington DC, 1990)

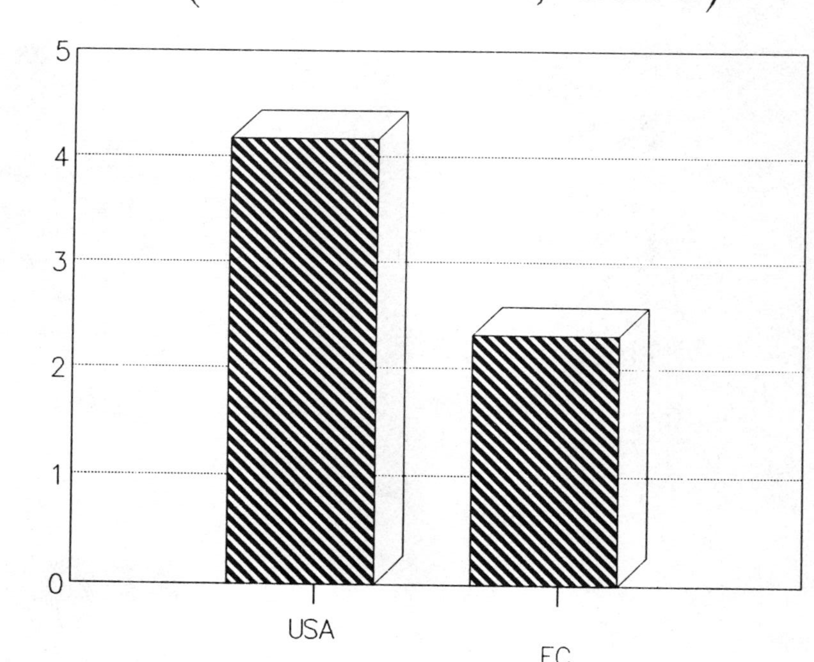

Figure 3 Net greenhouse gas emissions per capita, 1987 (Source: *World Resources 1990-1991*, World Resources Institute, Washington DC, 1990)

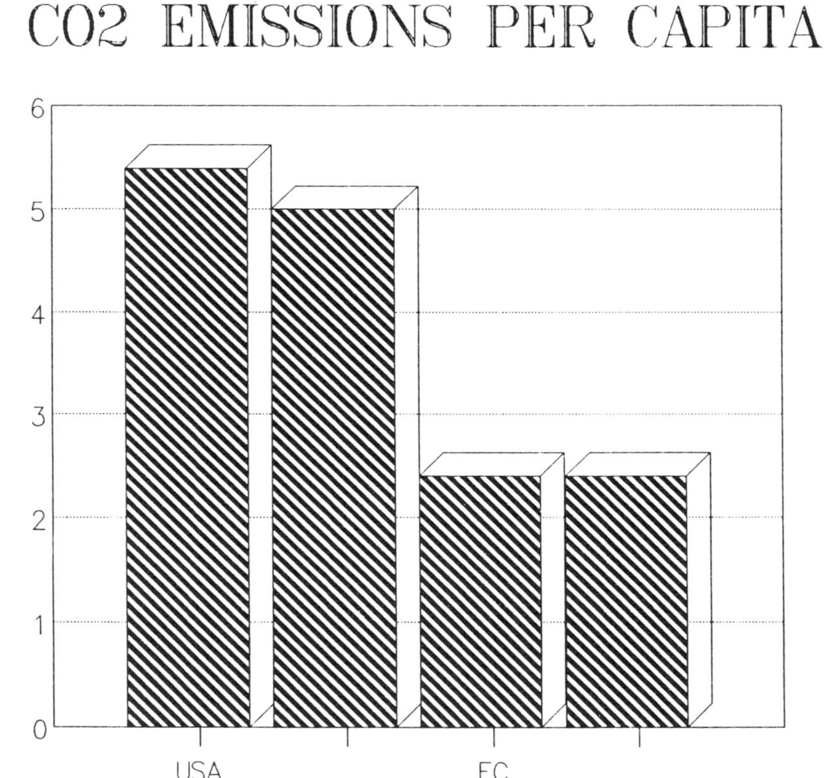

Figure 4 Carbon dioxide emissions per capita
(Source: *World Resources 1990-1991*,
World Resources Institute, Washington DC, 1990)

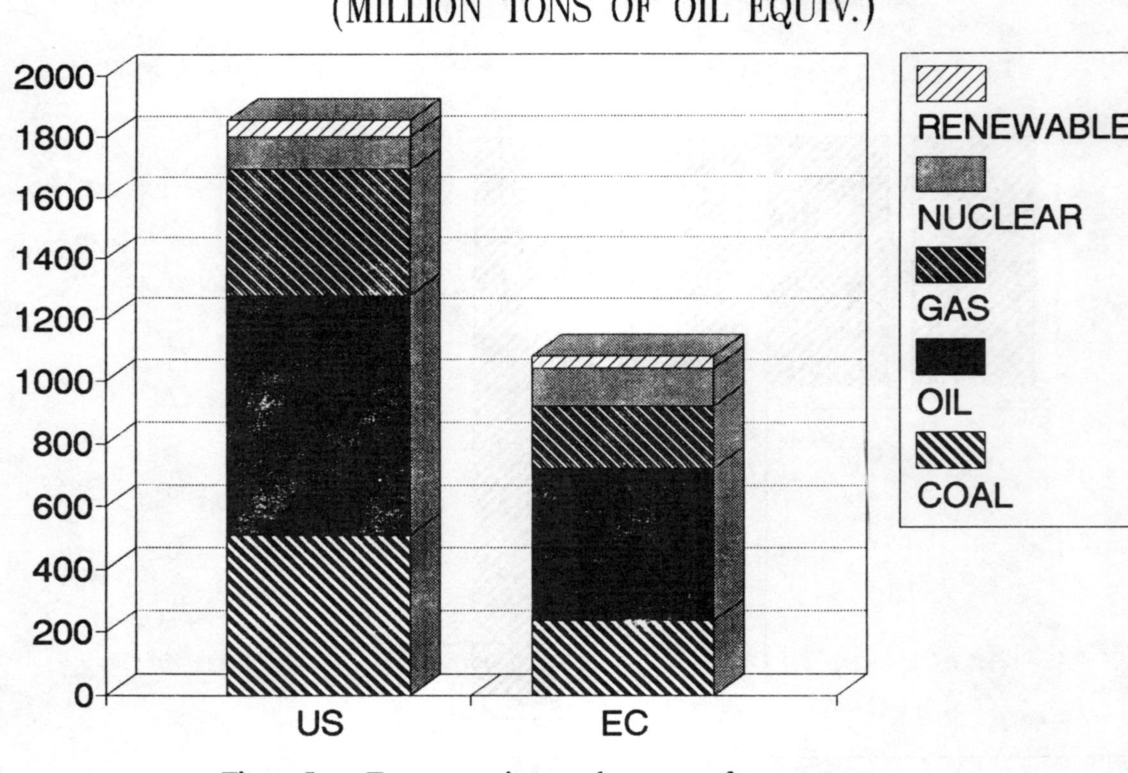

Figure 5 Energy requirement by source of energy (Source: *Environmental Data Compendium 1989*, OECD, Paris, 1989)

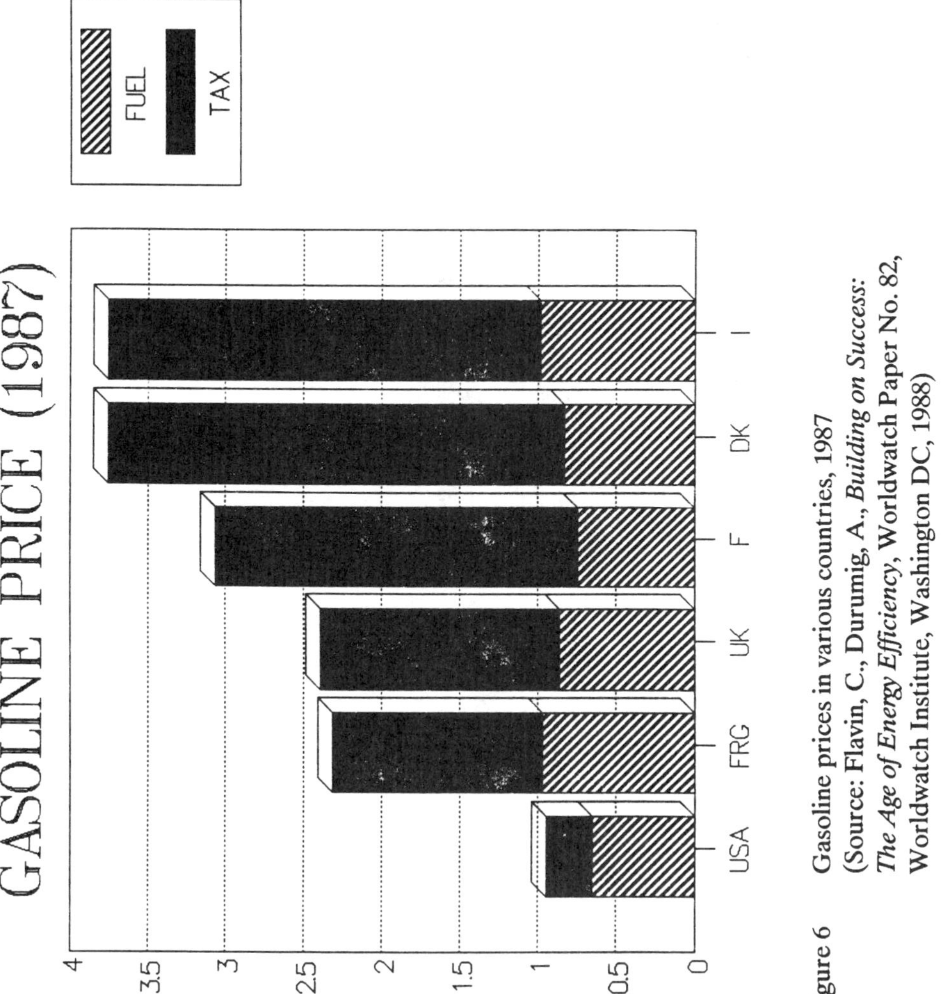

Figure 6 Gasoline prices in various countries, 1987
(Source: Flavin, C., Durumig, A., *Building on Success: The Age of Energy Efficiency*, Worldwatch Paper No. 82, Worldwatch Institute, Washington DC, 1988)

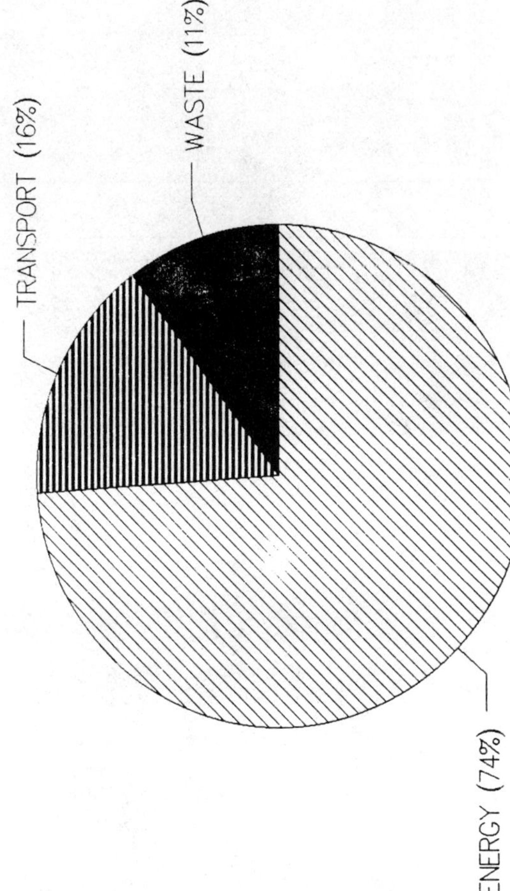

Figure 7 Reduction of carbon dioxide by sector in the Netherlands in the year 2000
(Source: *Memorandum on Energy Conservation*, Ministry of Economic Affairs, the Netherlands, SDU Publishers, The Hague, June 1990)

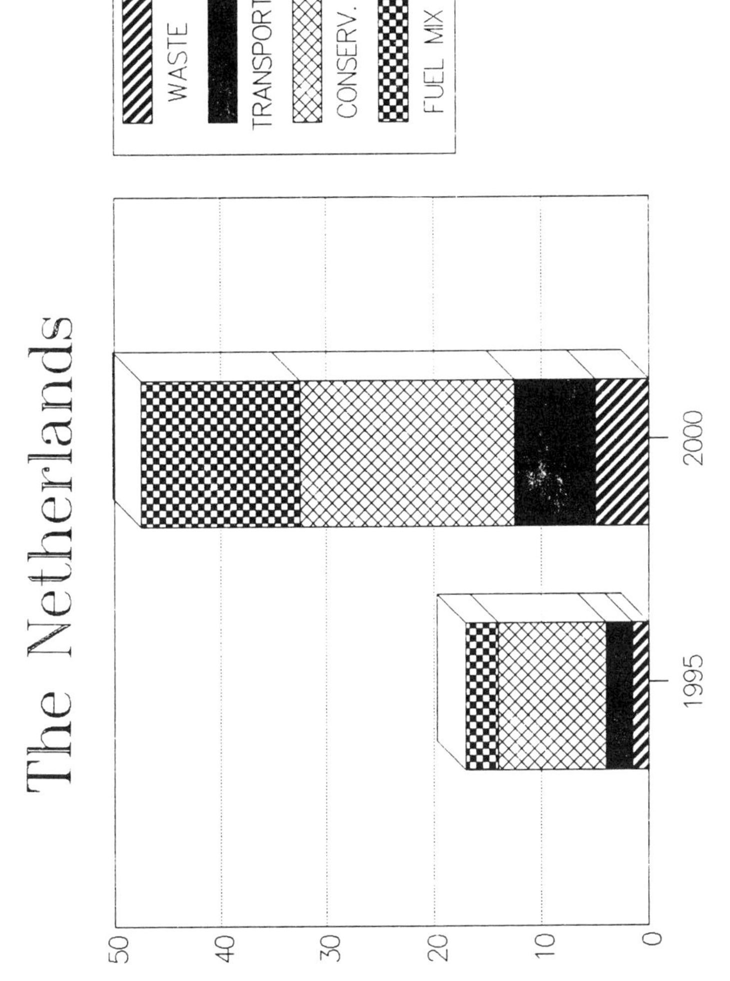

Figure 8 Reduction of carbon dioxide by sector in 1995 and 2000 (Source: *ibid.*, Figure 7)

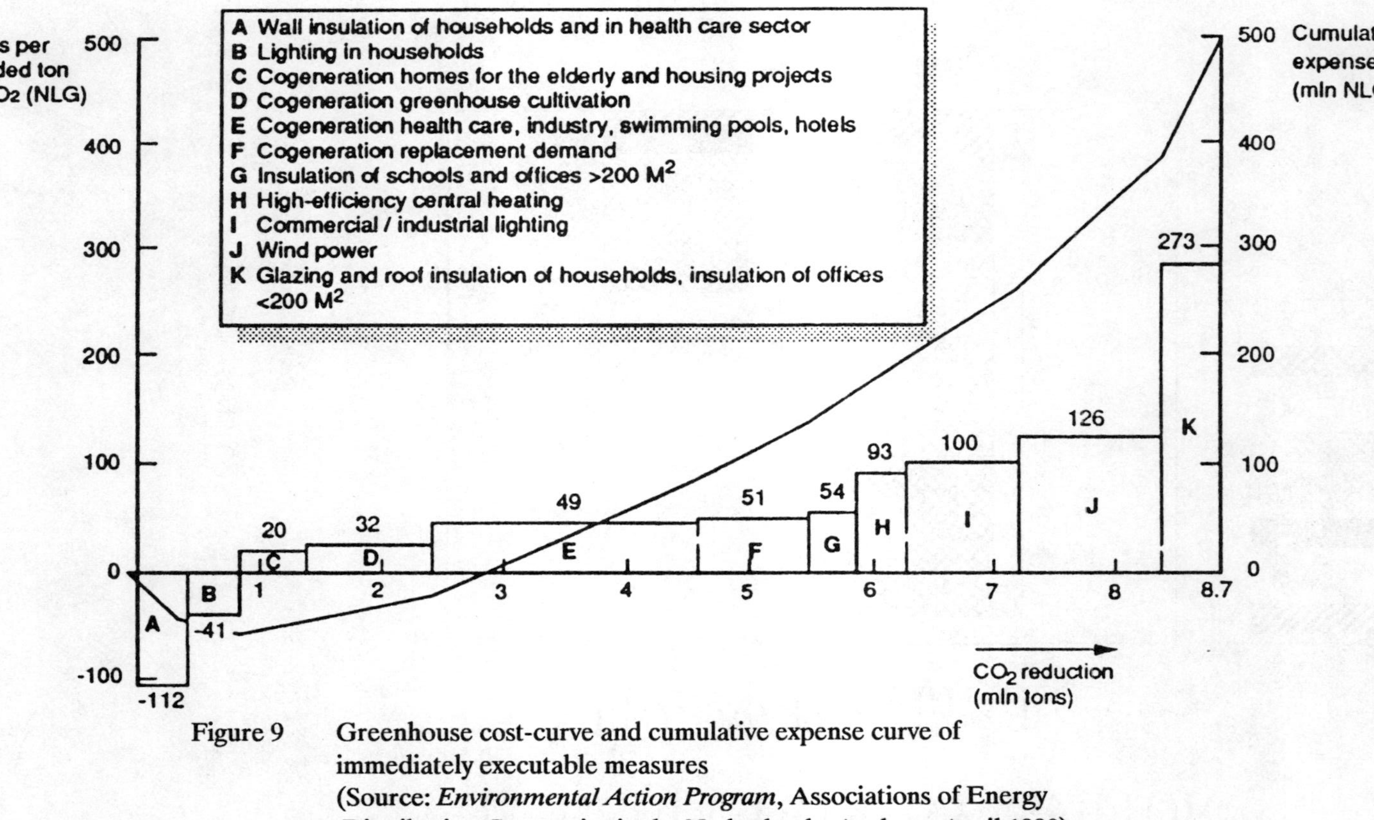

Figure 9 Greenhouse cost-curve and cumulative expense curve of immediately executable measures
(Source: *Environmental Action Program*, Associations of Energy Distribution Companies in the Netherlands, Arnhem, April 1990)

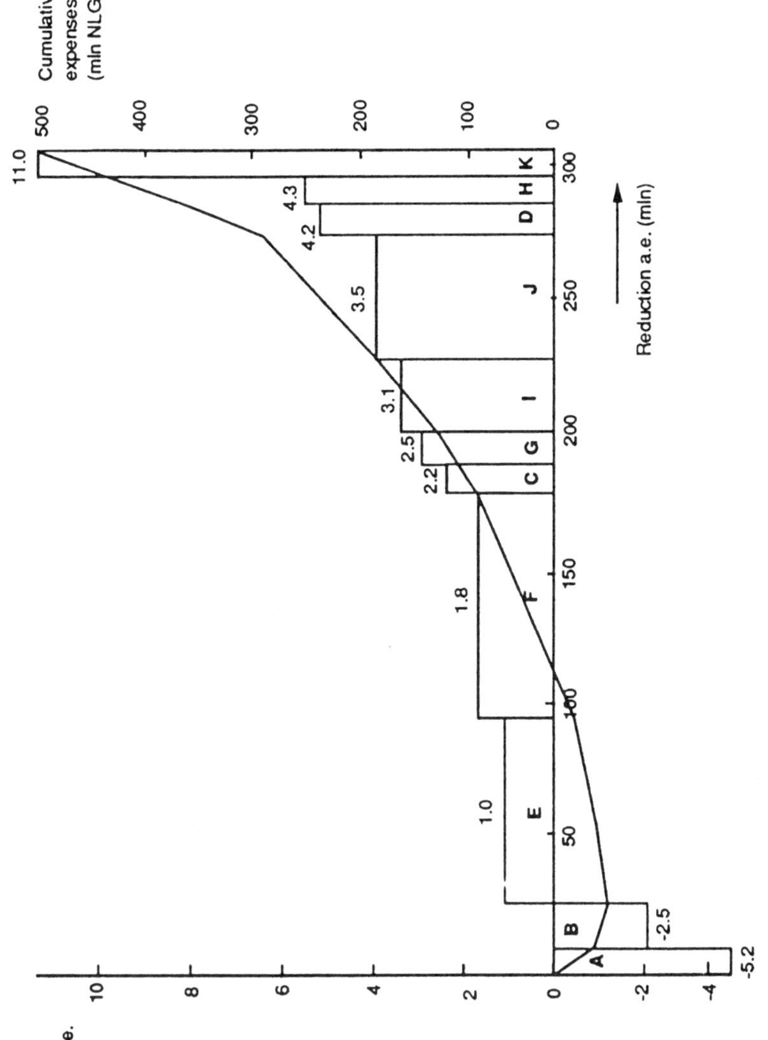

Figure 10 Acidification cost-curve and cumulative expense curve of immediately executable measures
(Source: *ibid.*, Figure 9)

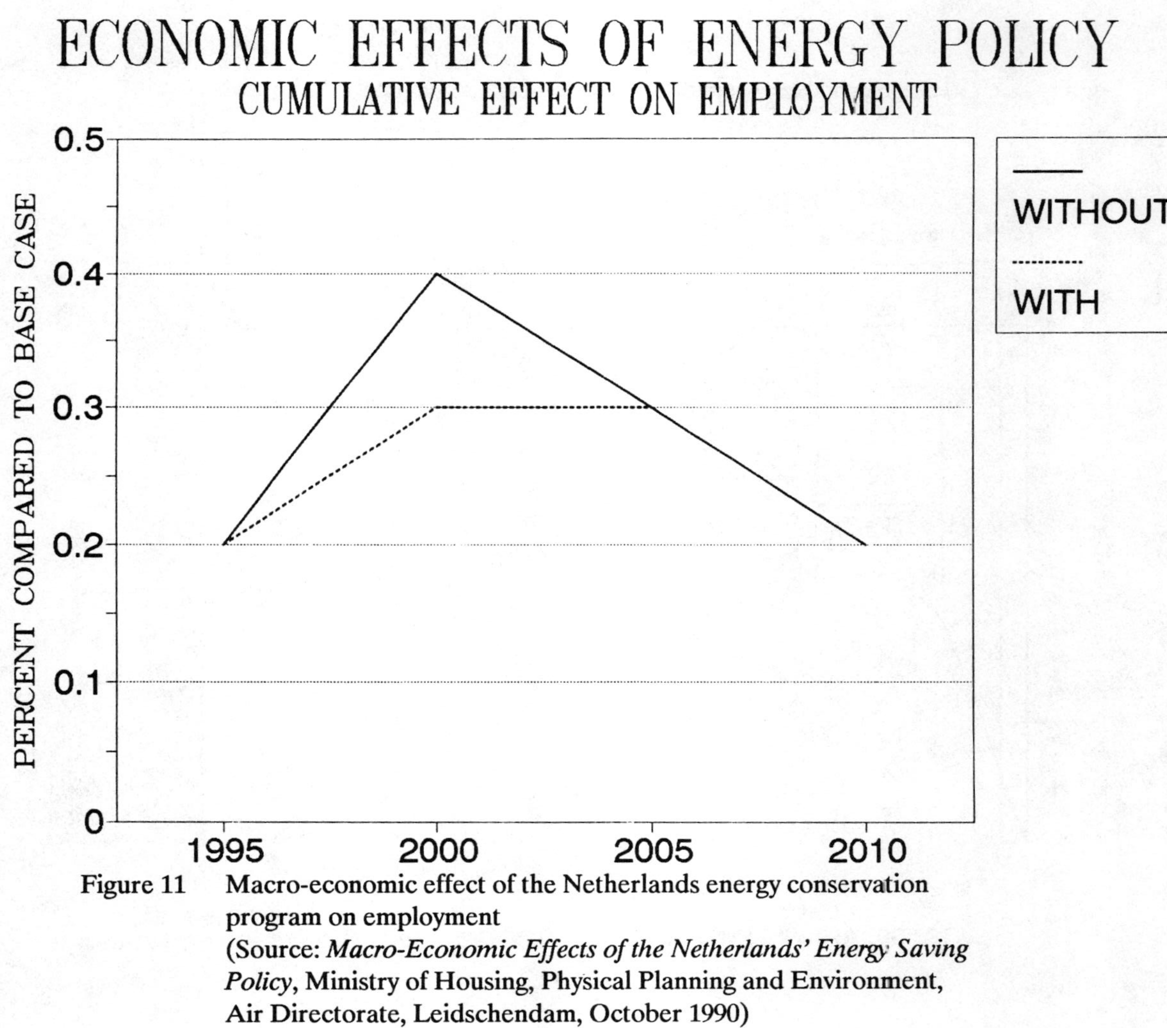

Figure 11 Macro-economic effect of the Netherlands energy conservation program on employment
(Source: *Macro-Economic Effects of the Netherlands' Energy Saving Policy*, Ministry of Housing, Physical Planning and Environment, Air Directorate, Leidschendam, October 1990)

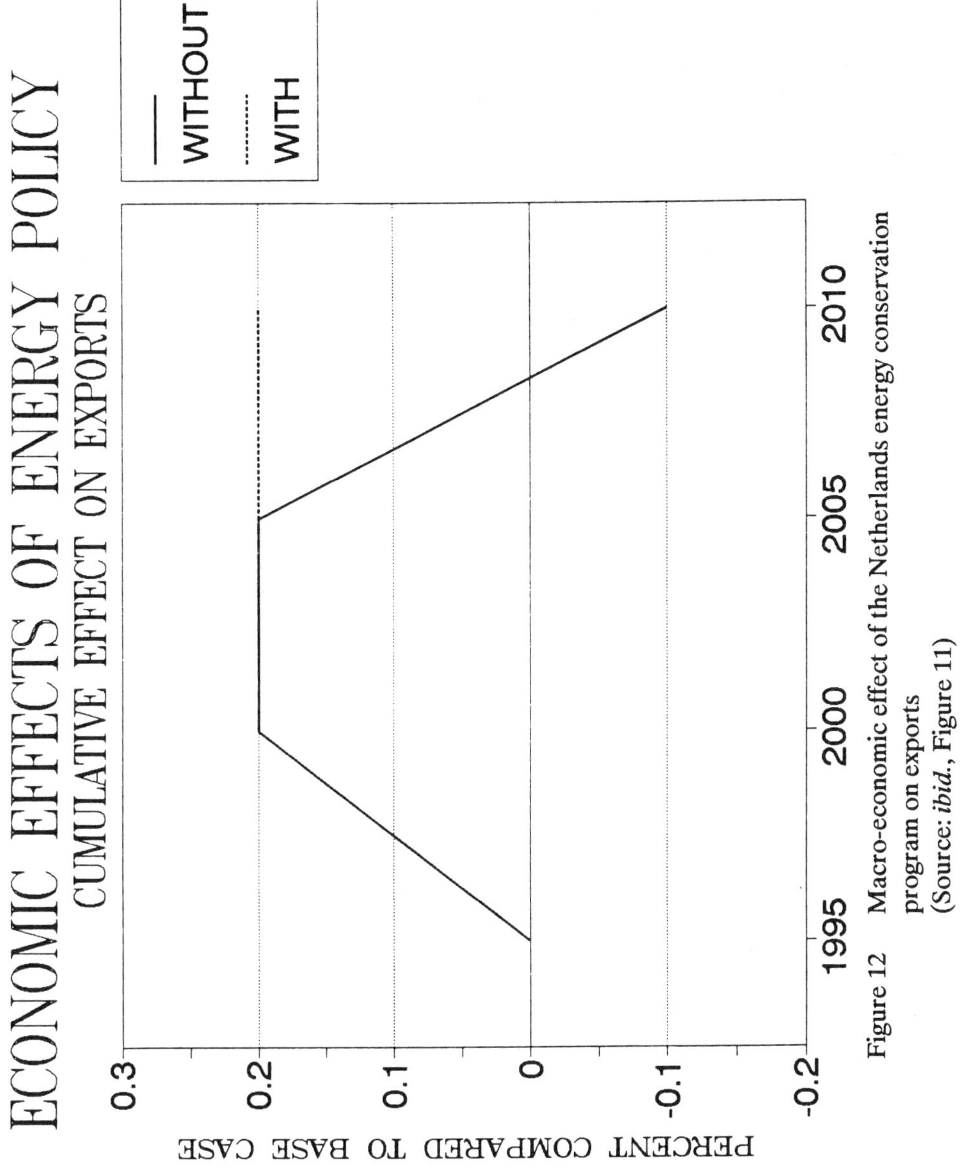

Figure 12 Macro-economic effect of the Netherlands energy conservation program on exports (Source: *ibid.*, Figure 11)

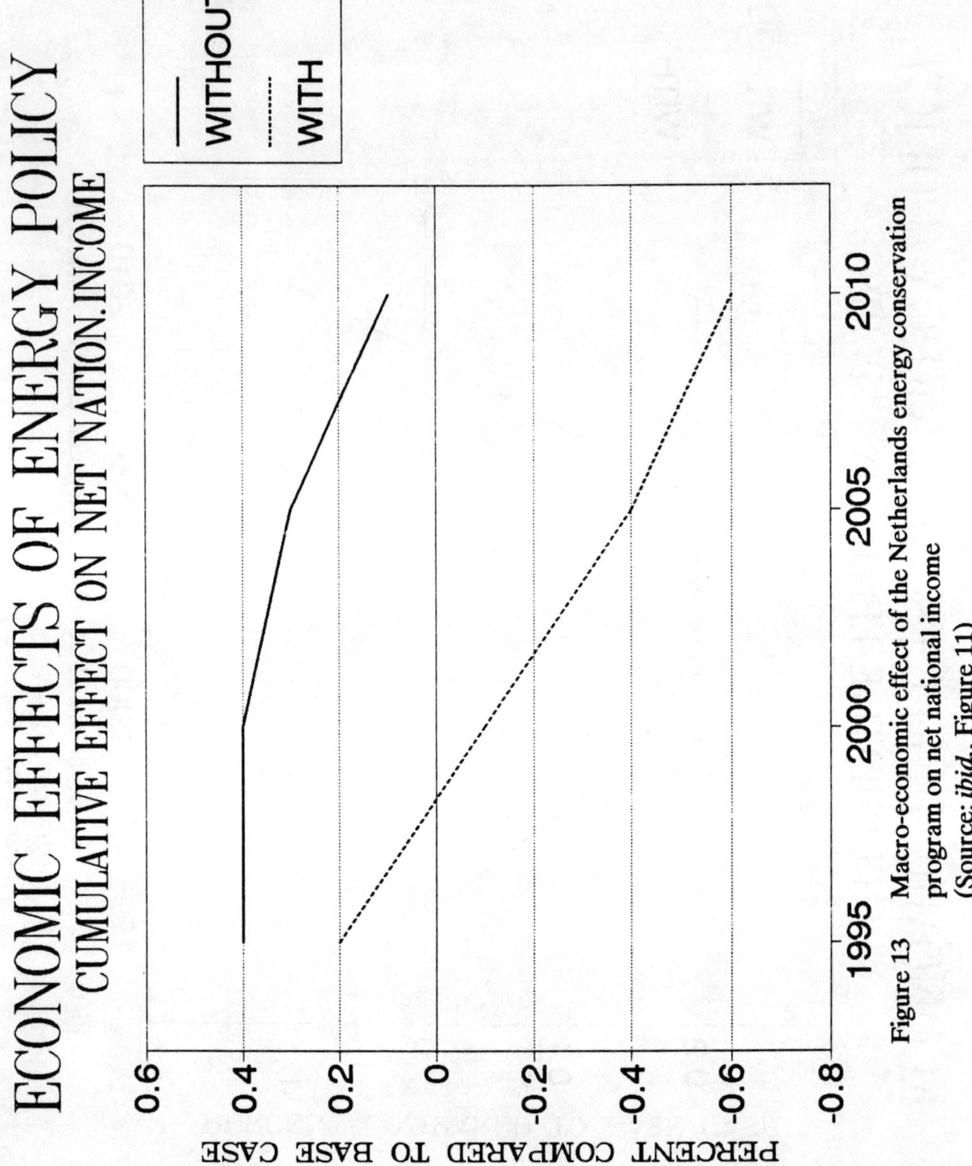

Figure 13 Macro-economic effect of the Netherlands energy conservation program on net national income (Source: *ibid.*, Figure 11)

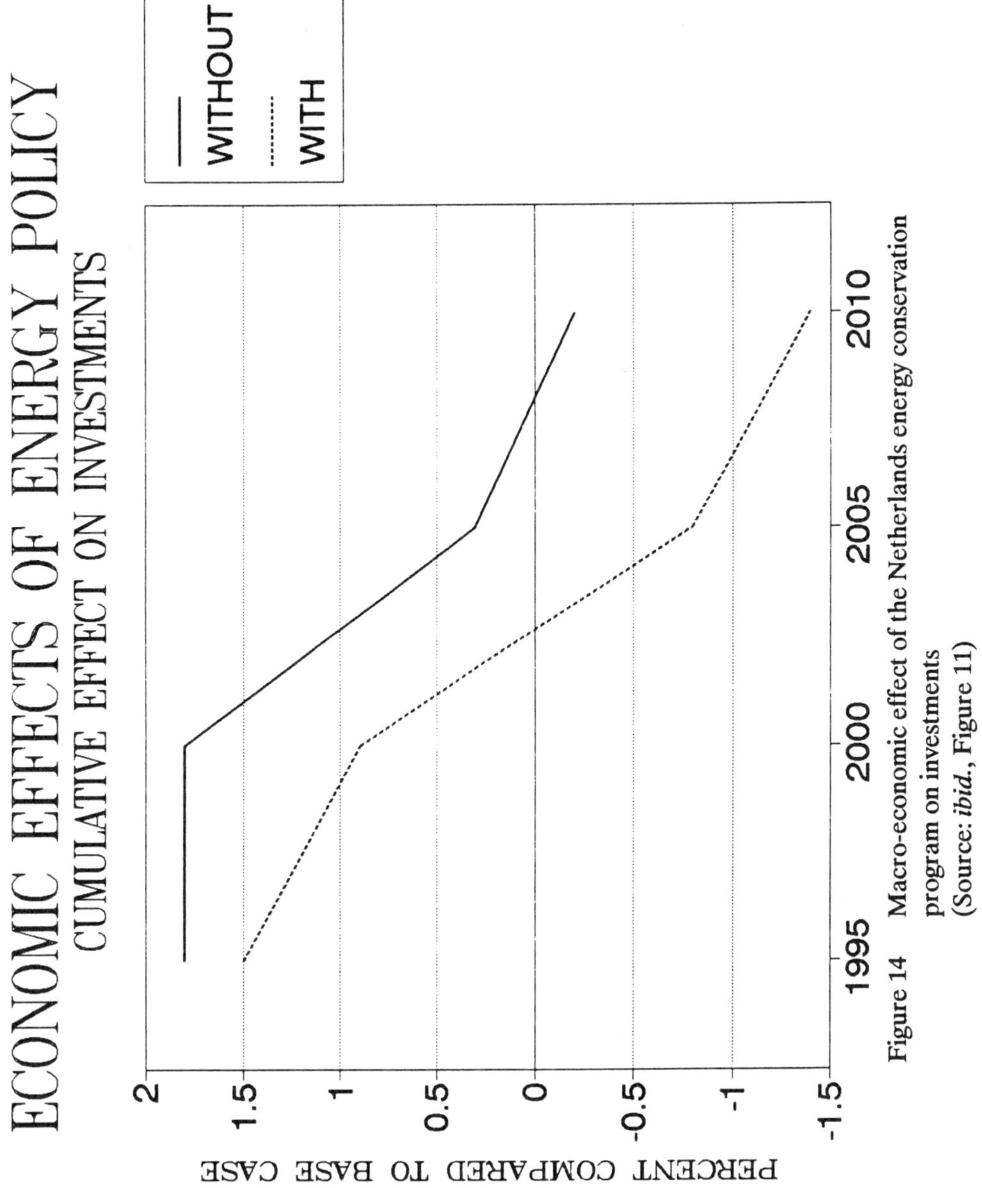

Figure 14 Macro-economic effect of the Netherlands energy conservation program on investments (Source: *ibid.*, Figure 11)

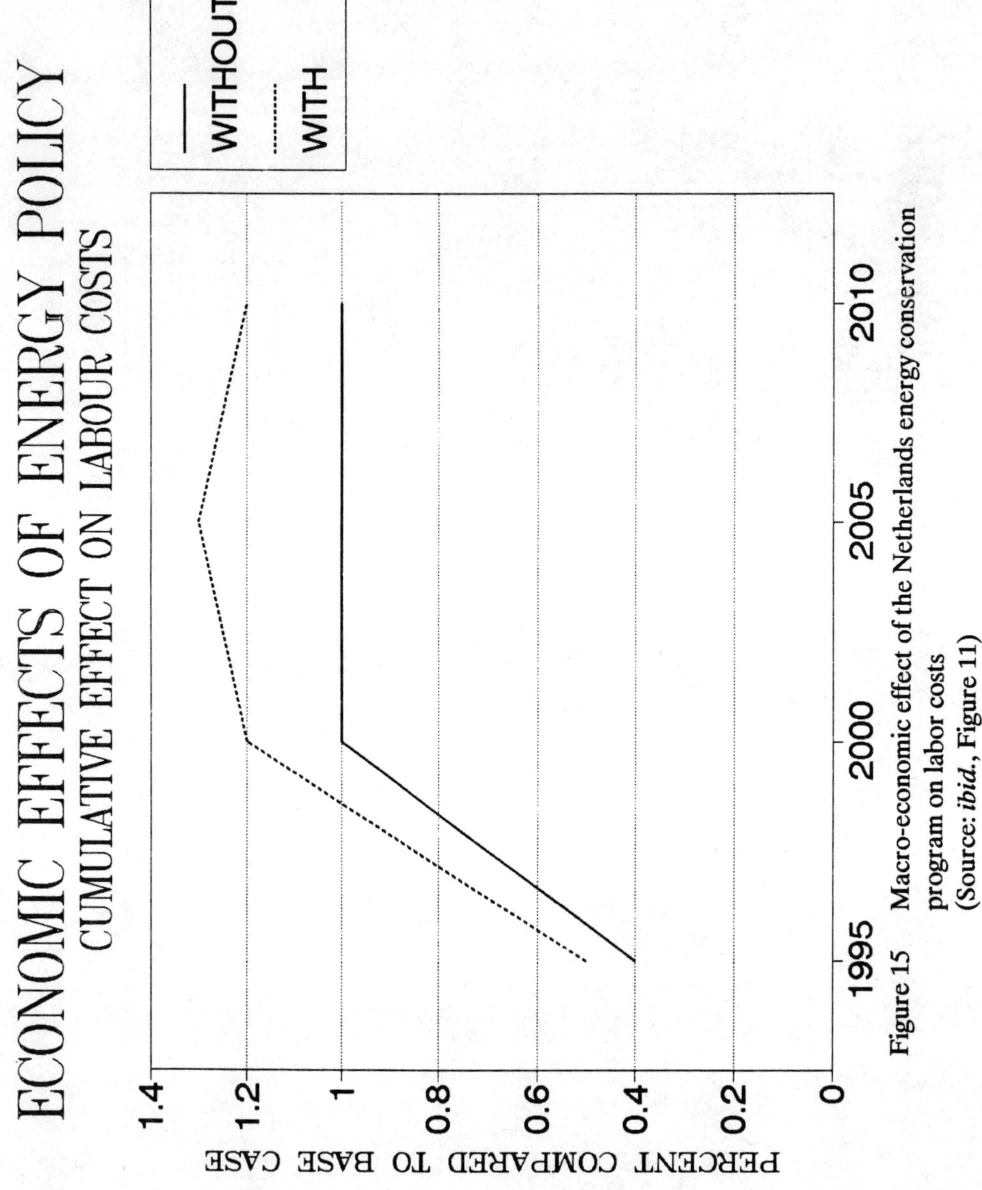

Figure 15 Macro-economic effect of the Netherlands energy conservation program on labor costs
(Source: *ibid.*, Figure 11)

References

1. European Commission, Communication of the Commission to the Council on Energy and Environment, COM(89) 369 def., Brussels, February 8, 1990.

2. *OECD Environmental Data-Compendium 1989*, OECD, Paris 1989.

3. *World Resources 1990-1991*, Oxford University Press, New York/Oxford 1990, p. 244 e.v.

4. Conclusions of the European Energy and Environment Council, Brussels, October 29, 1990.

5. Ministerial declaration of the Second World Climate Conference, Geneva, November 1990.

6. *National Environmental Policy Plan Plus*, Ministry of Housing, Physical Planning and Environment, The Hague, the Netherlands, June 1990.

7. *The Dutch Policy on Climate Change*, Ministry of Housing, Physical Planning and Environment, Air Directorate, Leidschendam, the Netherlands, November 1990.

8. *Memorandum on Energy Conservation – A Strategy for Energy Conservation and Renewable Energy Resources*, Ministry of Economic Affairs, SDU Publishers, The Hague, the Netherlands, June 1990.

9. *On the Road to 2015*, Ministry of Housing, Physical Planning and Environment, SDU Publishers, The Hague, the Netherlands, 1988.

10. *Environmental Action Program*, Associations of Energy Distribution Companies in the Netherlands, Arnhem, the Netherlands, April 1990.

11. *National Environmental Policy Plan*, Ministry of Housing, Physical Planning and Environment, SDU Publishers, The Hague, the Netherlands, May 1989.

12. *National Environmental Policy Plan Plus*, Ministry of Housing, Physical Planning and Environment, SDU Publishers, The Hague, the Netherlands, June 1990.

13. *Economic Effects of Three Scenarios for Environmental Policy in the Netherlands up to 2010*, Central Planning Bureau, Working Paper 29, The Hague, October 1989.

14. *Economic Aspects of Carbon Dioxide Reduction*, Ministry of Housing, Physical Planning and Environment, Air Directorate, Leidschendam, April 1990.

15. *Macro-Economic Effects of the Netherlands' Energy Saving Policy*, Ministry of Housing, Physical Planning and Environment, Air Directorate, Leidschendam, October 1990.

16. *Summary of Actions in Member Countries to Deal with Problem of Climate Change*, OECD, Committee of Energy Research and Development, Paris, October 1990.

JAPAN'S POLICY ON GLOBAL WARMING

Yasuhiro Shimizu

Embassy of Japan
2520 Massachusetts Avenue, N.W.
Washington, D.C. 20008

During the past few years, we have seen a growing awareness of the need for industrialized nations to take definite action on global warming and Japan is not an exception. On October 23, 1990, the Japanese government formulated the "Action Program to Arrest Global Warming," which set the targets of stabilizing greenhouse gas emissions. The main purpose of this paper is to explain some of the features and the background of the Action Program.

World emissions of CO_2 production from 1987 were 5.3 billion tons. After the U.S., U.S.S.R. and China, Japan was fourth with 4.7% of the world's production (Figure 1). If you consider CO_2 production in relation to gross domestic product (GDP), production per capita and land area, the results are most interesting. Table 1 shows that Japan has the lowest CO_2 emissions per GDP of all the industrialized countries. She also has one-third the emissions per capita when compared to the U.S. and ranks very well among all industrialized nations. However, Japan produces five times the U.S. production of CO_2 per unit of land area. These rankings are possible because Japan has allowed only minimal increases in CO_2 emissions since the oil crisis of the early 1970s (Figure 2). GNP in the same period has nearly doubled.

The Japanese government announced the "Action Program to Arrest Global Warming" in October 1990. The targets are to stabilize total and per capita CO_2 emissions by the year 2000. Concurrently methane, nitrous oxide and other greenhouse gas emissions should not be increased. Since the population is expected to increase several percentage points in that time period, it will require innovative procedures to achieve the desired goals.

The measures used to attain these levels of emissions are several. Forests in Japan and abroad should be developed and maintained. Buildings, both old and new, should be designed and, if necessary, remodeled to operate at high efficiency, with low energy demands. Cogeneration of electricity should be used wherever possible and biomass use is encouraged. Japan will emphasize the use of nuclear power as a non-air-polluting source of energy. Every effort will be made to increase energy savings in the manufacturing and transportation industries.

Copyright 1991 by Elsevier Science Publishing Company, Inc.
Global Climate Change: The Economic Costs of Mitigation and Adaptation
James C. White, Editor

At the same time, energy savings from changes in life style will be encouraged through public education in all media. Table 2 lists a number of measures which, if followed, would substantially decrease oil use in Japan. It is expected that most of the savings shown in the table will be accomplished.

A new organization, the Research Institute of Innovative Technology for the Earth (RITE), has been established in Japan with a $22 million budget for 1990. This institute will concentrate on chemical and biological CO_2 fixation technology, nonpolluting refrigerant materials to replace PCB-like compounds, biodegradable and recyclable plastics and environmentally sound manufacturing processes.

Figure 3 shows Japan's projected energy sources for the next twenty years. This program will maintain overall efficiency in Japan and lower its use of oil and coal. An important consideration is the reduction of oil dependency by almost one-half in the next twenty years.

After the approval and inception of the "Action Program to Arrest Global Warming," the Japanese government revised and increased its alternative energy supply target. There will be an increased demand for energy in the future with a growing GNP and an increasing population. Japan must be very innovative to achieve more energy with lower emissions but Japan hopes to accomplish her goals and maintain broad options to cope with global warming and still maintain its economy.

Table 1. Comparison of CO_2, GDP, and Land Area

Comparison of carbon dioxide emissions per GDP or land area (1987)

1987 Year

	CO_2 emission (100 million tons)		CO_2 emission per GDP	CO_2 emission per capita	CO_2 emission per land area
		(%)	(g/dollar)	(10kg/person)	(1,000tons/km^2)
WORLDWIDE*	52.25	100.0	–	106.3	39.0
ASIA*	12.78	24.5	–	44.6	46.4
Japan	2.44	4.7	102.6	199.7	645.2
China	5.29	10.1	1,802.6	49.5	55.3
Korea	0.42	0.8	349.4	100.7	432.5
NORTH AMERICA*	13.51	25.9	–	505.5	69.3
U.S.A.	12.38	23.7	275.2	507.7	132.1
Canada	1.13	2.2	302.5	436.5	11.3
CENTRAL AND SOUTH AMERICA*	2.27	4.3	–	55.1	10.9
WEST EUROPE*	7.80	14.9	–	220.2	217.0
Federal Republic of Germany	1.86	3.6	166.1	303.5	745.8
France	0.95	1.8	108.3	170.2	173.0
Italy	1.06	2.0	141.1	184.0	351.0
United Kingdom	1.59	3.0	276.5	279.8	649.8
Spain	0.44	0.8	152.7	113.4	87.1
Netherlands	0.52	1.0	242.0	353.0	1,402.3
EAST EUROPE*	4.02	7.7	–	289.8	315.0
Soviet Union	9.70	18.6	–	346.3	43.3
AFRICA*	1.45	2.8	–	25.4	4.8
OCEANIA*	0.72	1.4	–	287.2	8.4

(Note) 1. Carbon dioxide emission as a result of burning fossil fuel is calculated by carbon quantity.

2. Based on statistics provided by the United Nations, the World Bank and others.

3. For areas designated by *, 1986 population figures were used to calculate per capita carbon dioxide emissions.

Table 2. Potential Energy Savings by Individuals

Potential energy saving by individuals
(Calculated in terms of the amount of crude oil that can be saved.)

	Annual energy saving per household	National annual energy saving
Lowering temperature (heating) by 1°C.	Approx. 20ℓ	Approx. 800,000 kℓ
Raising temperature (air conditioning) by 1°C.	Approx. 7ℓ	Approx. 250,000 kℓ
Shortening each shower by 1 minute.	Approx. 25ℓ	Approx. 700,000 kℓ
Reducing frequency of reheating bath water by two times a week.	Approx. 12ℓ	Approx. 350,000 kℓ
Turning off a "pilot burner" when not in use.	Approx. 13ℓ	Approx. 250,000 kℓ
Reducing lighting time by 1 hour.	Approx. 30ℓ	Approx. 1,200,000 kℓ
Replacing a 60 Watt incandescent lamp with a 20 Watt fluorescent light of the same luminosity.	Approx. 20ℓ	Approx. 800,000 kℓ

(Note) Based on the Panel to Promote Conservation of Resource and Energy.

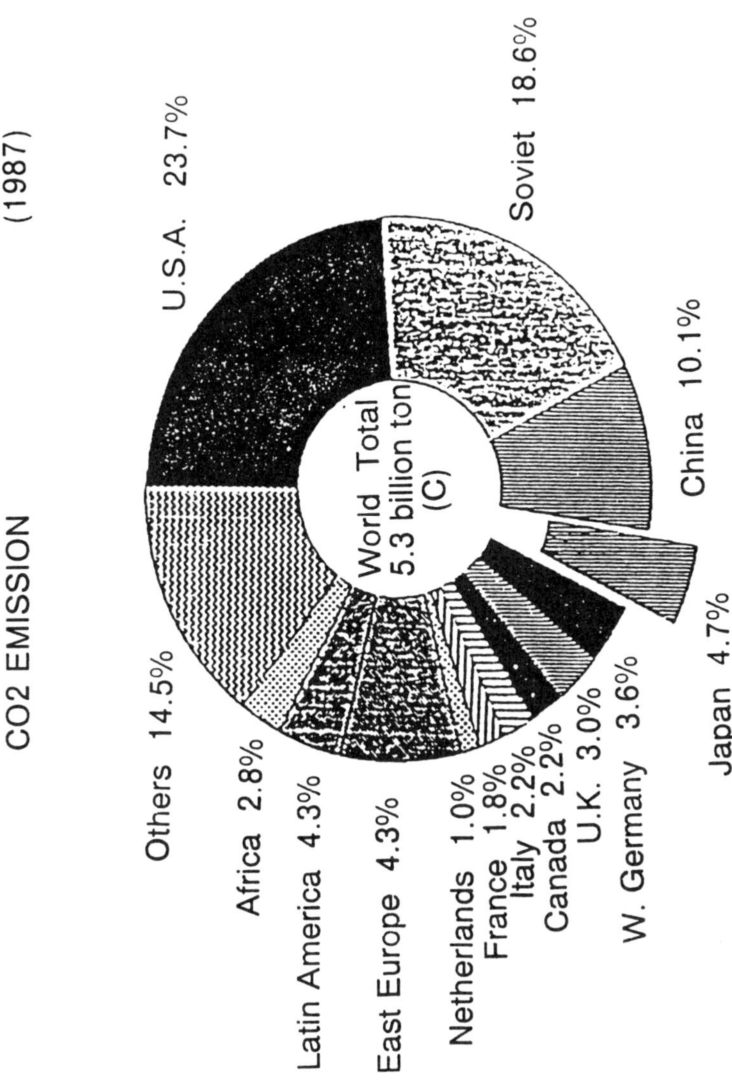

Source: U.N. Energy Statistics

Figure 1. CO_2 Emissions

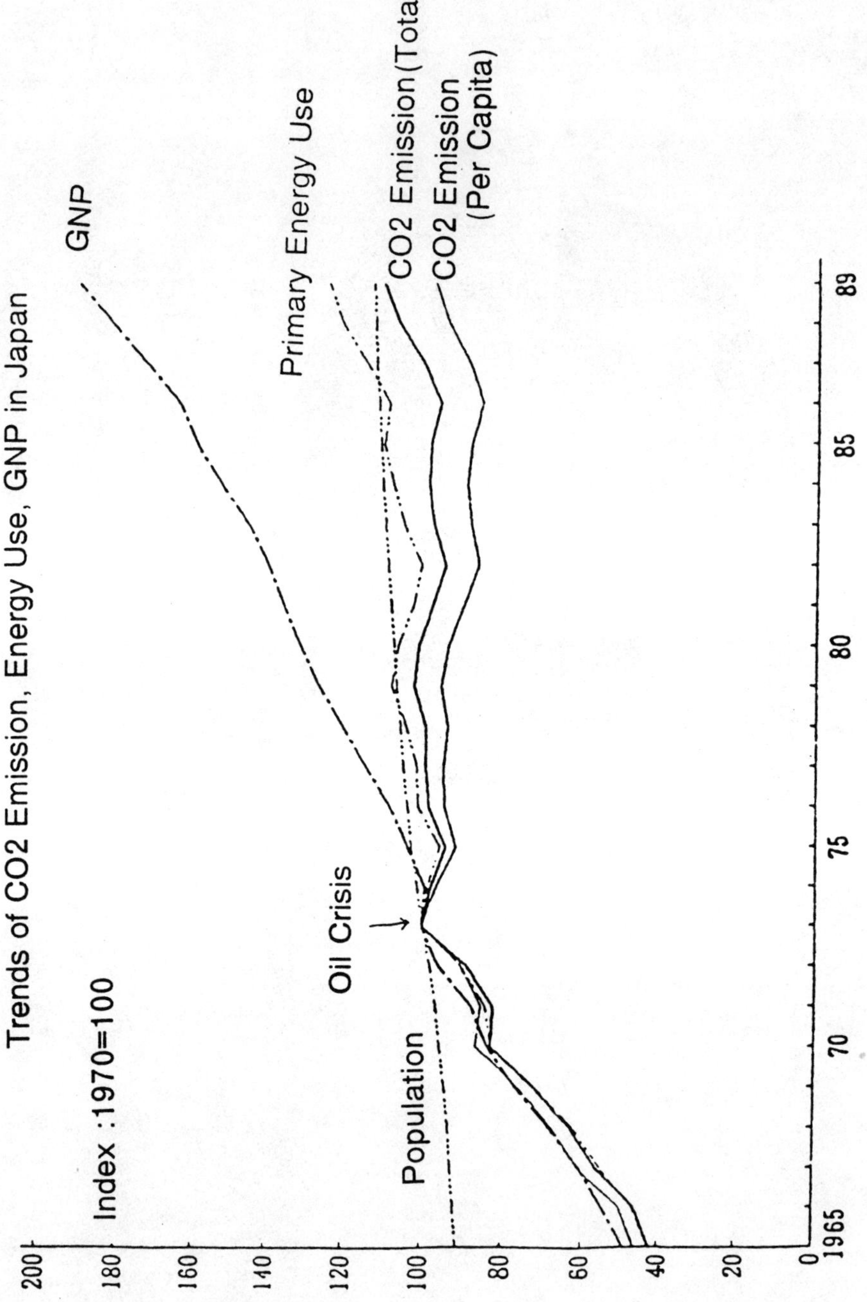

Figure 2. Trends of CO_2 Emissions, Energy Use and GNP

Long-term Alternative Energy Supply Goal
(October 30, 1990)

	1989	2000	2010
Total Energy Supply (million Kl)	499	594	657
New Energy *	1.3%	3.0%	5. 3%
Hydro	4.6%	3.7%	3.7%
Geothermal	0.1%	0.3%	0.9%
Nuclear	8.9%	13.3%	16.9%
Natural Gas	10.0%	10.9%	12.2%
Coal	17.3%	17.5%	15.7%
Oil	57.9%	51.3%	45.3%

* New Energy = Hydrogen, Solar, Wind, Wave, etc.

1. Promoting Non Fossil Fuel

	1989	2000	2010
Non Fossil Fuel Dependency *	14.8%	20.2%	26.8%

*Non Fossil Fuel = New Energy, Hydro, Geothermal & Nuclear

2. Discouraging Oil

	1973	1989	2000	2010
Oil Dependency	77.4%	57.9%	51.3%	45.3%

Figure 3. Long-Term Alternative Energy Supply Goals

SUMMATION

Paul R. Portney

Resources for the Future
1616 P Street, NW
Washington, DC 20036

The papers presented here are extraordinarily diverse and very interesting. We've heard points of view not only from economics but also from engineering, ecology and a very rich diversity of other points of view on how to approach the issue of global climate change. That makes things more interesting but it makes my task more difficult.

My role as the summarizer is in some sense appropriate because I want to throw a little bit of cold water on what has come before. By that I mean that, in spite of some very commendably clear and helpful presentations, some people who should have been clarifying confusion actually stirred up the mud a little bit. And there has been a tendency for economists and all non-economists – and I'll group them together for the time being – to keep passing like ships in the night. Not even passing closely, truth be told, rather like one set moving down the coast of the United States and the others passing up the coast of Europe.

In a sense, this has been a disservice to each of the panelists, whom I would have liked to have seen address each other more directly, and a disservice to the audience. But there certainly has been enough good that you all ought to walk out of this conference having learned, as I have, quite a bit about the subject, even if you brought a lot to it.

My observation, furthermore, is not necessarily economic chauvinism because I think some of my fellow economists have been as guilty of promoting confusion or being insensitive to other approaches as the non-economists have been.

What this conference has really been about in my view – and I hope not too many of you would disagree with me – is how to use economics to think about the issues involved in the control of greenhouse gases. On the conference program it says, "Global Climate Change: Economic Costs of Mitigation and Adaptation." We've talked not only about costs but also about the potentially substantial beneficial effects of taking a variety of different kinds of control measures to try to contain carbon dioxide and other greenhouse gases.

Since the purpose has been how to think about using economics to address the issue of global warming, what I would like to do is concentrate on the various roles that

Copyright 1991 by Elsevier Science Publishing Company, Inc.
Global Climate Change: The Economic Costs of Mitigation and Adaptation
James C. White, Editor

economics can play, possibly touching on some of the presentations that have been made at this conference.

Let me begin by saying that I think there are three basic roles that economics can bring to the analysis of policies to control greenhouse gases, whatever they may be. The first is the most controversial: it involves what goals we should aim at on a country basis, globally, regionally, et cetera. How big a set of emission reductions should we try to accomplish? That is, of course, a benefit-cost type question, and I'll say a word or two very soon about the appropriateness or inappropriateness of economics in goal selection.

A second set of questions has to do with how we can meet predetermined goals at the least cost. Can economics tell us how to meet a particular goal, even one that has been chosen in complete ignorance of economics? A final question is: Given that we've decided to meet a particular goal using a particular approach, who will bear the burden? I'll say less about this third question. Let me focus primarily on the role of economics in helping us choose the goals of any kind of climate policy and, also, the uses of economics in helping us determine how we can meet particular goals at the least cost to society.

On the question of using economics, benefit-cost analysis in particular, to help design climate mitigation or greenhouse gas control policies, I want to begin with a very frank concession. While I certainly cannot speak for the entire economics profession, I believe economists cannot now and never will be able to satisfactorily express in dollars all of the costs and benefits that will result from climate control measures. To me that follows as inevitably as day follows night.

Having said that, I want also to say that the previous speakers, economists and non-economists alike, who have given you the impression that benefits and costs relate inevitably to changes in gross national product have done a great disservice. GNP is to social welfare as fish is to bicycle. That is to say, there's no necessary one-to-one connection between the two.

I also want to say that, while social welfare cannot unambiguously be measured by gross national product, it is the case that one doesn't find many countries with high GNPs striving to change and become like countries that have very low GNPs. There must be something about the characteristics of countries with high gross national products that makes them attractive to countries with low gross national products because the latter try to design growth measures so that their standards of living will more closely approximate those of the former. If one looks at migration patterns in the world, moreover, one does not find hundreds of thousands of people each year leaving the United States, West Germany, the United Kingdom or Japan and going to Ethiopia or countries with low gross national products.

To come back to my assertion that there's no necessary connection between social welfare and gross national product, let me remind you that cost-benefit analysis was invented, was created for the purpose of rectifying the fact that national income and product accounts don't count some very important aspects of life. They don't take into account leisure, on which we all place a high value, nor do they reflect environmental values, since environmental services aren't traded in markets; I can't buy two more units of clean air, nor can I buy a clear vista. We have devised benefit-cost analysis because of inherent weaknesses in the concept of gross national product.

Thus, any economist who stands up before you and says that a particular policy is poor because it will cause gross national product to fall, or who says this is a good policy solely because it will cause gross national product to go up, doesn't deserve your attention. I can't emphasize that strongly enough.

Nevertheless, I want to say that it seems indisputable to me – and I hope to you – that economics at least has an important role to play in helping make decisions about climate change targets, about the goals of climate control policies. If we don't believe that the way to make solid policy on a country, regional or global basis is to say, "What will we gain by controlling carbon dioxide, chlorofluorocarbons, methane, nitrous oxide or whatever?" and "How do those gains compare with what we will sacrifice?" – if that's not the basic framework, if that's not a sensible way to make policy, then you should stop listening to me right now.

I take it as axiomatic that that is a useful way to think about making decisions with respect to global environmental issues. What will we get if we pursue a particular policy, and what will we have to give up? And of course, ultimately, is that exchange worthwhile? Not exchange in the market sense, but rather is that a trade-off that we want to make?

Furthermore, to those who have impugned this kind of balancing act – that is, this kind of qualitative benefit-cost approach – on the grounds that there's uncertainty, that the time scales are long, et cetera, I would say that entering into some kind of planetary compact or using some other approach to environmental policymaking doesn't exactly make these problems go away. It's no easier to make these decisions if one eschews benefit-cost analysis; the uncertainties are every bit as pervasive with respect to the values that any of us would attach to forestalling sea level rise or temperature change in mid-latitudes, et cetera. All of these uncertainties are there regardless of the approach we use to make these decisions.

Let me turn to a much less controversial use of economics in making decisions about global climate change. I refer here to what economists like to call cost-effectiveness analysis – the less controversial first cousin of benefit-cost analysis.

For a long time, my colleagues at Resources for the Future and scholars in universities and other research organizations around the world have urged decentralized approaches to environmental policy. But twenty-five years of economists preaching this gospel hasn't made much difference in U.S. environmental policy certainly, although it has influenced to some degree the content of European environmental policy.

But after this 25-year period and repeated assertions that there's a better way to protect the environment, coupled with a growing awareness that we may have reached the end of command and control's effectiveness in certain environmental policy areas, a change has occurred. It was hastened by the willingness of one environmental group, the

Environmental Defense Fund, to embrace a new regulatory approach (with special kudos to Dan Dudek). I refer to the recent amendments to the Clean Air Act, particularly those portions dealing with acid rain, which represent the first substantial legislatively condoned use of economic incentives to achieve environmental goals.

Expectations are that the incentive-based approach will reduce the costs of meeting a ten-million-ton reduction in SO_2 emissions in the United States by about 40 to 50 percent. If that seems like an inconsiderable amount (that's on the order of three or four billion dollars a year), think of the savings if we applied this decentralized approach to the control of carbon dioxide emissions on a global basis – we've heard variously that the costs could be on the order of $100 to $200 billion per year, and as much as $500 billion per year in some scenarios if we look out far enough.

Somewhat to my surprise we've heard relatively little about the various economic incentive approaches that one could use in dealing with greenhouse gases. I'm not going to go into them all in great detail here because my time is limited.

There are a number of decentralized approaches one can use to meet predetermined environmental targets. Obviously, I refer to effluent taxes or marketable permits. We've heard much over the course of these last two days about carbon taxes. John Shiller, Richard Richels, and David Montgomery have all talked about them, particularly about the size of the carbon tax that would be required to meet a given reduction in carbon emissions. They've also talked about the losses in GNP that would be associated with imposing taxes like that on ourselves.

We've heard much less about marketable permits, although people like Dan Dudek and Michael Grubb and others have presented compelling briefs that suggest we should attempt to meet global CO_2 limitations through a system of marketable permits. I'm going to say relatively little about the choice between taxes and permits, but I do want to make one observation.

In choosing between a tax approach or a marketable permits approach, there are many respects in which these are perfectly symmetrical. Under one approach, we fix the price and let the quantities fluctuate; under the other approach, we fix the quantity of emissions and let the marginal cost fluctuate.

One of the interesting distinctions between the two has to do with the political economy of each approach. If we use a marketable permits approach, twenty years from now we will have created valuable environmental assets that people will have invested money in protecting. If we should decide twenty years from now that global warming is not the serious problem that some believe it is today, it will be very difficult for governments around the world to print additional carbon permits because we will be devaluing assets which people hold.

That is, if science evolves in future decades to convince us that, although we were correct to regulate in 1990 or 1995, we are now slipping into an ice age, it's going to be much more difficult to change climate policy under a permits approach than under a tax approach. The disadvantage of the tax approach – particularly to those who wish to take action now – is that it's much more obvious and visible, and the costs are more immediate than those associated with a marketable permit approach.

Under a tax approach, each country (or all of the countries of the world acting together) would agree to set a carbon tax at a certain level. If, at some future point in time, policymakers decide that the problem is not as serious as we think it is today, they could always vote to reduce the size of that carbon tax. In doing so, they would not be devaluing anybody's assets. It would therefore be easier to become less stringent in our policies than if we take a marketable permits approach.

Both carbon taxes or marketable discharge permits have the advantage, of course, that they would help us minimize the costs associated with whatever target it is that we're aiming at.

But since I've raised the issue of costs, I want to address what to me is the fundamental question and, parenthetically, also the fundamental frustration I sense, having listened to the debate and discussion here.

We have heard Tom Schelling, John Shiller, and Rich Richels talk about the potential great expense associated with ambitious reductions in annual emissions of carbon dioxide in the United States. We've also heard from at least two people, Mary Beth Zimmerman and Howard Geller, who have talked about how much money we'll save in the process of reducing CO_2 emissions if only we summon the political will to do this.

Even before this conference, I have been profoundly frustrated by the fact that bright and intelligent people say, "We're going to get rich reducing CO_2 emissions," and another set of people, apparently just as bright, earnest and intelligent, say, "Reducing CO_2 emissions is going to cost us an arm and a leg." I would be surprised if those of you who have attended the last day and a half don't feel at least some of the same kind of frustration.

Well, isn't this a researchable issue? If the pessimists are correct, control measures could cost on the order of one to five percent of gross national product. Isn't it worth spending some money, perhaps an awful lot of money, to see whether efficient light bulbs and other conservation measures will really save as much energy as Howard Geller, Mary Beth Zimmerman and others claim they will? Or are these conservation measures bound to be dismal failures, as other people – often but not always economists – suggest?

We have a Department of Energy, a National Science Foundation, an Environmental Protection Agency, all of which have the responsibility to fund research. If I can think of a single issue which ought to attract considerable research funding, it ought to be this question: Are we going to make money or are we going to break ourselves trying to reduce CO_2 emissions?

It's not quite so simple to answer this question as I've suggested. I was talking about this question with Alex Cristofaro recently and he quite correctly pointed out that, in addition to pure cost considerations, there are other features associated with energy conservation. In the case of automobile transportation, there's the question of convenience.

Fuel-efficient cars do save you money, but if you try to climb into the back seat of my small car, you have a difficult time. That didn't use to be the case with cars.

I think we all admit that there's some inconvenience to smaller cars. I think we also recognize that down-sizing cars to improve fuel economy probably does subject us to larger risks. I have two children who drive and I'm very mindful of that. I would like them to have a car with better fuel economy, but at the same time, rightly or wrongly, I can't help but feel that if I put them in a bigger car with a bigger engine between them and an oncoming car, they've got a better chance to survive an accident.

Similarly, with light bulbs. Everybody seems to agree that there exist light bulbs that are more energy-conserving. Lester Lave's architect seems to believe that neither he nor others will like these bulbs because their color value is different than current light bulbs. Any of you who go into a washroom and look at yourself in a fluorescent-lighted mirror know that the light quality is not quite as good as it is under an ordinary incandescent light. How much that's worth I don't know. If I could find these miracle light bulbs I would have them in my home. And we certainly have switched to more energy-efficient light bulbs at the offices of Resources for the Future and in an adjacent building that we've recently constructed. We want to save money and we want to save energy.

To repeat, the energy-saving potential of our economy is a researchable issue. It's one to which intelligent men and women ought to turn their attention and on which intelligent governments ought to spend money.

Why do I think this question is so important? Well, we were talking about just how much money is at stake – one percent of GNP, two percent of GNP, five percent of GNP. While I agreed with much of what Tom Schelling said yesterday, if I understood him correctly, he said at one point that he wasn't particularly concerned about one percent of GNP because that was only $100 billion or so.

My reaction to that is that there are only 100 one-percent chunks in GNP. Since GNP in the United States is quite large, I get worried when intelligent people say, "One percent isn't enough to worry about." One percent of a very big number is an awful lot to worry about, and that's why I think it's worth spending a lot of money on research to help clarify just who's right in this debate about whether or not we're going to get rich or get poor saving energy.

Let me conclude my remarks today by making two observations about the future of this debate having to do with economics, costs and benefits in the discussions of what to do in the case of global climate change.

Let me direct my first set of remarks to my fellow economists and those who find the economic point of view congenial. We've got to wake up here. Economists have to understand that their point of view is not the only one that makes a difference in this debate, or that *should* make a difference in this debate. The intricacies and ins and outs of ecology, meteorology, agronomy, psychology, philosophy and a hundred other disciplines are as important as the inputs of economics in making decisions about what countries should do on global climate change.

Moreover, economists have to do more than learn about other disciplines and the roles that they can play in helping make decisions about climate change. Based on what I've heard prior to the last couple of days, and even a little of what I've heard during this time, economists have to better understand their own paradigm and the weaknesses of that paradigm if they're going to be taken seriously in climate change debates.

Having taken my fellow economists to task, I want also to say that those who look down so condescendingly on economics and economic analysis as an input to decisions about global climate change also have to answer a wake-up call. Anybody who thinks it's untoward for people to say, "How much is this policy going to cost me, and just what am I and the other citizens of this planet going to get for this?" just has to be dreaming. It's not only human nature, it's exactly the right way to think about these things. And since benefit-cost analysis, in its most common-sense version, answers exactly that question – what are we going to get and what are we going to have to sacrifice – it has to play a role in policymaking about global climate change.

Far from being unethical to conduct such analyses, I believe it's immoral *not* to ask what kind of protection various CO_2 measures are going to bring us and also what sacrifices we'll have to make both for the present and for future generations if we decide to expend resources on controlling carbon emissions or CFCs or anything else rather than investing in a whole host of other things that can help make both the present and the future a better one.

Having given my view and warned the sometimes warring camps in this debate that they have much to learn from each other, I want to say one final thing. If we can initiate some serious research on the profoundly important questions that lie before us, stop arguing over trivial matters, and think more carefully and interdisciplinarily about the problems associated with global climate change and the possible avenues open to address it, then subsequent conferences are going to advance the matter substantially.

CONFERENCE PARTICIPANTS

Richard Allen, The Dow Chemical Company, Houston, TX

Valter Angell, World Bank, Washington, DC

Priscilla Auchincloss, Rep. Louise Slaughter, Washington, DC

Louis Barbash, Smithsonian, Washington, DC

Barbara Bassuener, U S EPA, Washington, DC

Keith Belton, American Chemical Society, Washington, DC

Perry Bergman, U S Dept of Energy, Pittsburgh, Pa

J.E. Berry, Energy Tech Support Unit, Oxfordshire UK

Eugene Bierly, National Science Foundation, Washington, DC

Leslie Black, Senate Energy & Nat. Res. Comm., Washington, DC

Theodore Breton, ICF Resources, Inc., Fairfax, VA

Elizabeth Brownstein, Smithsonian, Washington, D.C.

Martin Buechi, Embassy of Switzerland, Washington, DC

Ralph Burr, Committee Ren. Energy Comm.&Trade, Washington, DC

Lynne Carter Hanson, Uni. of Rhode Island, Narragansett, RI

Esther Castro, L.A. Dept of Water & Power, Los Angeles, CA

Esther Castro, LA Dept of Water & power, Los Angeles, CA

Matthew Chadderdon, Carrier Corporation, Syracuse, NY

Walter Chan, Environment Ontario, Toronto,Ontario, CD

Edgar Chase, ALAPCO, Fairfax, VA

Richard Chastain, Southern Co. Svcs., Birmingham, AL

Paul Christensen, Cornell University, Ithaca, NY

Julie Clendenin, Edison Electric, Washington, DC

Doug Cogan, Investor Res. Research Ctr, Plainfield, NH

William Coleman, Electric Power Research Institute, Palo Alto CA

Bill Coleman, EPRI, Palo Alto, CA

Mitchell Colgan, College of Charleston, Charleston, SC

Lee Coplan, Environment Ontario, Toronto,Ontario CD

Walter Corson, Global Tomorrow Coalition, Washington, DC

Ann Cronin-Cossette, Canadian Embassy, Washington, DC

Christine D'Ambrosia, Rochester Gas & Electric, Rochester, NY

James Davis, Harvard University, Cambridge, MA

Richard Dennis, Southern Co. Svcs., Birmingham, AL

Bert Drake, Smithsonian Institute, Edgewater, MD

John Dysart, SYCOM Corporation, Washington, DC

Craig Ebert, ICF, Inc., Fairfax, VA

Lloyd Ernst, Nat'l Rural El. Co-op Asso., Washington, DC

Jerry Eyster, AT Massey Coal Co., Richmond, VA

R.L. Feder, Allied Signal, Morristown, NJ

Kevin Finneran, Issues in Science & Tech, Washington, DC

Arthur Fish, Argonne Nat'l Lab, Washington, DC

Kenneth Friedman, U S Dept. of Energy, Washington, DC

Diana Furchtgott-Roth, American Petroleum Ins., Washington, DC

Bernard Gelb, Congressional Res. Svc., Washington, DC

T.J. Glauthier, WWF/CF, Washington, DC

Indur Goklany, Dept. of the Interior, Washington, DC

Marvin Goldman, University of California, Davis, CA

Wilson Gonzalez, St. of Ct-Energy Div., Hartford, CT

F.Joseph Graham, Union Carbide, Danbury, CT

Gary Griffith, Chemical Manu. Assoc., Washington, DC

Erik Haites, Barakat & Chamberlin, Toronto, CD

Scott Hajoft, Env. Defense Fund, Washington, DC

Patricia Hanson, Poolesville Jr/Sr High School, Edgewater, MD

Nelson Hay, American Gas Association, Arlington, VA

Ruth Heikkiner, Uni. of Maryland, Laurel, MD

Udi Helman, Ctr for Strategic & Int'l Studies, Washington, DC

Elbert Herrick, American Chemical Society, Woodbine, MD

Ann Hirschman, Concern, Inc., Washington, DC

Nancy Hirsh, Energy Conservation Coalition, Washington, DC

David Hodas, Widener Uni. School of Law, Wilmington, DE

John Hoffman, U.S. EPA, Washington DC

John Hogan, Seattle Dept. Construction & Land Use, Seattle, WA

Chris Holly, Electric Utility Week, Washington, DC

Connie Holmes, National Coal Association, Washington, DC

Michael Hulme, Climatic Research Unit, Norwich UK

Hillard Huntington, Stanford University, Stanford, CA

Ken Jensen, Barakat & Chamberlin, Inc., Oakland, CA

Joan Jordan, National Science Foundation, Washington, DC

John Justus, Congressional Research Svc., Washington, DC

Sally Kane, Office of the Comptroller, Washington, DC

Margaret Kelley, American Petroleum Institute, Washington, DC

Mark Kemmer, GMC, Washington, DC

N.J. Kertamus, S. Ca. Edison Company, Rosemead, CA

Stephanie Kinney, Dept. of State, Washington, DC

Olva Kjorven, Norwegian Research Institute, Alexandria, VA

Adam Klinger, U.S. EPA, Washington DC

Charles Komanoff, Komanoff Energy Associates, New York, NY

Mike L'Ecuyer, U.S. EPA, Washington DC

Shonali Laha, Carnegie-Mellon, Pittsburgh, Pa.

Wendy Laird, World Wildlife Fund, Washington, DC

Alan Lamont, Lawrence Livermore Nat'l Lab, Livermore, CA

Tomas Larsson, Dept of Energy Conversion, Gothenburg, Swedn

C. Michael Lederer, University of Ca., Berkeley, CA

Bryan Lee, Air & Water Pollution Report, Silver Springs, MD

Paul Leiby, Oak Ridge Nat'l Labs, Oak Ridge, TN

Robert Lindgren, Center for Global Change, University of Maryland

Lori Megdal, City of Austin, Austin, TX

Terrill Meyer, Poolesville Jr/Sr High School, Poolesville, MD

Alden Meyer, Union of Concerned Scientists, Washington, DC

John Meyers, Washington Int'l Energy Group, Washington, DC

Robin Miles-McLean, EPA, Washington, DC

Russell Misheloff, US Agency for Int'l Dev, Washington, DC

Frank Moller, Center for Global Change, University of Maryland

Linda Moodie, NOAA, Int'l & Interagency Aff. Off., Washington, DC

Jane Moody, Inside EPA, Arlington, VA

Richard Moss, Princeton University, Princeton, NJ

Stanley Mumma, Penn State, University Park PA

Susan Murray, Montgomery Co. Public Schools, Poolesville, MD

Beth Nalker, Env. & Energy Study Institute, Washington, DC

Barbara Nash, Arlington Co. Dept. of Parks, Arlington, VA

Stephen Neal, Pacific Gas & Electric, San Ramon, CA

Beth Nelkar, Env. & Energy Study Institute, Washington, DC

Alan Nickels, WPCF, Alexandria, VA

Donna Nickerson, US EPA, Washington, DC

James J. Nipper, Department of Water & Power, Los Angeles, CA

John Ogle, Dow Chemical, Freeport, TX

Roland Paine, AMS newsletter, Washington, DC

Thomas Parker, Jr., Chemical Manu. Assoc., Washington, DC

Nancy Parks, Sierra Club, Aaronsburg, PA

Malka Pattison, Dept. of the Interior, Washington, DC

Ralph Perhac, Electric Power Research Institute, Palo Alto CA

Charles Peterson, Mantech Env. Tech, Corvallis, OR

Tom Pinkston, Tenneco Gas, Houston, TX

E.H.M. Price, Dept. of Energy, London UK

Tom Pugh, National Env. Dev. Assoc., Washington, DC

Harry Quarles, Columbia Gas System, Wilmington, DE

Chris Raymond, Chronicle of Higher Education, Washington, DC

Hakim Rehman, Env. Resources Ltd, London, UK

Daniel Reifsnyder, Dept. of State, Washington, DC

Hiram Reisner, Greenhouse Effect Report,

David Roberts, Carolina Power & Light Co., Raleigh, NC

Leonard Rodberg, Harrington Center Queens Colleg, New York, NY

Walter Rosenbaum, EPA-Ass't Adm. Office, OPPE, Washington, D.C.

Dale Rothman, Cornell University, Ithaca, NY

George Sagiyama, Pillsbury,Madison & Sutro, Washington, DC

Leddi Saunders-McGrath, Uni. of Pa., King of Prussia, PA

John Schiller, Ford Motor Company, Dearborn, MI

Bruce Schillo, EPA, Washington, DC

Jane Scully, ,

Cheryl Shanks, American Chemical Society, Washington DC

Kari Smith, CA Energy Commission, Sacramento, CA

Nancy Paige Smith, St. Mary's College, St. Mary's City, MD

David South, Argonne Nat'l Labs, Chicago, IL

Philip Squair, A/C & Refrigeration Institute, Arlington, VA

W. Ross Stevens, III, DuPont Company, Wilmington, DE

Dave Stirpee, Alliance for Responsible CFC Policy, Arlington, VA

Vicky Sullivan, Southern Company Services, Birmingham, AL

Kazu Takemoto, The World Bank, Washington, DC

Jonathan Temple, British Embassy, Washington, DC

Michael Tinkleman, EPRI, Washington, DC

George Tomlinson, , Ile Perrot,Quebec CD

Russell Tucker, Edison Electric Institute, Washington, DC

Jan Vlcek, Sutherland, Asbill & Brennan, Washington, DC

F.A. Vogelsberg,Jr., E I DuPont de Nemours Co., Wilmington, DE

Rodney Weiher, Office of the Comptroller, Washington, DC

Clas-Otto Wene, Dept. of Energy Conversion, Gothenburg, Swedn

Robert Wenger, Southern Company Services, Inc., Atlanta, GA

David Yaden, Oregon Dept of Energy, Salem, OR

Kaoru Yamaguchi, U.S. - Japan Information Exchange, Washington, DC

Cathy Zoi, EPA, Washington, DC

Pamela Zurer, Chemical & Engineering News, Washington, DC

CONFERENCE PROGRAM

Program — Tuesday, December 4, 1990

7:45 a.m. Registration and Continental Breakfast

8:45 Welcome

Elizabeth Thorndike, Center for Environmental Information, Inc.

9:00 Keynote Address - Adaptive Planetary Management

William C. Clark, Kennedy School of Government, Harvard University

9:30 How Should We Address Economic Costs of Climate Change?

Different perspectives on addressing the question of economic costs will give conference participants insights about the range of viewpoints relevant to the issue.

Moderator: **J. Christopher Bernabo**, Science and Policy Associates, Inc.

Nancy Birdsall, World Bank

Peter G. Brown, School of Public Affairs, University of Maryland

Daniel J. Dudek, Environmental Defense Fund, Inc.

Richard Schmalensee, President's Council of Economic Advisors

John Shiller, Ford Motor Company

10:45 Break

11:00 The Benefits of Mitigation and Adaptation

This session will assess what can be predicted about benefits of mitigation and adaptation, given the scientific uncertainties.

Mary Beth Zimmerman, Alliance to Save Energy

Thomas Schelling, Kennedy School of Government, Harvard University

11:45 Methods of Analysis

Designed for the non-specialist in economics, this talk will describe basic concepts necessary to understand economic analysis.

William R. Cline, Institute for International Economics

12:15 p.m. Luncheon

1:30 Current Analyses of Economic Costs

Several major studies have recently analyzed economic costs of global climate change. These studies have been sponsored by federal agencies, research institutes and private sector interests.

Moderator: **Lester B. Lave**, Graduate School of Industrial Administration, Carnegie-Mellon University

"The Cost of Reducing CO_2 Emissions in the United States"

Alex Cristofaro, Air & Energy Studies, U.S. Environmental Protection Agency

"Saving Money and Reducing the Risk of Climate Change Through Greater Energy Efficiency"

Howard Geller, American Council for an Energy Efficient Economy

"Global CO_2 Emission Reductions — The Impacts of Rising Energy Costs"

Alan S. Manne, Stanford University, and **Richard Richels**, Electric Power Research Institute (presenter)

"Carbon Sequestration vs. Fossil Fuel Substitution: Alternative Roles for Biomass in Coping with Greenhouse Warming"

Robert H. Williams, Center for Energy and Environmental Studies

3:30 Break

4:00 Analyses continued

5:30 Adjourn

Program — Wednesday, December 5, 1990

9:00 a.m. Respondents Panel

Members of this panel will respond to the papers presented in the previous day's session.

Moderator: **Rosina Bierbaum**, U.S. Congress Office of Technology Assessment

Dale Rothman and **Duane Chapman** (presenter), Department of Agricultural Economics, Cornell University

Michael Gluckman, Electric Power Research Institute

James K. Hambright, STAPPA-ALAPCO

David Montgomery, Congressional Budget Office

John Reilly, Economic Research Service, U.S. Department of Agriculture

10:15 Break

10:45 Questions/Discussion

An opportunity for a dialogue between the panel and conference participants.

11:45 Luncheon

1:00 p.m. International Panel

The economic analysis studies have focused on impacts or effects in the United States. What do the studies signify for other nations, including the developing countries?

Keynote and Moderator: **James P. Bruce**, Chairman, Canadian Climate Program Board

Erik Helland-Hansen, United Nations Development Programme

Ying Chong Lu, Institute for Techno-Economics and Energy Systems Analysis, Bejing

Mohan Munasinghe, World Bank

Bert Metz, Counselor for Health and Environment, Royal Netherlands Embassy

Yasuhiro Shimizu, Environmental Attache, Embassy of Japan

2:30 Break

3:00 Questions

3:30 Summation

Paul R. Portney, Resources for the Future

4:30 Adjourn